Textbooks in Telecommunication Engineering

Series Editor
Tarek S. El-Bawab,
Professor and Dean,
School of Engineering,
American University of Nigeria

Telecommunications have evolved to embrace almost all aspects of our everyday life, including education, research, health care, business, banking, entertainment, space, remote sensing, meteorology, defense, homeland security, and social media, among others. With such progress in Telecom, it became evident that specialized telecommunication engineering education programs are necessary to accelerate the pace of advancement in this field. These programs will focus on network science and engineering; have curricula, labs, and textbooks of their own; and should prepare future engineers and researchers for several emerging challenges.

The IEEE Communications Society's Telecommunication Engineering Education (TEE) movement, led by Tarek S. El-Bawab, resulted in recognition of this field by the Accreditation Board for Engineering and Technology (ABET), November 1, 2014. The Springer's Series Textbooks in Telecommunication Engineering capitalizes on this milestone, and aims at designing, developing, and promoting high-quality textbooks to fulfill the teaching and research needs of this discipline, and those of related university curricula. The goal is to do so at both the undergraduate and graduate levels, and globally. The new series will supplement today's literature with modern and innovative telecommunication engineering textbooks and will make inroads in areas of network science and engineering where textbooks have been largely missing. The series aims at producing high-quality volumes featuring interactive content; innovative presentation media; classroom materials for students and professors; and dedicated websites.

Book proposals are solicited in all topics of telecommunication engineering including, but not limited to: network architecture and protocols; traffic engineering; telecommunication signaling and control; network availability, reliability, protection, and restoration; network management; network security; network design, measurements, and modeling; broadband access; MSO/cable networks; VoIP and IPTV; transmission media and systems; switching and routing (from legacy to next-generation paradigms); telecommunication software; wireless communication systems; wireless, cellular and personal networks; satellite and space communications and networks; optical communications and networks; free-space optical communications; cognitive communications and networks; green communications and networks; heterogeneous networks; dynamic networks; storage networks; ad hoc and sensor networks; social networks; software defined networks; interactive and multimedia communications and networks; network applications and services; e-health; e-business; big data; Internet of things; telecom economics and business; telecom regulation and standardization; and telecommunication labs of all kinds. Proposals of interest should suggest textbooks that can be used to design university courses, either in full or in part. They should focus on recent advances in the field while capturing legacy principles that are necessary for students to understand the bases of the discipline and appreciate its evolution trends. Books in this series will provide high-quality illustrations, examples, problems and case studies.

For further information, please contact: Dr. Tarek S. El-Bawab, Series Editor, Professor and Dean of Engineering, American University of Nigeria, telbawab@ieee.org; or Mary James, Senior Editor, Springer, mary.james@springer.com

More information about this series at http://www.springer.com/series/13835

Yi Lou • Niaz Ahmed

Underwater Communications and Networks

 Springer

Yi Lou
Harbin Engineering University
Heilongjiang
China

Niaz Ahmed
Harbin Engineering University
Heilongjiang
China

ISSN 2524-4345 ISSN 2524-4353 (electronic)
Textbooks in Telecommunication Engineering
ISBN 978-3-030-86651-8 ISBN 978-3-030-86649-5 (eBook)
https://doi.org/10.1007/978-3-030-86649-5

This Springer imprint is published by the registered company Springer Nature Switzerland AG
The registered company address is: Gewerbestrasse 11, 6330 Cham, Switzerland

This book is devoted to our families.

Foreword

Underwater Communication is an important field of study to build the capacity of exploring the 70% part of the earth. Conventional underwater communication is achieved with the help of acoustic waves and is known as underwater acoustic communication. Because of the physical behavior and complexity of underwater channels, research in underwater acoustic communication is far behind the research in terrestrial communication. Researchers have, therefore, proposed the use of new technologies such as optical communication and magneto-inductive communication to boost the research for underwater applications. To teach about underwater communication, it is thus important to realize the students' detailed understanding and the potential of all these technologies. Despite the fact that there are tons of underwater applications, there are very few universities that offer underwater communication studies. The few universities that do offer courses on underwater communication mainly focus on the details of underwater acoustic communication only, and there is a lack of a single book that offers students with knowledge of all the basic technologies used in underwater communication. This book serves this much-desired purpose and presents a detailed review of all the three main technologies used in underwater communication. I firmly believe that this book can replace the introductory courses already offered in lieu of underwater communication. Some of the unique features offered in this book are as follows:

- The authors of this book have put significant effort to put all the three main technologies of underwater communication, such as underwater acoustic communication, underwater optical communication, and underwater magneto-inductive communication, in one place. All these technologies offer different flavors of underwater communication and play an important role in the advancement of underwater research.
- Apart from presenting the three underwater communication technologies, another feature of this book is its simplistic approach toward all the topics. The simplistic approach makes this book attractive for students of different grades. This book is an introductory level book that can be used for both undergraduate and graduate-level students.

Keeping in view the above-mentioned unique features, I firmly believe that this book can be very useful for readers interested in underwater communication. This single book can help to build a detailed understanding of all the different domains of underwater communication.

Harbin, China   Gang Qiao
April 2021

Preface

During the graduate studies, it was realized that there is much more to underwater communication; however, the literature available is only focused on underwater acoustic communication. Comparing to the research articles, there are not enough books available that could serve as a syllabus for a single complete course on underwater communication. Each course, however, does rely on some recommended books, but a single book that covers all the fundamentals is really missing. This always made us think about coming up with a book that can go through all the fundamental topics related to underwater communication.

Underwater communication is a really vast field, and most commonly, the study is linked with acoustic communication only. It is true that underwater acoustic communication is the only technology that allows long-range communication underwater, but that comes at the cost of low data rates and large delays. To speed up the communication and achieve better data rates, optical and magneto-inductive communications have been studied and presented for underwater applications. Similarly, a lot of material and content is available in books for underwater acoustic communication; however, the content related to underwater optical and underwater magneto-inductive communication is not easily available, and thus it required a lot of effort to think through the contents and structure of this book. After a careful and detailed thought process, basic fundamentals have been gathered from research articles, so that this book can serve the purpose of a single platform where readers can learn about each aspect of underwater communication. This book has been designed with three main parts: underwater acoustic communication, underwater optical communication, and underwater magneto-inductive communication.

This book is written on three different levels. First, it may be read by people with limited mathematics and engineering background. The simplistic approach adopted in this book is adopted to keep such an audience in mind. Some of the specific terminologies may be hard to understand at first, but going through the text in general and the figures, the reader will find it easy to follow the book content. Another important feature of this book is that the three parts are written independently to each other, and a reader can easily choose to read any part without the knowledge of other parts.

The second group of audience of this book are students and teachers with basic communication backgrounds. Students with terrestrial and underwater backgrounds will find this book easy to follow and read. Keeping this vast audience, the book is written with basic communication terminologies only. Parts of the book that required specific underwater communication knowledge have been explained in detail, so that readers can relate to their field of studies.

The third group of the audience are professionals from both academia and industry involved in underwater communication. We, however, do not expect all the experts to agree with the content presented in this book. However, we do hope that the content and information presented in this book will give a deep insight into the fundamentals of underwater communication.

Harbin, China   Yi Lou

Harbin, China   Niaz Ahmed

April 2021

Acknowledgments

It is a pleasure for us to thank Prof. *Julian Cheng* for valuable recommendations. At the same time, we would like to thank the dean of our college and our mentor Prof. *Gang Qiao*, who provided us a working environment where we could easily focus and spend our efforts on writing this book. Without his support, it would not be possible for us to accomplish this project.

We also extend our gratitude to our lab mates who encouraged us and shared their useful comments that helped us in shaping and writing this book.

A special thanks to our students: Ruofan Sun, Muhammad Muzzamil, Xinhao Qu, Yinheng Lu, Xinxin Yu, Yu Bi, Chenlu Yang, and Xueqin Xia, who assisted us during the entire process of compiling and writing this book. It was their effort and hard work that made it possible to finish writing in a timely manner.

Last but not the least, we would like to thank our family members (wives and kids) whose share of time was spent on the book. It was their patience and support that kept us going to complete this book.

Contents

Acronyms

2PSK	Binary phase-shift keying
AODV	Ad hoc on-demand distance vector routing
AOP	Apparent optical properties
AOW	Adaptive on-demand weighting algorithm
APD	Avalanche photo diode
ASK	Amplitude-shift keying
BCH	Bose–Chaudhuri–Hocquenghem
CDM	Code division multiplexing
CDMA	Code division multiple access
CH	Cluster head
CIR	Channel impulse response
CMA	Constant-mode blind equalization
COFDM	Coded OFDM
CP	Cyclic prefix
CRC	Cyclic redundancy check
CSMA	Carrier sense multiple access
DAPIM	Differential amplitude pulse interval modulation
DFE	Decision feedback equalizer
DFT	Discrete Fourier transform
DHPIM	Dual-header PIM
DPC	Dirty paper coding
DPIM	Digital pulse interval modulation
DPPM	Differential pulse position modulation
DPSK	Differential phase-shift keying
DPSSL	Diode pumped solid state laser
FDM	Frequency division multiplexing
FDMA	Frequency division multiple access
FH	Frequency hopping
FPFD	Flash lamp-pumped, frequency-doubled
FSK	Frequency-shift keying
FSO	Free space optical

FZ	Forced-zero
GFDM	Generalized frequency division multiplexing
HDTV	High-definition television
HROV	Hybrid remotely operated vehicle
IDFT	Inverse discrete Fourier transform
IOP	Intrinsic optical property
ISI	Intersymbol interference
LD	Laser diode
LDPC	Low-density parity check
LED	Light emitting diode
LiPaR	Light path routing protocol
LMS	Least mean square
LOS	Line-of-sight
LT	Luby Transform
MAC	Media access control
MACA	Multiple access with collision avoidance
MANET	Mobile ad hoc network
MI	Magnetic induction
MIMO	Multiple-input multiple-output
MMSE	Minimum mean square error
MRT	Maximum ratio transmission
MSE	Mean squared error
NLOS	Non-line of sight
NOMA	Non-orthogonal multiple access
NRZ	Non-return to zero
OFDM	Orthogonal frequency division multiplexing
OFDMA	Orthogonal frequency division multiplexing multiple access
OLSR	Optimized link state routing
OOK	On/off key
OSI	Open systems interconnection
OTF	Optical transfer function
PAPM	Pulse amplitude and position modulation
PAPR	Peak-to-average power ratio
PDM	Polarization division multiplexing
PIN	Positive intrinsic negative
PN	Pseudo-random
PolSK	Polarization shift keying
PPM	Pulse position modulation
PSK	Phase-shift keying
PMT	Photo multiplier tube
PWM	Pulse width modulation
RF	Radio frequency
RLS	Recursive least mean square
RS	Reed Solomon
RSC	Recursive systematic convolutional

RTE	Radiative transfer equation
RZ	Return-zero
SC-FDE	Single-carrier frequency-domain equalization
SCM	Sub-carrier multiplexing
SIM	Sub-carrier intensity modulation
TCP/IP	Transmission control protocol/internet protocol
TDM	Time division multiplexing
TDMA	Time division multiple access
TDoA	Time difference of arrival
UACN	Underwater acoustic communications network
UOT	Underwater optical turbulence
UWA	Underwater acoustic
UWAC	UWA communication
UWC	Underwater wireless communication
UWN	Underwater wireless network
UWOC	Underwater wireless optical communication
VSF	Volume scattering function
WDM	Wavelength division multiplexing
WKB	Wentzel, Kramer, Brillouin

Part I
Underwater Acoustic Communications and Networks

Chapter 1
Basic Principles of Underwater Acoustic Communication

1.1 Introduction

With the recent concept of smart ocean and underwater internet of things (IOT), there requires higher technical specifications of underwater acoustic communications and networks, such as communication rate, reliability, and delay. In order to meet the high demands, an understanding of the basic guidelines for underwater acoustic communication (UWAC) is needed, and therefore the propagation mechanism of underwater acoustic (UWA) signals along with the characteristics of an UWA channel is introduced in this chapter.

The propagation of vibrations generated by a sound source through the air or other mediums is called an acoustic wave. Acoustic waves can travel over long distances in water and are the main carrier of information in ocean. Marine mammals, like whales and dolphins, also use these acoustic waves for ranging, feeding, and communication.

Acoustic waves are currently the most feasible means of underwater detection and communication [1]. With the development of maritime military and commercial activities, UWAC technology has also developed rapidly in recent years. UWAC technology is thus mainly used in submarines, deep-sea manned submersibles, underwater cordless robots, remote control, data acquisition of seabed-based equipment, etc.

1.2 Theory of UWA Waves Propagation

1.2.1 Speed of Underwater Sound

The most significant acoustic parameter is sound speed, which is one of the fundamental physical quantities affecting the propagation of sound waves in water.

© The Author(s), under exclusive license to Springer Nature Switzerland AG 2022
Y. Lou, N. Ahmed, *Underwater Communications and Networks*, Textbooks in Telecommunication Engineering, https://doi.org/10.1007/978-3-030-86649-5_1

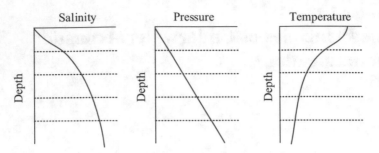

Fig. 1.1 Variation of environmental parameters with ocean depth

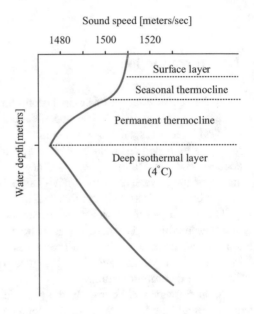

Fig. 1.2 Typical sound velocity stratification profile of deep sea

Sound (acoustic) waves are elastic longitudinal waves in water with the propagation speed given by the formula $c^2 = (\partial p/\partial \rho)_s = 1/(\rho K_a)$, where ρ is density of the medium, K_a is the adiabatic compression coefficient. Studies have shown that ρ and K_a are functions of temperature, salinity, and static pressure and therefore sound speed in seawater is mainly affected by these factors with temperature being the most obvious. In general, an increase in temperature, salinity, and static pressure eventually leads to an increase in sound speed. The variation in the ocean environment parameters with increasing ocean depth is shown in Fig. 1.1 [2].

A typical sound speed profile as a function of water depth is shown in Fig. 1.2. According to Fig. 1.2, there are four layers in the profile: surface layer, deep isothermal layer, seasonal and permanent thermocline layers. In the surface layer, water depth is usually several tens of meters. Due to the mixing effect of wind, both temperature and salinity in this layer tend to be homogeneous, which leads to constant sound speed. This layer is also called the mixed layer. Next, in the deep isothermal layer, the water temperature is always maintained at about $4\,^{\circ}\mathrm{C}$. However, water pressure is not the same and thus the main cause of sound speed variation, resulting in a positive gradient of sound velocity.

The increase in water depth causes the water temperature to drop in the thermocline layer, which is illustrated in Fig. 1.1. In this layer, increase in pressure and salinity cannot compensate the effects of temperature drops. Therefore, the sound velocity profile shows a negative gradient in depth. The negative gradient also varies with seasons in a seasonal thermocline layer. While the seasonal variation is not obvious in the permanent thermocline layer. The relationship of sound speed with temperature, salinity, and static pressure can be expressed by

$$c = 1449.22 + 4.6T - 0.055T^2 + (1.34 - 0.010T)(S - 35) + 0.016z, \qquad (1.1)$$

where c is the sound speed in water (m/s), T is the temperature ($^{\circ}$C), ρ is the salinity (‰), z is the depth (m). Moreover, for $-4\,^{\circ}\mathrm{C} \leq T \leq 30\,^{\circ}\mathrm{C}, 0 \leq S \leq 37$‰, and $1\,\mathrm{kg/cm}^2 \leq P \leq 1000\,\mathrm{kg/cm}^2$, Wilson formula is applied which is given by

$$c = 1449.14 + c_t + c_p + c_s + c_{stp}, \qquad (1.2)$$

where the expressions for c_t, c_p, c_s, and c_{stp} are provided in Eq. (1.3).

$$c_t = 4.5721T - 4.4532 \times 10^{-2}T^2 - 2.6045 \times 10^{-4}T^3 + 7.9851 \times 10^{-6}T^4,$$

$$c_p = 1.60272 \times 10^{-1}P + 1.0268 \times 10^{-5}P^2 + 3.5216 \times 10^{-9}P^3$$
$$- 3.3603 \times 10^{-12}P^4,$$

$$c_s = 1.39799(S - 35) + 1.69202 \times 10^{-3}(S - 35)^2,$$

$$c_{stp} = (S - 35)\left(-1.1244 \times 10^{-2}T + 7.7711 \times 10^{-7}T^2\right)$$
$$+ 7.7016 \times 10^{-5}P - 1.2943 \times 10^{-7}P^2 + 3.1580 \times 10^{-8}PT \qquad (1.3)$$
$$+ 1.5790 \times 10^{-9}PT^2 + P\left(-1.8607 \times 10^{-4}T\right.$$
$$\left. + 7.4812 \times 10^{-6}T^2 + 4.5283 \times 10^{-8}T^3\right)$$
$$+ P^2\left(-2.5294 \times 10^{-7}T + 1.8563 \times 10^{-9}T^2\right)$$
$$+ P^3\left(-1.9646 \times 10^{-10}T\right).$$

The sound speed is a very important physical property and plays an essential role in sonar equipment performance. Hence, the measurement of sound speed is an integral part of sea trials. Sound speed measurements are often carried out by a sound velocity meter, where the basic principle is the so-called cyclic method, as shown in Fig. 1.3.

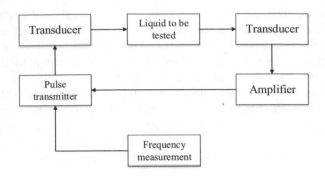

Fig. 1.3 Principle diagram of sound speed measurement

The probe of the sound velocity meter is placed in the desired region. A series of high-frequency pulses are emitted and then propagated in a known identified path L. Once the repetition period T of the pulse has been measured, the sound speed c is then determined by $c = L/T$. The accuracy of this method generally reaches 0.1%.

1.2.2 Underwater Sound Transmission Loss

When sound waves propagate in water, the sound intensity gradually decreases as the propagation distance increases. This phenomenon is influenced by several factors, which are discussed in this subsection.

The reduction of sound intensity is roughly divided into two categories: acoustic absorption, and extended losses. Acoustic absorption is the process of converting acoustic energy into thermal energy caused by physical effects such as thermal conductivity, viscosity, and molecular relaxation of seawater. Extended losses on the other hand, refer to the change in the spatial distribution of acoustic energy during sound propagation. The influencing factors include geometric expansion of the wavefront surface, reflections from the seafloor and surface, sound scattering, and sound propagation path bending due to various inhomogeneities in the seawater medium. The reduction of acoustic energy can be described by *transmission loss*, which is defined as [3]

$$TL = 10 \lg \frac{I_1}{I_r}, \tag{1.4}$$

where I_1 is the sound intensity at 1 m from the equivalent sound center of the source and I_r is the sound intensity at the distance r (m) from the source. The transmission loss is composed of both expansion and absorption losses and can be expressed as

$$TL = TL_1 + TL_2, \tag{1.5}$$

where TL_1 and TL_2 represent expansion loss and absorption loss, respectively.

When calculating the expansion loss of a plane wave, the sound pressure in the propagation direction of the harmonic plane wave can be written as

$$p = p_0\, e^{j(wt-kx)}, \tag{1.6}$$

where p_0 is the amplitude of the plane wave sound pressure, ω is the angular frequency, k is the wave number. It is known that the sound intensity of a plane wave is proportional to p^2 and does not vary with x. Therefore, we have $I_1 = I_x$. Here, I_1 is the sound intensity at 1 m from the source equivalent sound center, I_x is the sound intensity at x meters from the source equivalent sound center. According to the definition of transmission loss, TL_1 is expressed as

$$TL_1 = 10\lg \frac{I_1}{I_x} = 0 \ (\mathrm{dB}). \tag{1.7}$$

This is because the plane wavefront does not expand with distance. That is, the propagation loss will not be caused by wavefront expansion.

For the calculation of the expansion loss of a simple harmonic uniform spherical wave along the sagittal diameter r, the sound pressure can first be expressed as

$$p = \frac{p_0}{r} e^{j(wt-kr)}. \tag{1.8}$$

In Eq. (1.8), p_0/r is the amplitude of the spherical wave sound pressure. Because the amplitude and the distance r are inversely proportional, the sound intensity I_r is inversely proportional to r^2, and the expansion loss of the spherical wave is

$$TL_1 = 10\lg \frac{I_1}{I_r} = 20\lg r \ (\mathrm{dB}). \tag{1.9}$$

For typical sound propagation expansion losses, in general, the propagation losses due to expansion can be written as

$$TL_1 = n10\lg r \ (\mathrm{dB}), \tag{1.10}$$

where r is the propagation distance and n is a constant which takes on different values for different propagation conditions. In more detail:

- $n = 0$: This situation applies to plane wave propagation with no expansion loss.
- $n = 1$: This situation applies to columnar wave propagation, where the wavefront surface expands according to a cylindrical lateral pattern. For example, sound propagates in an ideal waveguide consisting of a fully reflective seafloor and a fully reflective sea surface.

- $n = 3/2$: Shallow sea sound propagates in the presence of submarine sound absorption. At this time, $TL_1 = 15 \lg r$, this is an amendment for the column propagation loss by accounting for interfacial acoustic absorption.
- $n = 2$: This situation applies to spherical wave propagation, where the wavefront expands in a spherical plane.
- $n = 3$: This applies to the loss of sound propagation after the sound waves have passed through the negative leapfrog layer in shallow seas.
- $n = 4$: After accounting for the acoustic reflection interference effects of a flat sea surface, the loss of sound propagation in the far-field region can be represented as $TL_1 = 40 \lg r$. It is an amendment for spherical propagation losses after accounting for multipath interference. This law also applies to the sound intensity attenuation in the far-field of a dipole source field.

In terms of expansion laws, the actual expansion losses are less likely to be as simple as mentioned above due to the actual sea area's inhomogeneity, where the acoustic signal is refracted, scattered, and reflected. The shape of the extended wavefront may be more complex, but it is always approximated by spherical or columnar waves for computational reasons.

Furthermore, the expansion loss discussed above is under the assumption that the medium is ideal. However, since seawater is an inhomogeneous medium in practice, plane waves in the propagation process still produce a certain acoustic energy loss. That is, the sound energy is converted into other forms of energy. This loss of sound energy during propagation is called sound wave absorption. Like the expansion loss, sound wave absorption also produces the phenomenon that the sound intensity gradually weakens as the propagation distance increases.

Under the assumption of no expansion loss, as the distance increases, the differential expression of the sound intensity change dI can be expressed as

$$dI = -\beta I dx, \tag{1.11}$$

where β is related to the sound absorption capacity of the medium.

From Eq. (1.11), it can be seen that when a plane wave propagates in an absorbing medium, the sound intensity decays with increase in distance x following the exponential law. Take the logarithm of Eq. (1.11) with a base of 10 and multiply by 10, we have

$$10 \log I_2 - 10 \log I_1 = -\alpha(x_2 - x_1), \tag{1.12}$$

where

$$\alpha = -\frac{10 \log I_2 - 10 \log I_1}{x_2 - x_1}, \tag{1.13}$$

where α refers to the logarithmic absorption coefficient (with a logarithmic base of 10) in decibels per meter, which means that the sound intensity is attenuated by α decibel per unit distance traveled due to absorption by the medium.

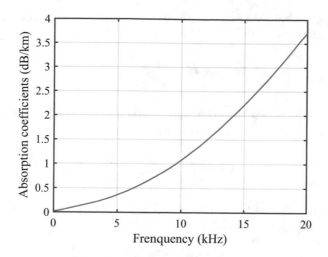

Fig. 1.4 The relationship between absorption coefficient and frequency

By definition, it can be seen that the propagation losses due to absorption are

$$TL_2 = \alpha(x_2 - x_1). \tag{1.14}$$

It can be seen that the propagation loss caused by seawater absorption is equal to the absorption coefficient multiplied by the propagation distance. If x is written as r and combined with Eq. (1.10), we obtain the total propagation loss. It is equal to the expansion loss plus the absorption loss, i.e.,

$$TL = n \cdot 10 \lg r + \alpha(r_2 - r_1). \tag{1.15}$$

The above equation's absorption coefficient can be obtained by consulting Fig. 1.4 between the absorption coefficient and frequency. Equation (1.15) is a common formula for calculating propagation losses and has very important engineering and theory applications.

1.3 Acoustic Field Model

In underwater acoustics, the acoustic field model is usually used to describe the propagation law of sound waves. The propagation process of sound waves is a wave process [4] and can be characterized by the wave equation. The wave equation is the theoretical basis of all mathematical models describing the propagation of the acoustic wave. And the acoustic field analysis is to solve the wave equation under the conditions of a given solution so as to obtain the general laws of the wave process. In a homogeneous ideal fluid medium, the three-dimensional wave equation of small amplitude sound waves is

$$\nabla^2 p = \frac{1}{c^2} \frac{\partial^2 p}{\partial t^2}, \tag{1.16}$$

where p, c, and ∇^2 denote sound pressure, sound velocity, and the Laplace operator, respectively, and the Laplace operator can be expressed as

$$\nabla^2 = \frac{\partial^2}{\partial x^2} + \frac{\partial^2}{\partial y^2} + \frac{\partial^2}{\partial z^2}. \tag{1.17}$$

In a steady-state acoustic field generated by a simple harmonic source, the wave equation (see Eq. (1.16)) simplifies to

$$\nabla^2 p + k^2 p = 0, \tag{1.18}$$

where $k = \omega/c$ is the wave number, $\omega = 2\pi f$ represents the angular frequency of the sound source.

Equation (1.18), also known as the Helmholtz equation, is the expression of the wave equation in the frequency domain. In a general macroscopic nonhomogeneous medium, both density ρ and sound velocity c are functions related to spatial orientation. At this point, the wave equation cannot be obtained directly, but the wave function can be introduced by

$$\psi = \frac{1}{\sqrt{\rho}} p. \tag{1.19}$$

From this, the wave equation for ψ can be obtained as

$$\nabla^2 \psi + K^2(x, y, z) = 0, \tag{1.20}$$

where

$$K^2 = k^2 + \frac{1}{2\rho} \nabla^2 \rho - \frac{3}{4} \left(\frac{1}{\rho} \nabla \rho \right)^2. \tag{1.21}$$

Solving Eq. (1.20) or Eq. (1.18) is complicated or even impossible for general non-uniform media spaces. Even if a solution is available under certain conditions, it is often in the form of an advanced transcendental function, which does not give a very intuitive physical picture. Therefore, the actual ocean channel is often divided into several special types based on its physical and geometric characteristics to obtain approximate solutions. Different forms of solutions result in different sound field models. The following five types of sound field models are commonly used: ray theory models, normal models, multipath expansion models, fast field models, and parabolic equation models. These five models can be further subdivided into distance-dependent and distance-independent models. Distance-dependent means that the model assumes columnar symmetry, i.e., the channel is a horizontal hierarchical channel, and the acoustic propagation properties are only a function of depth. Distance-dependent implies that some properties of the seawater medium are not only a function of depth but also a function of distance and grazing angle. Distance-dependent is also subdivided into two-dimensional (2D) models related to distance and depth and three-dimensional (3D) models related to distance, depth, and grazing angle.

1.3.1 Ray Theory Models

Classical ray acoustics considers the transmission of sound energy by rays (sound lines) in the acoustic field. There are two basic equations in the field of ray acoustics: (a) the Eikonal equation to determine the travel law of the sound lines, and (b) the intensity equation to determine the intensity of the sound lines. These two equations can be obtained under certain approximation conditions.

In ray theory, the approximate solution of the wave equation is the product of the amplitude function $A(x, y, z)$ and the phase function $P(x, y, z)$ of the sound pressure, i.e.,

$$\psi(x, y, z) = A(x, y, z)e^{iP(x,y,z)}. \tag{1.22}$$

Substituting the above equation into the Helmholtz equation and separating the real and imaginary parts yields

$$\frac{1}{A}\nabla^2 A - [\nabla P]^2 + k^2 = 0, \tag{1.23}$$

$$2[\nabla A \cdot \nabla P] + A\nabla^2 P = 0. \tag{1.24}$$

For high-frequency sound sources, we have

$$\frac{1}{A}\nabla^2 A \ll k^2. \tag{1.25}$$

That is, A is approximately constant within a wavelength, and Eq. (1.23) can then be simplified to

$$[\nabla P]^2 = k^2. \tag{1.26}$$

This is the Eikonal equation and can help to determine the trajectory of the sound line. Similarly, Eq. (1.24) is the intensity equation from which the intensity of the sound line can be determined.

In a layered medium, the ray acoustic field can be expressed as

$$\psi(x, z) = \sqrt{\frac{W \cos \theta_0}{R \left(\frac{\partial R}{\partial \theta}\right)_{\theta_0} \sin \theta_z}} e^{ikx + ik \int_0^z \sqrt{n^2(z) - \xi^2}\, dz}, \tag{1.27}$$

where W is unit stereo angle radiation power, θ_0 is the grazing angle of the sound line at the source, θ_z is the grazing angle at any depth z, ξ is the separation constant, and R is the distance.

The most important practical result in ray theory is the Snell law. It describes the law of refraction of sound lines in a medium with varying sound speed. Snell states that in a medium consisting of equal velocity layers, the grazing angle $\theta_1, \theta_2, \ldots$, of the sound lines is related to the sound speed $c_1, c_2, \ldots$, in the layers. The relation is

$$\frac{\cos \theta_1}{c_1} = \frac{\cos \theta_2}{c_2} = \cdots = \text{const}. \tag{1.28}$$

Ray theory is an approximation theory of geometric acoustics. In ray theory, we do not consider the attenuation of sound energy in transmission. The sound line diagram can give a visual image of the acoustic field and is an important method for solving the acoustic field. However, this method does not give a reliable image of the acoustic field if the sound lines bend or change in intensity over a range of wavelengths. Therefore, this method is only suitable for high-frequency sound transmissions, and Etter gives an approximate method for determining high frequencies, i.e.,

$$f > 10c/H, \tag{1.29}$$

where f is frequency, H represents layer height, and c represents sound speed.

Although the ray model can be theoretically applied to distance-related problems, three-dimensional (3D) models are seldom used in practice due to the complexity of implementation. Therefore, people more often use one- or two-dimensional models. There are three solutions to the distance-related problems in 2D models:

(1) The whole distance is divided into discrete distance segments. In each segment, the environmental parameters are kept constant. First, the sound lines are tracked in the first distance segment, and then some sound lines are selected and mapped to the second distance segment. The above process is repeated for the remaining distance segments until a complete sound ray trace is obtained for the whole distance segment.
(2) The sound velocity curve is expressed as a segmented linear function of depth, and the distance-depth plane between these profiles is divided into triangular

sectors. In each sector, the sound velocity varies linearly with distance and depth. The sound trajectory corresponds to an arc in the sector.

(3) When the environmental parameters vary slowly with distance, the sound velocity curves are fitted with a cubic spline function. Similar to the first method, the sound ray trajectories are obtained for different distance segments, but the trajectories of adjacent distance segments are obtained by linear interpolation.

When dealing with distance-dependent ray tracing models, the main problem encountered is how to ensure a smooth transition of the sound velocity curve at the adjacent measurement points of different distance segments. In practice, this is reflected in two problems: First, the sound velocity profile in the transition region obtained by interpolation should have a reliable physical meaning. Second, the resulting sound trajectory must be computable.

1.3.2 Normal Wave Models

A normal model solution is an exact integral solution of the wave equation. It describes the sound propagation in terms of normal models (eigenfunctions). Each eigenfunction is a solution to the wave equation. The normal model solution is obtained by superimposing the normal waves and satisfying the boundary and source conditions. For a point source acoustic field in a layered medium with no horizontal gradient, the normal model field can be expressed as the product of the depth and distance functions in column coordinates, i.e.,

$$\psi(r, z) = U(z)\varphi(r).$$ (1.30)

Substituting the above equation into Eq. (1.18) and performing the separation of variables yields

$$\frac{\mathrm{d}^2 U}{\mathrm{d}z^2} + \left(k^2 - \xi^2\right)\varphi = 0,$$ (1.31)

$$\frac{\mathrm{d}^2\varphi}{\mathrm{d}r^2} + \frac{1}{r}\frac{\mathrm{d}\varphi}{\mathrm{d}r} + \xi^2\varphi = 0.$$ (1.32)

Equation (1.31) is the depth equation, i.e., the normal wave equation. It gives the solution of standing wave form of the sound wave in the depth direction. Equation (1.32) is the distance equation, which gives the solution of traveling wave form in the distance direction and describes the transmission characteristics of the acoustic wave.

The normal wave equation (see Eq. (1.31)) can be reduced to an eigenvalue problem. Its solution is called the Green function. The distance equation (see

Eq. (1.32)) is a zero-order Bessel equation whose solution is a zero-order Hankel function.

Under the far-field approximation, the normal model solution can be expressed as

$$\psi(r, z) = g(r, p) \sum \frac{u_m(z_s) u_m(z)}{\sqrt{k_m}} e^{i(k_m r - \delta_m r \frac{\pi}{4})}, \tag{1.33}$$

where $g(r, \rho)$ is a function of the distance r and the density ρ and represents the geometric distribution of the acoustic waves, u_m is the mth normal wave relative to the spatial frequency k_m, and u_m and k_m are the eigenvectors and eigenvalues of the normal model equation (see Eq. (1.31)), respectively, and δ_m denotes the attenuation coefficient, which is the sum of various plane wave attenuation, dielectric attenuation, seawater absorption, and boundary loss.

The normal model is an important tool for the analysis of shallow ocean' sound fields. It gives a complete picture of the sound propagation properties determined by the inherent oceanic simplex mode, especially when considering the influence of seafloor parameters. The normal model is suitable for point source acoustic fields in stratified media. It ignores the interactions between the normal waves and the continuous spectral structure of the model. Generally speaking, for a certain acoustic signal frequency, only a limited number of normal waves can be propagated in the channel. The higher the acoustic signal frequency, the higher is the order of the propagation of the normal waves. The lower the frequency of the acoustic signal, the smaller is the number of normal waves, and the corresponding model calculation is also less. Therefore, the normal model has the advantages of high accuracy and low computational effort in low-frequency applications.

One advantage of the normal models over the ray theory models is that the propagation loss can be calculated for all depths and distances for a given frequency and source depth. In the ray theory models, the propagation loss needs to be recalculated when the transmitting depth or receiving depth changes. Whereas, the normal models require information about the seafloor structure for its calculation.

1.3.3   Multipath Expansion Models

In the multipath expansion model, the integral solution of the wave equation is expressed as a sum of a finite number of partial integrals. Each partial integral corresponds to a specific sound path. The multipath expansion model is also known as the WKB (Wentzel, Kramer, Brillouin) method because it uses the WKB approximation in solving the normal model equation. The WKB approximation simplifies the solution of the normal models by assuming that the sound velocity varies slowly with depth and each normal wave has a sound line corresponding to it. The WKB method considers only a certain number of normal modes to express the sound pressure field as a finite integral sum. This method is suitable for simulation of deep-sea, medium-and high-frequency sound propagation.

The multipath expansion model is a bridge between the normal model and the ray theory model. It has accuracy of the normal model and simplicity of the ray model. The multipath expansion model is quite similar to the ray model, however, multipath expansion model takes into account the first-order bypass and focal dispersion cases. Thus, it can be used to deal with the problems of focal dispersion and shadow areas that cannot be processed by ray theory model.

1.3.4   Fast Field Models

The fast field models are applicable to horizontally stratified nonhomogeneous media. In the fast field models, $c(r) = c(z)$, that is $k(r) = k(z)$. The sound field can be decomposed into an infinite superposition of horizontally propagating waves, i.e.,

$$\psi(r, z) = \int F(\eta, z)e^{j2\pi nr} \, d\eta, \tag{1.34}$$

where r is the distance in column coordinate, z denotes the vertical coordinate, $F(\eta, z)$ is the solution of the following equation for a given solution condition.

$$\frac{d^2 F(\eta, z)}{dz^2} + 4\pi^2 \left[k^2(z) - \eta^2\right] F(\eta, z) = 0. \tag{1.35}$$

From $k^2(z)$ and the known boundary conditions, the numerical solution of the Eq. (1.35) yields $F(\eta, z)$, and the FFT method is then used to determine $\psi(r, z)$. This result is accurate for all frequencies.

The operational simplification of the sound velocity distribution into exponential form can greatly simplify the calculation, but it makes the treatment of the sound velocity distribution difficult in general channels, hence require more application experience.

1.3.5   Parabolic Equation Models

In wave propagation problems, the parabolic approximation has been used since 1940s. It was first applied to the propagation of long-range tropospheric radio waves, then to microwave waveguides, laser propagation, etc. For UWA propagation problems, it was first applied in 1973.

In parabolic equation model, the wave equation is expressed in column coordinates. The key to this model is to replace the elliptic equation in Eq. (1.32) with a parabolic equation, assuming that the energy propagation velocity is similar to the transverse or longitudinal sound velocity.

$$\frac{\partial^2 \varphi}{\partial r^2} + \frac{1}{r}\frac{\partial \varphi}{\partial r} + \frac{\partial^2 \varphi}{\partial z^2} + k^2 n^2 \varphi = 0. \tag{1.36}$$

Assume that φ takes the following form:

$$\varphi(r, z) = F(r, z)R(r). \tag{1.37}$$

The separation of variables with k^2 as the separation constant yields

$$\frac{\partial^2 R}{\partial r^2} + \frac{1}{r}\frac{\partial R}{dr} + k^2 R = 0, \tag{1.38}$$

$$\frac{\partial^2 F}{\partial r^2} + \frac{\partial^2 F}{\partial z^2} + \left(\frac{1}{r} + \frac{2}{R}\frac{\partial R}{\partial r}\right)\frac{\partial F}{\partial r} + k^2 n^2 F - k^2 F = 0. \tag{1.39}$$

Equation (1.38) is a zero-order Bessel equation whose solution is a first-class zero-order Hankel function:

$$R = H_0^{(1)}(kr). \tag{1.40}$$

In the far-field condition, take its asymptotic form as

$$R = \sqrt{\frac{2}{\pi kr}}e^{i(kr-\pi/4)}. \tag{1.41}$$

Substitute it back into Eq. (1.39), and assume that

$$\frac{\partial^2 F}{\partial r^2} \ll 2k\left(\frac{\partial F}{\partial r}\right). \tag{1.42}$$

We can get

$$\frac{\partial^2 F}{\partial r^2} + 2ik\frac{\partial F}{\partial r} + k^2\left(n^2 - 1\right)F = 0. \tag{1.43}$$

This is the parabolic equation, where n is a function of depth z, distance r, and grazing angle θ.

The parabolic equation model can be solved recursively in numerical calculations. The normal model provides the sound field state at the initial position, and the new sound field state can be solved recursively at subsequent distances. This model is particularly suitable for the analysis of far-field propagation characteristics of horizontal inhomogeneous media.

1.4   Structure and Performance of UWAC System

1.4.1   Basic Structure of UWAC System

UWAC systems in general are mostly digital communication systems. The main components include transmitter and receiver transducers, encoder, decoder, modulator, demodulator, and UWA channel, as shown in Fig. 1.5. The following is a brief description of the basic functions of each part of the system [5].

1.4.1.1   Channel and Noise

A channel is a signal path in any transmission medium. Abstractly, it is a band of frequencies through which a signal is allowed to pass. Similarly, the band may restrict or degrade the signal too. The UWA channel is a wireless transmission channel in water that uses sound waves as the carrier of information. The water medium and the boundary conditions such as the surface, and the seafloor have complex and variable transmission characteristics so that the UWA channel has a more serious impact on the transmission of signals. On the other hand, noise is usually a collective term for the noise and interference inherent in various devices and channels in a communication system.

1.4.1.2   Encoder and Decoder

The coder and decoder usually include both the source encoder/decoder, and the channel encoder/decoder. In a digital communication system, the analog or digital signal output from the source needs to be converted into a binary digital signal for transmission. The process of efficiently converting the analog or digital source output into a binary digital sequence is called source coding. Similarly, the process of converting the binary digital sequence into an analog or digital source output is called source decoding. Moreover, the device that performs these functions is called a source encoder/transcoder. In some digital communication systems, like image

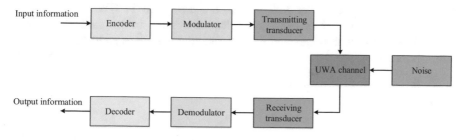

Fig. 1.5 Structure of a UWA digital communication system

transmission systems, in addition to analog-to-digital conversion, source coding is also required to perform the task of data compression to transmit a large amount of data required for dynamic image display in a limited amount of time.

Due to the imperfect performance of the communication system and the influence of noise and interference, errors may occur in the transmission of digital signals, resulting in degradation of the quality of transmitted information. In order to limit the errors, error control techniques are needed. One such technique is error correction coding/decoding which is also known as error control coding/decoding, and it belongs to the category of channel coding. The channel coding technology mainly studies the concept of error detection, error correction, and its basic implementation method. The encoder generates supervisory coded elements or redundant coded elements based on the input to achieve error control. The decoder then performs error detection or error correction based on these coded elements.

In addition, if other coding techniques are used in the communication system to improve the system performance, the corresponding encoder/decoder is required in the system. For example, in UWA multiple-input multiple-output (MIMO) systems, spatial diversity is usually obtained using space–time coding, so the system will also include a space–time encoder/decoder.

1.4.1.3   Modulators and Demodulators

The signal coming out of the encoder is a digital baseband signal. If this baseband signal is transmitted directly, it is called a digital baseband transmission. Baseband transmission uses a wired channel and has a limited transmission distance. The transmission of wireless signals over longer distances requires the use of a carrier wave. The process of modulating a digital baseband signal on a carrier to form a digital carrier signal is called modulation. The function of modulation is to transform a baseband digital signal into a specific frequency band suitable for channel transmission. Common digital modulation methods include amplitude-shift keying (ASK), frequency-shift keying (FSK), and phase-shift keying (PSK), where the latter two are commonly applied in UWAC systems. Similarly, on the receive side, the process of separating a baseband signal from the modulated carrier signal is called demodulation.

In addition to the basic carrier modulation techniques, other modulation techniques, such as multi-carrier modulation, are also used in UWAC systems to improve the performance of the communication system. In this case, the system should include the appropriate modulator/demodulator, such as multi-carrier modulator/demodulator.

1.4.1.4   Synchronization

Synchronization is an essential part of a digital communication system. Synchronization is to make the communication system's transceiver and transmitter at the same pace in time and frequency. Synchronization usually includes carrier synchronization, bit synchronization, frame synchronization, etc. Synchronization is necessary for digital communication and greatly affects the performance of a communication system.

1.4.2 Performance Indicators of UWAC System

1.4.2.1 Transmission Rate

(1) Symbol transmission rate
The code rate is the number of codes or symbols transmitted by a communication system per unit time, denoted by R_B, and is measured in B (baud). The code transmission rate is also known as the transmission rate or baud rate. Digital signals have multi-decimal and binary. The code rate is related to the transmission time T, that is [6]

$$R_B = \frac{1}{T} \text{ (B)}. \tag{1.44}$$

(2) Information rate
The information rate is the amount of information transmitted by a communication system per unit time, denoted by R_b, and its unit is b/s (bits per seconds). According to the definition of information, each coded element or symbol contains a certain amount of information in bits. Therefore, there is a clear relationship between the code rate and the information rate, which is given by

$$R_b = R_B H \text{ (b/s)}, \tag{1.45}$$

where H denotes the average amount of information contained in each code in the source. When each code is transmitted with equal probability, the information volume has the maximum value ($1bM$). At this time, the information rate reaches the maximum, i.e.,

$$R_b = R_B 1bM \text{ (b/s)}, \tag{1.46}$$

where M is the binary number of the code. Communication signals with different transmission rates are affected by different channel transmission characteristics. For example, in a shallow sea with horizontal transmission, the multipath effect will cause a high transmission rate in the signal out of the inter-code interference, resulting in serious signal distortion. While in the deep sea and vertical transmission channel, this signal distortion will be much less.

Therefore, the transmission rate that a UWAC system can achieve is also related to the channel environment.

(3) Frequency band utilization

The transmission rate of the UWA channel is related to the application environment of the communication system. Therefore, when comparing the effectiveness of different communication systems, their transmission rates and occupied bandwidths needs to be taken into account. Therefore, the real measure of the transmission efficiency of digital communication systems should be the frequency band utilization, that is, the transmission rate of codes allowed in each unit of the frequency band. It is denoted as η and can be expressed as

$$\eta = \frac{R_B}{W} \ (\text{B/Hz}), \tag{1.47}$$

where W is the system bandwidth occupied by the digital signal. It depends on the code rate R_B.

Most UWA channels are severely limited in terms of available bandwidth due to the communication distance and used frequency. Therefore, in theory, communication systems with high-frequency band utilization are more suitable for application in UWA channels.

1.4.2.2   Error Probability

A measure of the reliability of digital communication systems is the error probability. Error probability has several definitions with different meanings.

1. Code error rate
   The code error rate is the ratio of the number of codes with errors to the total number of codes transmitted in a communication system.
2. Bit error rate (BER)
   The bit error rate is the ratio of the number of information (bits) in error to the total number of information (bits) transmitted in a communication system.

There are two factors affecting the BER: one is the imperfect design of the system, and the other is the presence of various kinds of interference and noise. The former effect can theoretically be eliminated with proper design, while the latter effect cannot be avoided. Communication systems usually use error control techniques to reduce the BER and improve reliability.

1.5 Characteristics of the UWA Channel

In the early stage of UWAC studies, scholars usually used relatively simple UWA channel models. As more and more researches have been carried out on UWA channels, the understanding of UWA channels has become more in-depth [7].

1.5.1 Multipath Effects

Like radio communications, UWACs are also subject to multipath effects. A schematic diagram of multipath signal generation in the process of acoustic communications is shown in Fig. 1.6.

In Fig. 1.6, the seafloor and the sea surface are the upper and lower boundaries of the ocean, at which sound waves are reflected and refracted. In reality, the actual situation is much more complex than the one shown in Fig. 1.7, because neither the seafloor nor the sea surface is an ideal interface. For instance, the sea surface is not flat, and many waves on its surface are in constant motion. The seafloor is also irregular. Although the seafloor is fixed compared to the surface, the seafloor can also allow sound waves to penetrate to some extent. The phenomenon of multipath transmission is particularly prominent in shallow horizontal channels. The acoustic signal from the source is reflected, scattered, and refracted several times at different

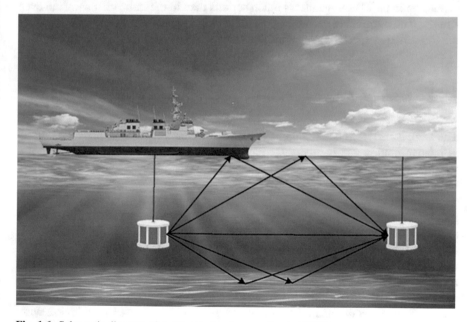

Fig. 1.6 Schematic diagram of multipath signal generation in UWAC

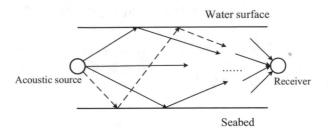

Fig. 1.7 Schematic diagram of surface-bottom reflection multipath

times along different paths before it reaches the receiver. The phenomenon causes fading and phase shifting of the transmitted signal. The elimination of interference from multipath phenomena has always been a primary challenge for UWACs.

Generally, multipath phenomena are also a serious problem in electromagnetic environments, but not as severe as in underwater sound propagation environments. There are two main reasons for this:

(1) There are generally no signals in the electromagnetic environment that are reflected multiple times between two interfaces.
(2) The propagation speed of UWA signals is only about 1500 m/s, which is 1/200000 of the propagation speed of free space electromagnetic waves.

Because of the above two reasons (especially the second one), the multipath expansion in the UWA channel is much serious than in the radio channel. And even a minor path differences can cause considerable multipath delay expansion, limiting the transmission speed and transmission quality of UWAC.

1.5.2 Doppler Effect

The Doppler effect is a common physical phenomenon that exists between objects in relative motion and is represented as a change in frequency or wavelength of a wave. The wave can either be a mechanical wave (such as an acoustic wave), or an electromagnetic wave.

The main cause of Doppler is the relative motion between receiver and transmitter, as the ratio of the relative velocity of transmitter/receiver to that of sound speed is often a few parts per thousand. This ratio is five orders of magnitude larger than the ratio in a radio environment. Thus, the Doppler effect on the system is much more severe in UWACs than in the radio communications environment. In addition to the relative motion between the receiver and transmitter, both surface wave motion and surface turbulence can also cause Doppler shifts. Sea surface wave motion is the most dominant factor and increases with increase in sea breeze/wind level. The Doppler rate expansion is defined as the maximum difference in Doppler rates along the propagation path and is expressed in Eq. (1.48):

$$D_d = \max \left\{ \frac{|v_p - v_q|}{c} \right\}, \forall p, q, \tag{1.48}$$

where v_p is the Doppler rate of the pth path, i.e., the change rate of pth channel propagation length, c is the speed of underwater sound.

1.5.3 Transmission Loss

UWA signals in seawater lose energy as they propagate due to the phenomenon of acoustic absorption in seawater. The nature of acoustic absorption is attributed to the shear viscosity, volumetric viscosity, and relaxation phenomena of the seawater medium.

This frequency-dependent absorption loss can limit the UWA channel bandwidth and limit the communication distance of the UWAC system.

Another type of transmission loss is the expansion loss, which is similar to the free space propagation loss in an electromagnetic environment. However, given the relatively limited spatial extent of the underwater environment, expansion losses are generally divided into columnar and spherical expansions. Columnar expansion losses are proportional to the expansion distance, and spherical expansion losses are proportional to the square of expansion distance.

Besides, the presence of sediment, air bubbles, and organisms in the submerged medium causes inhomogeneity that leads to scattering, also results in transmission losses. These factors play a secondary role in transmission loss, while the aforementioned acoustic absorption and expansion losses are the main factors.

1.5.4 Environmental Noise

There are various background sound sources in a marine environment, from both natural environment, and from human activities. The noise caused by human activities is mainly mechanical noise from navigation. Normal UWACs are likely to be interrupted by the presence of a moving vessel in the vicinity. Noise from the natural environment mainly comes from hydrodynamic noise, marine biological noise, underwater earthquakes, the sound of wind, rain, and waves on the water surface. The presence of noise limits the communication quality and range of UWAC systems. Theoretically, increasing the transmitting power can mitigate the effect of noise, but increasing the transmitting power is not always feasible. When calculating system performance indexes for UWAC systems, the calculation of deep-

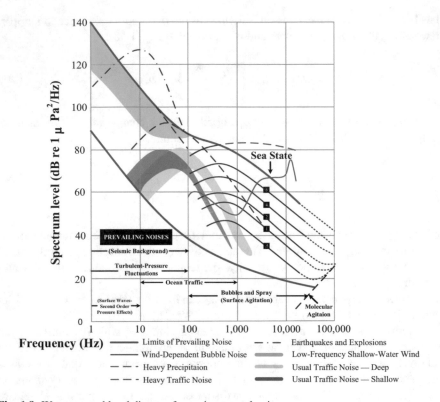

Fig. 1.8 Wenz spectral level diagram for environmental noise

sea environmental noise is usually obtained by looking up the main parameters in a table. The deep-sea environmental noise spectral level is shown in Fig. 1.8.

Generally speaking, in the frequency band between 500 Hz and several tens of kHz, marine ambient noise decreases by approximately 5 dB per octave. In the frequency band below 500 Hz, ship navigation noise is an important source of the noise. Particularly in and around harbors, ship noise and noise from coastal industrial activities are the dominant noises. This noise is a shallow ambient noise, and this mixing varies with time and location.

1.5.5   Channel Time Variation

The speed at which underwater sound waves travel in the ocean is not constant. It is mainly depth and distance-dependent, while somewhat dependent on geographical location and weather. The temperature and salinity of the seawater also have an effect on the speed of sound. Together with the fact that seawater is dynamic in terms of organisms, currents, and waves, all these factors make the channel behave as time-varying.

The relationship between sound velocity and temperature T (°C), depth z (m) and salinity S (‰) is given in the first subsection of this chapter and is specified in Eq. (1.1).

Simultaneously, internal waves, vortices, temperature gradients, density stratification, and refractive diffraction effects together cause the channel environment to be dependent on time, space, and frequency fluctuations.

The Doppler effect and multipath propagation also cause time variation in the channel, which is similar to the radio environment.

1.5.6 Propagation Time Delay

Propagation delay is the time taken by a signal/wave to travel a certain distance through a transmission medium, i.e., the total time elapsed from the time the data is sent by the sender until it is received by the receiver (or from the time the receiver sends an acknowledgment frame until the sender receives it).

The channel delay spread is defined as the maximal difference in the times-of-arrival of channel paths:

$$D_\tau = \max\left\{\left|\tau_p - \tau_q\right|\right\}, \ \forall p, q. \tag{1.49}$$

In UWAC, the slow speed of sound waves and the significant multipath phenomenon results in a large channel delay extension. For example, a difference of 15 m between the arrival positions of two objects results in a 10 ms difference in arrival time (here we assume that sound travels at 1500 m/s). In shallow water, the typical delay spread is around a few tens of milliseconds; in deeper water, the delay spread can be measured in seconds. For UWACs, large delay spreads lead to severe inter-code interference caused by waveform time dispersion (also known as time propagation).

1.6 Classification of UWA Channels

The channel is a general term for the transmission medium between the sending and receiving ends. From the perspective of studying message transmission, the channel can be divided into a narrow channel and a wide channel. The narrow channel is the physical transmission medium, while the generalized channel includes the sending and receiving devices in addition to the narrow channel, and the various parts between the source and the host are seen as channels, which simplifies the analytical model. However, regardless of the type of generalized channel, the narrow channel is the most important and fundamental component of the generalized channel.

And its characteristics are the main factor affecting the quality of communication and the main issue to be considered in designing the communication system and determining the engineering implementation plan [8].

From the communication theory point of view, the ocean is the acoustic channel, that carry out long-distance information transmission. It is much more complex than a radio channel. The distribution of sound velocity in ocean, seabed, and sea surface are the main factors affecting acoustic propagation. The ocean channel belongs to a non-uniform, double-interface, random, inhomogeneous medium channel and moreover, a time dispersed slow fading channel. The increase in propagation distance and signal frequency will cause energy loss. The available bandwidth is only a few kilohertz as UWA channel capacity is less. The propagation process of time-varying space change while multipath effect is also serious. There are two main effects of acoustic channels on UWAC systems: firstly, the way in which sound propagates in the ocean and the average loss of energy propagation leading to signal attenuation at the receiving end; secondly, the transformations carried out on the signal, deterministic transformations leading to distortions in the received waveform and random transformations leading to information loss.

1.6.1 Coherent Multipath Channels

The oceanic channel is a space-varying, frequency-varying, time-varying, random channel that transforms the energy of the target signal (acoustic propagation loss) while also transforming the emitted waveform of the source. Thus the acoustic channel can be regarded as a time-varying, space-varying random filter that transforms the emitted waveform. However, if the observation or processing time is short, the acoustic channel can be considered as a time-invariant filter in most applications, and the acoustic channel is experimentally proven to be a slowly time-varying coherent multipath channel. The acoustic signal arrives at the receiver point along different paths in the ray acoustic perspective, and the total received signal is an interferometric superposition of the signals transmitted by all the sound lines passing through the receiver point. The coherent multipath channel model is one in which the medium and boundaries are time-invariant, and the source and receiver locations are determined. The signals from the source arrive at the receiver point along with different pathways, which interfere with each other and superimpose, resulting in complex spatial interference patterns and complex filtering characteristics. The sound field's intensity and the received waveform in a multipath channel are determined by the eigenwave parameters. An eigenwave is a collection of all the sound lines that reach the reception point and make a significant contribution to the sound field.

Sound waves are micro-amplitude waves that satisfy the superposition theorem, so it is rational to assume that the coherent multipath channel is linear and can be described by a linear time-invariant filter.

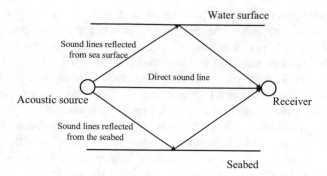

Fig. 1.9 Schematic diagram of the multipath channel model

In a coherent multipath channel, the acoustic signal arrives at the reception point at different moments along different paths of the sound lines, and the total received signal is an interferometric superposition of the signals transmitted by all the individual sound lines passing through the reception point, producing a complex spatial filtering characteristic, as shown in Fig. 1.9. Marine environment and signal frequency affect the formation of the multipath effect. And the multipath effect is formed by the following mechanisms: in shallow water, the reflected energy at the interface is the main cause; in deep water, the bending of sound lines with different grazing angles is the common cause. The multipath effect causes a temporal expansion of the signal, sometimes up to several hundred milliseconds in shallow channels, and from tens of milliseconds to several seconds in deep-sea SOFAR (sound fixing and ranging) channels.

Channel Impulse Response (CIR) function $h(t)$ for a multipath channel is

$$h(t) = \sum_{i=1}^{N} A_i \delta\left(t - \tau_{0i}\right), \tag{1.50}$$

where N is the number of intrinsic sound lines passing through the reception point, A_i and τ_{0i} are the signal amplitude and signal time delay of the ith route to the reception point, respectively.

Performing the Fourier transform on both sides of Eq. (1.50) gives

$$H(f) = \sum_{i=1}^{N} A_i e^{-j2\pi f \tau_{0i}}. \tag{1.51}$$

If the sea depth, the profile shape of the seabed, the sound velocity profile, and the relative geometry of the source and receiver points are known, the specific form of the coherent multipath channel system function can be derived from Eq. (1.50) or Eq. (1.51) by simply calculating the sound line parameters.

The interference of the multipath effect may not only cause the frequency characteristics of the UWA channel to appear as interphase "passbands" and "stopbands," much like a "comb filter," but may also lead to a certain degree of interference. This can also lead to interference in the phase extinction or phase length of the signal waveform at a certain frequency, resulting in a significant loss of energy at that frequency and thus affecting the judgment of the information. In addition, because the sound field at the receiver point is the result of an interfering superposition of multiple arrivals, the system function of the channel is sensitive to changes in environmental parameters and the relative position of the source and receiver. The order of sensitivity of the channel's system function to environmental parameters is leading position variation, water layer thickness variation, horizontal position variation, and sound velocity variation in the water layer.

The unique multipath expansion of the UWA channel can open up new opportunities if it is exploited. The sensitivity of the channel system function to the environmental parameters' changes and the relative position of the source and receiver is an important guide to the study of single-array multi-user UWACs. At the same time, mastering the time-varying and space-varying properties of the channel system function is also of great significance for the realization of high-quality mobile UWACs.

1.6.2 Shallow Sea Acoustic Channels

A shallow sea implies a horizontal propagation distance of several times the depth of the seawater. In a geographical sense, shallow seas are inland seas such as harbors and bays and offshore on the continental shelf, which often extends outwards to the continental shelf's edge. Most of the country's waters are shallow seas, and the shallow sea depths of the continental shelf are all within 200 m. When sound travels at a certain depth in the ocean, the sound waves are reflected many times by the surface and the seafloor and are confined between the upper and lower boundaries of the ocean. The acoustic properties of the surface and the seafloor have an important influence on the sound field. The shallow sea acoustic propagation conditions are more severe than those of the deep sea.

In shallow seas, many physical parameters of the sea surface, seawater medium, and sea bed determine the sound propagation loss. UWA channels are greatly affected by boundary conditions (sea surface and bottom) and sea temperature distribution. Among them, the seabed has a particularly strong influence on sound propagation. The fluctuation of sound signals is one of the unique characteristics of ocean propagation. Even if the same signal is sent at the same location, the signal arriving at a fixed receiving point will change with time. The first reason is that the seawater itself is uneven, the temperature is uneven, and the seawater is in turbulent motion, which causes time-varying interference effects between various propagation paths. Secondly, because of the surface reflection sound fluctuates greatly at close range. However, as the distance increases, the fluctuation tends to decrease. This is

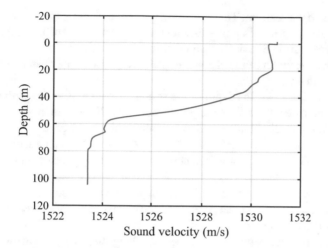

Fig. 1.10 Distribution of sound velocities in shallow waters

because, as the glancing angle on the surface becomes small, the sea surface's role tends to become more and more unstable, and the influence of the sea surface tends to increase, and it tends to be more and more comprehensive. On the other hand, shallow seabed's acoustic characteristics are also very complex, usually stratified by density or sound speed, which seriously affects the propagation loss.

Figure 1.10 gives the measured sound velocity distribution for a shallow sea with a sea depth of about 105 m.

1.6.3 Deep-Sea Acoustic Channels

The deep-sea acoustic channel is characterized by the distribution of sound velocities in the deep sea, which is mainly characterized by the presence of sound velocity profiles that refract the sound lines upwards. If the effect of the seafloor on sound propagation is negligible, the sea area under consideration is the theoretical deep sea, where refraction-refraction sound lines and refraction-surface reflection sound lines are important to sound paths. In the deep ocean, there are three main effective ways to achieve long-range propagation: deep ocean sound channel (SOFAR sound channel) propagation, convergence zone propagation, and surface waveguide propagation.

The sound velocity distribution in the deep sea has a minimum value, and the water layer in which it is located is called the sound channel axis. The refraction effect determines that the sound lines tend to bend towards the water layer where the sound velocity is lower during propagation, so that in the deep-sea sound channel (also known as the SOFAR channel), part of the sound lines that start at the source is retained in the channel without the loss of sound energy caused by reflections from the surface and the seafloor. The area along the channel axis is the convergence zone, and its propagation loss is small. Therefore, especially when the sound source is located near the channel axis, the sound signal can travel far along the channel axis. And the effective width of the channel impulse response in the channel convergence area is very small.

In different sea areas, the depth of the sound channel axis varies with seasons. In the South China Sea, the average depth of more than 2000 m, there are typical deep-sea sound velocity distribution and year-round, the sound channel axis is located in 1000–1200 m. In the Pacific Ocean, the Atlantic Ocean, and other deep-sea areas, the sound velocity profile in a certain depth bending results in refraction and reflection of sound waves. The spread of acoustic energy is also concentrated in a relatively small area (convergence zone) when the waveguide also has an invisible boundary. In the Arctic, the deep-sea acoustic channel axis is located at or near the surface of the ice cover.

In current practical applications, the ultra-long-distance propagation of sound waves in the acoustic channel transmits distress signals from wrecked aircraft and ships and monitors earthquakes, volcanic eruptions, and tsunamis underwater. Therefore, if communication nodes are laid on the sound channel axis, waves can be transmitted thousands of kilometers away in that channel in a manner similar to the propagation of light waves in optical channels, enabling long-range UWA.

In the deep-sea acoustic channel, the impulse response function of the channel is very stable. The acoustic signal has small fluctuations and can be transmitted over long distances with high intensity and low distortion. For UWAC, if the transceiver and transmitter nodes are placed near the acoustic channel axis, on the one hand, the inter-code interference caused by multipath expansion is small, and on the other hand, the propagation loss is small. The use of deep-sea channel characteristics will facilitate the realization of long-range, high-quality UWACs.

In addition to deep ocean acoustic path propagation, convergence zone propagation, and surface waveguide propagation are also effective ways to achieve long-range communication in the deep ocean.

The convergence zone can only be formed if the depth of the sea is large enough for deep refraction paths to existing in the ocean. The necessary condition is that the speed of sound near the seafloor is greater than or equal to the speed of sound at the source, in which case sonar equipment near the surface can take advantage of the convergence zone effect. It is called convergence zone propagation because the sound waves emitted from the source near the sea surface form a downward beam, and the sound energy propagates along a refracted sound line that neither touches the sea surface nor the seafloor. The distance between the high-intensity zones is called the convergence zone distance, and it is the most important parameter for the use of sonar, and this distance varies greatly with geographical location.

Sea surface waveguide propagation uses the surface sound channel formed by the isothermal layer of the sea surface for sound propagation. The isothermal layer shows a positive gradient sound velocity distribution because the static pressure makes the sound velocity increase slightly with depth, and there are good conditions for sound propagation when the layer thickness is large. The isothermal layer is usually produced in a mild and windy ocean and is maintained by the agitation of the sea by wind and waves. The isothermal layer can extend deeper after a major storm and disappear after a long period of calm wind. This shows that the surface layer varies significantly with the weather, sea state, day and night, and season and is not a very stable waveguide.

In July 2010, the "Jiaolong" manned submersible dived to a depth of 3759 m. And in July 2011, it dived to a depth of 5182 m. In July 2011, at a depth of 5182 m, it successfully placed the logo of the China Oceanic Association and a carved wooden Chinese dragon on the bottom. And subsequent tests will be carried out to a depth of 7000 m. In addition, the development of marine resources also requires the support of deep-sea communications. Therefore, making full use of the deep-sea acoustic channel's characteristics for acoustic communication will become a promising way of communication to the dive and can provide information transmission services for deep-sea resource development and marine research.

1.7 History of UWACs

Like other communication technologies, UWAC technology has undergone a process of development from analog to digital [9]. Early analog UWAC technology is represented by the hydrophone, which generally uses Single Sideband (SSB) modulation technology and generally operates at 8–11 kHz. As it is analog technology, it requires a considerable amount of transmitting power, in the order of a hundred watts. Hydrophone can provide analog voice communication about two kilometers away and has not yet formed a network.

Advances in semiconductor technology, represented by ultra-large-scale integrated circuits, have dramatically changed the pace of communications technology. Many algorithms that were considered too complex and unrealistic in the past have been gradually applied with advances in semiconductor technology, with digital UWAC technology gradually replacing analog UWAC technology.

The theory of UWAC technology is relatively mature, but there are still some problems in its implementation. The UWAC technology has made great progress, to achieve a faster communication rate, higher communication quality, better communication network, but there is still need to progress in MIMO technology, coding technology, spread spectrum technology.

With the development of marine development and national defense construction, the use of UWAC technology to transmit information has greatly increased the demand. In the early twentieth century, UWA communication systems were used for the transmission of control information between surface control platforms and underwater divers. In 1945, the United States developed the first well-functioning underwater wireless telephone system for communication between submarines. Until the 1980s, UWAC was mainly used in the military sector, and its technical research and application were limited in scope. Today, however, UWAC has been widely used from military to commercial fields, and research on UWAC has attracted more and more attention. From coherent digital modulation, spread spectrum technology, adaptive equalization to multi-carrier modulation, multiple inputs and multiple-output systems, and underwater communication network technology, various communication technologies have all become research areas for UWAC technology.

From the perspective of communication, UWAC systems and radio communication systems have many similarities in terms of channel characteristics or system composition, but in terms of performance indicators and technical implementation of many aspects, the two are very different. The primarily reason is that the UWA channel has complex and variable transmission characteristics, presenting many challenges for the high data rate and high reliability of UWAC. It also makes the development of UWAC technology to lag behind radio communication technology. Since 1980s, with the emergence of the commercial value of UWAC systems and the advancement of UWAC technology, especially after the successful application of coherent detection communication technology, UWAC systems have been rapidly developed. And many new research topics have emerged in terms of channel characteristics and communication technology. The latest results of these research topics have, in turn, contributed to the development of UWAC systems and expanded their applications, making UWAC a new research discipline that integrates UWA physics and wireless communication technology.

? Questions

1. Try to explain what factors are involved in the variation of sound speed in seawater in UWACs and how they affect sound speed variation.
2. What are the main differences between UWAC and wireless communication?
3. What are the characteristics of UWA channels? What are the categories into which they can be divided? What are the reasons for their formation?
4. Use Wilson's formula to calculate the sound speed in the following cases: (1) The temperature is $20\,^\circ\text{C}$, the salinity is $20\%_{00}$, and the depth is $1\,\text{m}$. (2) The temperature is $15\,^\circ\text{C}$, the salinity is $20\%_{00}$, and the depth is $20\,\text{m}$.
5. At $5\,\text{kHz}$ ($\alpha = 0.3\,\text{dB/km}$) and $20\,\text{kHz}$ ($\alpha = 3\,\text{dB/km}$), if the unidirectional propagation loss is $80\,\text{dB}$. Consider the following cases:

   (1) spherical expansion plus absorption loss;
   (2) column expansion plus absorption loss.

   What is the detection distance in each case?

References

1. D.B. Kilfoyle, A.B. Baggeroer, The state of the art in underwater acoustic telemetry. IEEE J. Ocean. Eng. **25**(1), 4–27 (2000)
2. M. Stojanovic, Recent advances in high-speed underwater acoustic communications. IEEE J. Ocean. Eng. **21**(2), 125–136 (1996)
3. M. Stojanovic, J.A. Catipovic, J.G. Proakis, Phase-coherent digital communications for underwater acoustic channels. IEEE J. Ocean. Eng. **19**(1), 100–111 (1994)
4. P.C. Etter, *Underwater Acoustic Modeling: Principles, Techniques and Applications* (CRC Press, Boca Raton, 1995)
5. M. Stojanovic, J. Catipovic, J.G. Proakis, Adaptive multichannel combining and equalization for underwater acoustic communications. J. Acoust. Soc. Am. **94**(3), 1621–1631 (1993)
6. J. Catipovic, A. Baggeroer, K. Von Der Heydt et al., Design and performance analysis of a digital acoustic telemetry system for the short range underwater channel. IEEE J. Ocean. Eng. **9**(4), 242–252 (1984)
7. M. Chitre, S. Shahabudeen, L. Freitag et al., Recent advances in underwater acoustic communications & networking, in *OCEANS 2008* (IEEE, Piscataway, 2008), pp. 1–10
8. L. Freitag, M. Stojanovic, S. Singh et al., Analysis of channel effects on direct-sequence and frequency-hopped spread-spectrum acoustic communication. IEEE J. Ocean. Eng. **26**(4), 586–593 (2001)
9. B. Li, J. Huang, S. Zhou et al., MIMO-OFDM for high-rate underwater acoustic communications. IEEE J. Ocean. Eng. **34**(4), 634–644 (2009)

Chapter 2
Modulation Technology in UWAC System

2.1 Introduction

Modulation is the process of making the specific parameters of the carrier signal change in response to the baseband signal, which is essentially a spectrum shift. The roles and purposes of modulation include: (a) converting the modulated signal(baseband signal) into a modulated signal (frequency band signal) suitable for channel transmission; (b) realizing channel multiplexing and improving channel utilization; (c) reducing interference and improving system immunity; and (d) realizing interchange between transmission bandwidth and signal-to-noise ratio [1, 2].

Since the acoustic signal is low frequency, the modulated bandpass sampling signal is usually generated directly in $x[n]$. From the perspective of continuous modulation or discrete modulation for the parameters of the carrier signal, modulation technology can be divided into analog modulation and digital modulation.

Compared with space radio communication, the research of UWA started relatively late. The research progress of its various technologies thus also lags behind radio communication. The modulation techniques used in UWAC system are similar to those used in radio communication. Compared with radio communication, the biggest difference in UWAC is the channel. The previous chapter has introduced the characteristics of the UWA channel. It is highly noisy, strongly reverberant, narrow channel bandwidth, severe multipath effects, time-varying, space-varying, and frequency-varying channel. When wireless communication modulation technology is applied to a UWA channel, its complexity will change accordingly. Early UWAC mainly used analog frequency modulation to meet simple data communication. The UWAC has not been further developed well until the 1970s.

As the UWAC detection area extends into deep water, new requirements are placed on the detection depth and detection distance of UWA equipment. This requires high data transmission rate for UWAC. The UWAC modulation method then transforms from the analog domain to the digital domain.

Digital communication has few advantages such as, robust to interference, the capability of using regenerative relays, encryption, switching, easy equipment integration, etc. The development of digital technology also promotes the development of various digital modulation technologies. Digital modulation technologies mainly include ASK, FSK, PSK, and other modulation methods derived from them.

2.2　Amplitude Shift Keying Modulation

Amplitude shift keying (ASK) is a modulation technology that uses digital baseband signals to control carrier amplitude for information transmission. Due to the impact of amplitude fluctuation of UWA signal, it is difficult for multistage ASK to set appropriate detection threshold during demodulation. Therefore, ASK in UWAC mostly uses on–off keying (OOK), which is the simplest amplitude keying. The OOK signal can be expressed as

$$
S_{OOK}(t) = \begin{cases} s(t)\cos(2\pi f_c t + \theta) & \text{(send 1)} \\ 0 & \text{(send 0)} , \end{cases} \tag{2.1}
$$

where $s(t)$ is the digital baseband signal, and f_c is the carrier frequency.

The modulation schematic diagram and waveform diagram are shown in Fig. 2.1. Figure 2.1a is the OOK signal obtained by analog multiplication, Fig. 2.1b is the OOK signal obtained by digital keying, and Fig. 2.1c is the modulated time-domain waveform, which uses unipolar non-return to zero codes to control the carrier amplitude. When the symbol is 1, the carrier is sent out, and when the symbol is 0,

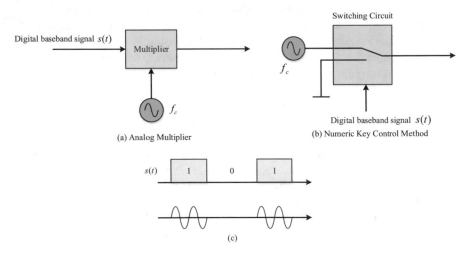

Fig. 2.1 OOK modulation scheme

the carrier is not sent out. On–off keying mode is adopted in the telegraph working mode of 660 communication sonar in China. But the power utilization rate of this method is very low, and the ability to resist multipath interference is very weak, so it is only suitable for occasions where the required communication rate is not too high.

There are two types of demodulation methods. One is the non-coherent demodulation method (i.e., envelope detection method), which recovers the original modulated signal directly from the amplitude of the modulated wave. The other is the coherent demodulation method (i.e., synchronous detection method), which is applied to linear modulated signals. The critical aspect of implementing coherent demodulation is to restore a coherent carrier at the receiver that is strictly synchronized with the modulating carrier. The coherent reference signal is then multiplied by the carrier frequency using a multiplier. The performance of the recovered carrier is directly related to the demodulation performance of the receiver. The corresponding demodulation principle block diagram is shown in Fig. 2.2, where (a) is the non-coherent demodulation method and (b) is the coherent demodulation method.

Because of the low power utilization rate of ASK, most modern digital UWAC systems use FSK or PSK modulation mode. These two common modulations are further introduced in Sects. 2.3 and 2.4.

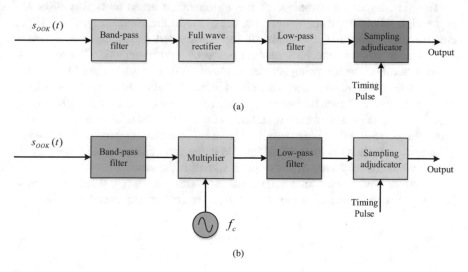

Fig. 2.2 Block diagram of demodulation principle. (**a**) Non-coherent demodulation. (**b**) Coherent demodulation

2.3 Frequency Shift Keying Modulation

Frequency shift keying (FSK) modulation is a non-coherent modulation. This modulation is highly adaptable to the time and frequency expansion of the UWA channel. A typical multi-frequency keyed UWAC system is given in a paper published by J. Catipovic et al. in 1989. The carrier band of the system is 20–30 kHz, and the maximum data transmission rate is 5 Kbps. The frequency band of the system is divided into 16 subbands. In each subband, a 4FSK signal is transmitted. Thus, out of the total of 64 channels, 16 of them are used simultaneously to transmit 32 bits of information in parallel. The system can successfully transmit data in the horizontal shallow channel of 4000 m and the deep vertical channel of 3000 m with a false bit rate in the order of 0.01–0.1. However, the shortcomings of FSK modulation include wide bandwidth, low band utilization, and the requirement of a high signal-to-noise ratio. Moreover, in case of Doppler shift, a certain degree of frequency redundancy must be set so that the limited bandwidth of the UWA channel cannot be fully utilized. In addition, although FSK modulation does not need to face the problem of carrier phase recovery, it does not solve the inter-code interference caused by multiple channels. Some systems insert a certain interval between consecutive code elements to eliminate inter-code interference, causing a reduction in communication rate.

The most important advantage of the FSK method based on energy detection is the high reliability of the communication. Due to the complexity of the UWA channel, the phase and amplitude of the transmitted signal can cause severe distortion. In contrast, the FSK method based on energy detection is immune to phase distortion. The following is an example of 2FSK to explain the FSK signal.

FSK uses different frequency carriers to transmit digital information. Similarly, 2FSK uses two different frequency carriers to represent two levels in the digital signal. That is, the message corresponding to the levels 1 and 0 loads at two different signal frequencies. When the transmitted level is 1, the carrier signal with frequency f_1 is sent, and when the transmitted level is 0, the carrier signal with frequency f_2 is sent, while the amplitude of both carrier signals is the same.

Therefore, the carrier frequency changes with the binary baseband signal between the two frequency points f_1 and f_2, which can be expressed as

$$s(t) = \begin{cases} A \cos(2\pi f_1 t + \theta_1) & \text{(send 1)} \\ A \cos(2\pi f_2 t + \theta_2) & \text{(send 0)}. \end{cases} \tag{2.2}$$

The generation of 2FSK signal is shown in Fig. 2.3a, and the waveform of 2FSK signal is shown in Fig. 2.3b.

The binary FSK signal can be decomposed into two 2ASK signals where each 2ASK signal uses a unique carrier frequency. Therefore, the time-domain representation of the 2FSK signal has the following form:

$$s(t) = s_1(t) \cos(2\pi f_1 t + \theta_1) + s_2(t) \cos(2\pi f_2 t + \theta_2). \tag{2.3}$$

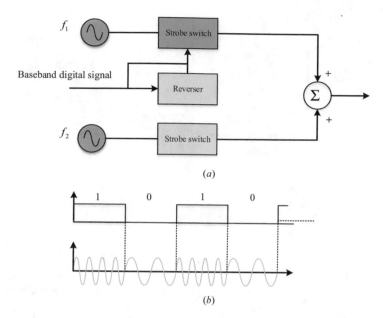

Fig. 2.3 2FSK signal generation block diagram and waveform diagram

The demodulation mode of the FSK signal can be divided into coherent demodulation and non-coherent demodulation. In practice, non-coherent demodulation is the main method. The demodulation principle block diagram is shown in Fig. 2.4, where (a) is the non-coherent demodulation mode, and (b) is the coherent demodulation.

Figure 2.4a shows the principle block diagram of non-coherent demodulation. It can be seen that it is actually composed of two ASK receivers. The principle is to use a pair of bandpass filters to distinguish the central frequencies of f_1 and f_2 signals. When ignoring the influence of noise, assuming a carrier pulse of frequency f_1 is received, the envelope detector I outputs a non-zero envelope $\rho_1(t)$ to the sampler decision, and $\rho_2(t)$ is approximately equal to 0. When a carrier pulse of frequency f_2 is received, the envelope detector II outputs a non-zero envelope, and $\rho_2(t)$ is approximately equal to 0. The detected envelopes $\rho_1(t)$ and $\rho_2(t)$ are decided by the decision generator at the sampling decision time t_0. If $\rho_1(t) - \rho_2(t) > 0$, it is judged as 1; otherwise, it is judged as 0. Thus, the decision circuit recovers the baseband digital signal according to the positive and negative polarity of the sampling value at time t_0. In the presence of noise, the above acceptance process is the same, except that the influence of noise may lead to a wrong decision.

Figure 2.4b is the principle block diagram of coherent demodulation. By comparing (a) and (b), it can be seen that the two envelop detectors in (a) are replaced by two coherent demodulation modems in (b). Obviously, their working principle is similar to that of non-coherent demodulation. Non-coherent demodulation systems are used more commonly as compared to coherent demodulation systems because

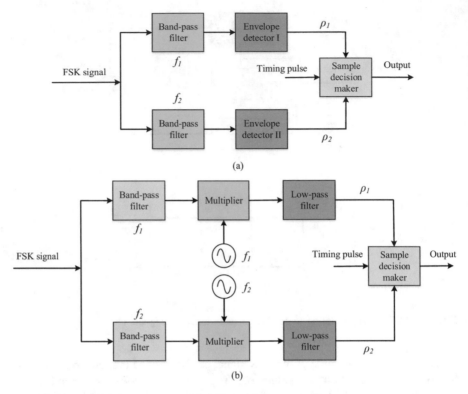

Fig. 2.4 FSK demodulation principle block diagram. (**a**) Non-coherent demodulation. (**b**) Coherent demodulation

of the simplicity of non-coherent demodulation equipment. Coherent demodulation generates two local carriers with reference to the transmitted FSK signal carriers, which in practice is more complex.

2.4   PSK Modulation

Phase-shift keying (PSK) is a modulation technique that represents the input signal information in terms of the carrier phase. PSK modulation has been used in UWAC since the 1980s. It can be divided into absolute phase-shift keying modulation and differential phase-shift keying (DPSK) modulation, which is one of the fastest developing and most successful fields. PSK is used to directly represent digital information during transmission. Its performance of frequency band utilization and bit error rate in white noise is the best among digital modulation techniques. However, correct phase reference should be obtained when the PSK signal is received. Otherwise, serious errors of the demodulation signal will be caused.

DPSK does not require carrier extraction and is superior to PSK in resisting frequency drift, multipath, and slow phase jitter. Therefore, in its initial use, coherent communication relies mainly on DPSK. Compared with FSK, PSK has higher energy and spectral efficiency, but it faces the problem of carrier recovery due to the multipath phenomenon. Therefore, the PSK modulation communication technology is limited to the deep-water vertical channel. For the UWA channel with multipath effect such as near horizontal channel, the PSK technology becomes more and more unstable. Binary absolute phase-shift keying and binary differential phase-shift keying are introduced in the following sections.

2.4.1   Binary Phase-Shift Keying Modulation

For an ideal binary phase-shift keying (2PSK) modulation signal, the carrier of the initial phase 0 and π is usually used to represent the binary 1 and 0 in the digital baseband signal, respectively. So its expression is

$$S_{2PSK}(t) = \begin{cases} A\cos(2\pi f_c t + 0) = A\cos 2\pi f_c t \\ A\cos(2\pi f_c t + \pi) = -A\cos 2\pi f_c t. \end{cases} \tag{2.4}$$

The modulation principle block diagram of 2PSK signal is shown in Fig. 2.5a, b. Wherein, $s(t)$ is the digital baseband signal. Since the signal waveform representing

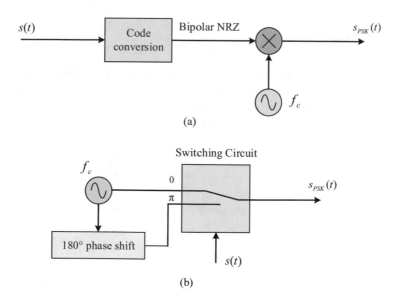

Fig. 2.5 Modulation principle block diagram and time waveform of 2PSK signal. (**a**) Analog method. (**b**) Key control method

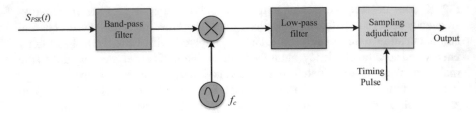

Fig. 2.6 Principle block diagram of 2PSK coherent demodulation

the two codes is the same, and the polarity is opposite, the time expression of the 2PSK signal can be written as

$$S_{2PSK}(t) = s(t) \cos 2\pi f_c t. \tag{2.5}$$

In Eq. (2.5), $s(t)$ is a bipolar non-return to zero signal. The demodulation mode of PSK signal is coherent demodulation method, and its demodulation principle block diagram is shown in Fig. 2.6. A similar working principle to that of FSK coherent demodulation can be seen in Fig. 2.6. However, when applying this method for demodulation, it should be noted that this method has the phenomenon of "inverted work." Because in the process of 2PSK signal carrier recovery, there is 180° phase ambiguity. In practice, we often use binary differential phase-shift keying to overcome this phenomenon.

2.4.2 DPSK Signal

Binary differential phase-shift keying (2DPSK), also called relative phase-shift keying, uses the relative phase changes of the adjacent codes' carriers to transmit digital information. Assuming that the initial phase difference between two adjacent symbols is $\Delta\varphi$, the relation between digital information and $\Delta\varphi$ can be defined as

$$\Delta\varphi = \varphi_k - \varphi_{k-1} = \begin{cases} 0 & \text{(defined 0)} \\ \pi & \text{(defined 1).} \end{cases} \tag{2.6}$$

To illustrate this relationship, use the following example:

Binary digital information:		1	1	0	1	0	0	1	1	0;
2DPSK signal phase:	(0)	π	0	0	π	π	π	0	π	π;
or	(π)	0	π	π	0	0	0	π	0	0.

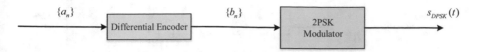

Fig. 2.7 2DPSK modulation principle block diagram

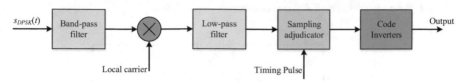

Fig. 2.8 Block diagram of coherent demodulation principle

The generation method of 2DPSK signal is to carry out code transformation (differential coding) of absolute code $\{a_n\}$ to get relative code $\{b_n\}$ (differential code) and then carry out absolute phase modulation to generate 2DPSK signal, as shown in Fig. 2.7.

In Fig. 2.7, the differential encoder converts absolute code into relative code. And its coding rule is $b_n = a_n \oplus b_{n-1}$, where $\oplus$ is binary addition, and b_{n-1} is the preceding code of b_n. The initial b_{n-1} can be set by itself.

The demodulation method of DPSK signals can also be categorized into coherent demodulation and non-coherent demodulation. Coherent demodulation, also called code inversion method, mainly compares the polarity of code elements. The demodulation principle block diagram is shown in Fig. 2.8.

According to Fig. 2.8, the demodulation principle of 2DPSK is to perform coherent demodulation and recover the relative code. Then, it is reduced to the relative code by a code inverse converter to recover the sent binary digital information. Due to the ambiguities of carrier phase, the demodulated relative code may have 1 and 0 inversions. Since the absolute code obtained by differential decoding does not have any inversion, thus the problem caused due to the carrier phase ambiguity does not arise here.

Another non-coherent demodulation method for 2DPSK is the differential coherent demodulation method, also known as the phase comparison method. Its demodulation principle block diagram is shown in Fig. 2.9. Compared with the coherent demodulation method, this demodulation method does not need a particular coherent carrier. In contrast, the coherent demodulation method requires accurate carrier synchronization, especially in the severe multipath delay UWA channel. As this synchronization is more difficult to achieve, the non-coherent demodulation method is more common in practical applications. In this demodulation mode, the multiplier can be considered as a phase comparator, and its result reflects the phase difference between adjacent codes. The original digital information can be recovered directly after processing by low-pass filtering and sampling decision, so the code reverse converter is not needed.

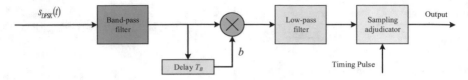

Fig. 2.9 Schematic diagram of differential coherent demodulation

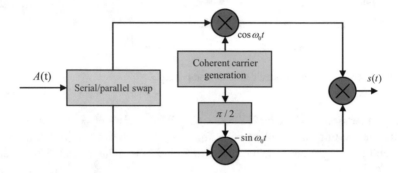

Fig. 2.10 Block diagram of QPSK modulation principle

2.4.3 QPSK Signal

QPSK is commonly used in UWA coherent communication. When using this modulation method to communicate, the receiver must first restore the carrier, that is, to realize carrier synchronization and then carry out orthogonal demodulation to move the signal to the baseband for processing. QPSK is a further extension of the PSK concept where the number of phases modulated is not limited to two. The carrier can carry any amount of phase information, and the received signal is demodulated by multiplying it by a sine wave of the same frequency to obtain the phase-shift information. Using QPSK, the carrier can have four different phase shifts (four chips). Each chip can represent two binary bytes. This modulation method enables the same carrier to transmit two bits of information instead of the original one, which doubles the carrier's band utilization. QPSK signal can be expressed as

$$
S_{QPSK}(t) = \begin{cases} A\cos(2\pi f_c t + \frac{\pi}{4}) & 11 \\ A\cos(2\pi f_c t + \frac{3\pi}{4}) & 01 \\ A\cos(2\pi f_c t - \frac{3\pi}{4}) & 00 \\ A\cos(2\pi f_c t - \frac{\pi}{4}) & 10. \end{cases} \tag{2.7}
$$

The principle block diagram of QPSK modulation is shown in Fig. 2.10. First, the input binary data sequence is converted to polar form by a non-return to zero (NRZ) level coder. The binary waveform is partitioned by a divider into two independent binary waveforms consisting of the odd and even bits of the input sequence for modulating a pair of quadrature carriers. This results in a pair of binary

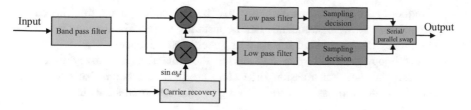

Fig. 2.11 Block diagram of QPSK demodulation principle

PSK signals. The orthogonality of $\phi_1(t)$ and $\phi_2(t)$ allows these two signals to be detected independently. In the final stage, the two binary PSK signals are summed to obtain the QPSK signal.

Because QPSK can be viewed as a composition of two orthogonal 2PSK signals, it can be demodulated using a demodulation method similar to that of 2PSK signals. In other words, it consists of two coherent demodulators of 2PSK signals (Fig. 2.11).

2.5 Spread Spectrum Communication Technology

Spread spectrum communication technology, optical fiber communication technology, and satellite communication technology are called three high-tech communication transmission modes after entering the information age. The main characteristic of spread spectrum communication is that the signal bandwidth used to transmit information is greater than the bandwidth of the information itself to obtain spread spectrum gain to resist all kinds of interference in the transmission process. Spread spectrum communication technology is a kind of communication technology that can effectively resist UWA interference. It has three main characteristics. First, the required frequency bandwidth of the signal is much larger than the minimum bandwidth needed for information transmission. Second, the expansion of the frequency band is achieved by an independent code order. The code order is realized by means of coding and modulation, independent of the transmitted information data. Third, the same code is applied at the receiver side for simultaneous reception, expansion, and recovery of the transmitted information data [3].

In terms of modulation, it differs little from conventional data communication, except that a spreading processing step is added before modulation. It expands the symbols to be transmitted with a feature code, and the expanded symbols are called code chips. At the receiving end, a despread process is also added to restore the N code slices to one symbol. Spectrum expansion is achieved in communication systems in the following ways:

(1) Direct sequence spreading: a high code rate direct sequence pseudo-random (PN) code is used to spread the low-speed data to be transmitted at the transmitter end, and the same PN code is used to despread at the receiver end.

(2) Frequency hopping spreading: the use of PN codes to control the carrier frequency in a wider frequency band constantly varies. The essence is a multi-frequency frequency-shift keying with specific pseudocodes.

(3) Time hopping spread spectrum: divide the time axis into many time slots so that the data transmission time slots are pseudo-random. This technique can be understood as time-shift keying of multiple time slices using specific pseudocode control.

This book will focus on direct sequence spread spectrum technology and frequency hopping spread spectrum technology for the introduction.

2.5.1 Processing Gain and Anti-interference Tolerance of Spread Spectrum Technology

Spread spectrum technology reduces the SNR of the receiver input by extending the signal spectrum at the transmitting end and increases the SNR by despreading. The improvement of SNR is often described by processing gain. The processing gain is usually defined as the ratio of the spread spectrum signal bandwidth B to the information bandwidth B_{in}, which is often expressed in decibels in engineering, i.e.,

$$G_P = 10 \lg \frac{B}{B_{in}}. \tag{2.8}$$

In general, the greater the spread spectrum gain, the stronger the anti-interference ability, but the relation is not that simple. The system also needs to ensure that the output has a certain SNR to work properly (such as CDMA cellular mobile communication system for 7 dB). And to deduct the system internal SNR loss L, we must introduce the concept of anti-interference tolerance M, which is defined as

$$M = G_P - \left[\left(\frac{S}{N} \right) + L \right]. \tag{2.9}$$

In the formula, G_p is the processing gain, and S/N is the signal-to-noise ratio.

2.5.1.1 Common Spread Spectrum Technology

(1) Direct Sequence Spread Spectrum Technology
In UWAC, the schematic diagram of direct sequence spread spectrum technology is shown in Fig. 2.12.

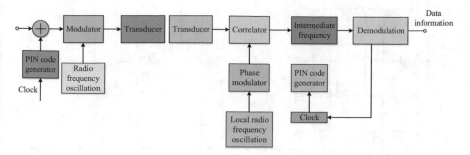

Fig. 2.12 Schematic diagram of electromagnetic wave wireless direct spread communication system

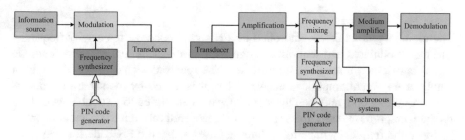

Fig. 2.13 Block diagram of FH communication technology

Direct sequence spread spectrum is the bandwidth extension of the low-speed data transmitted by a PN code with a high code rate. 2PSK (binary phase-shift keying modulation) is a commonly applied modulation method of direct sequence spread spectrum modulation in UWAC. Direct sequence spread spectrum technology can effectively counteract the multipath effect in UWAC. But there are problems to be noted: the data bits are transmitted at a very low rate. "Near and far effect" will interfere with multiple users when networking. The requirement of synchronization accuracy is higher than that of the frequency hopping (FH) spread spectrum communication, and the robustness of the direct sequence spread spectrum is not as good as that of FH spread spectrum communication.

(2) FH Spread Spectrum Communication Technology
The generation of FH signals is mainly composed of a frequency synthesizer and frequency hopping instruction generator (used to design a specific PN code). It has strong anti-interference and anti-multipath performance. In the FH system, the key technologies affecting the system are the design of FH patterns, generation of PN codes, fast, accurate, reliable synchronization capture technology, and multi-user detection technology. The principle block diagram of FH communication technology applied in UWAC is shown in Fig. 2.13.

At the transmitter, the source generates the transmit signal. After analog-to-digital conversion, the signal sent by the source is converted into a digital signal.

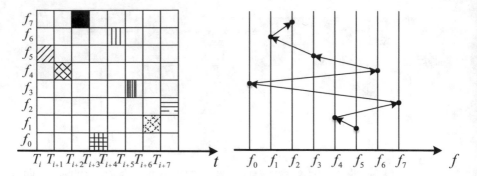

Fig. 2.14 Frequency hopping pattern of FH communication system in a certain time period

The digital signal is firstly controlled by PN code to change the carrier frequency and then modulated and sent out through the transducer. At the receiving end, the signal is received through the transducer. The received signal passes through an amplifier for signal amplification. After that, it is mixed by the same PN code as the transmitting end, and then the signal power is amplified to a suitable value by a medium amplifier. Finally, the sent signal is recovered through the demodulator.

Since the frequency hopping technology uses a specific PN code for frequency spreading and then combines the information code sequence generated by the source to form the control parameters of the frequency synthesizer, it avoids the problems encountered in the direct sequence spread spectrum system in UWAC. In the frequency hopping system, the frequency hopping is determined by the hopping pattern. The design of the frequency hopping pattern is also a key technology. A well-designed hopping pattern design is not only beneficial to anti-jamming and avoiding tracking but also necessary to meet the requirements of multiple users sharing the same frequency band. Figure 2.14 shows the frequency hopping patterns corresponding to eight frequency points (f_5, f_4, f_7, f_0, f_6, f_3, f_1, f_2) in a time slot of the frequency hopping communication system.

(3) Sweep-Spread Carrier Modulation
In all previous modulation schemes, the carrier frequency remains constant throughout the data transmission period or at least for each symbol interval. To further improve the multipath resolution, frequency oscillations can be introduced artificially, i.e., sweep-spread carrier modulation, thus converting the time delay distinction of different multipaths into frequency intervals. Therefore, this method can be considered as a special form of spread spectrum communication, which implicitly has the ability to reduce noise.

2.5.2 Advantages of Spread Spectrum Technology

(1) Strong anti-interference and low bit error rate: the wider the spectrum extended by spread spectrum communication technology, the higher the processing gain and the better anti-interference capability.

(2) Easy to use in the same frequency leading to an improved wireless spectrum utilization.

(3) Strong anti-fading and anti-multipath interference performance: Since the signal band after spreading becomes wide (much larger than the relevant bandwidth), the spectral fading during transmission is also reduced. Besides, the effective received signal in a spread spectrum system needs to fall on the right time slot. Multipath signals passing through different paths will generate arrival time differences. When the time difference exceeds the width of one code slice, it does not have much effect on the demodulation of the previously arrived signals that have been synchronized. Therefore, this will enhance the performance of the system against multipath interference.

(4) Spread spectrum communication has an encryption function, and its confidentiality is strong. Since the signal is extended to a very wide band, the power per unit band becomes very small, i.e., the power spectral density of the signal is very low, so its confidentiality is enhanced.

(5) The low power spectrum density of spread spectrum communication technology makes it possible to interfere very little with the various narrowband communication systems that already exist.

(6) With multiple access capability, it can realize code division multiple access. Using the coded orthogonal divisibility of spread spectrum sequences, people can transmit multiple user signals in the same medium, at the same frequency, and at the same time, thus realizing code division multiple access. Compared with frequency division multiple access and time division multiple access, code division multiple access is more suitable for UWAC considering the narrow bandwidth and large propagation time delay of the UWA channel.

2.6 Orthogonal Frequency Division Multiplexing (OFDM) Technology

Early wireless networks or mobile communication systems used single-carrier modulation, modulating the signal (voice or data) to be transmitted onto a single carrier and then transmitted by an antenna. If the signal is hidden in the carrier's amplitude, there are AM and ASK modulation systems. If the signal is hidden in the carrier frequency, there are FM and FSK modulation systems. If the signal is hidden in the carrier phase, there are PM and PSK modulation systems. These are single-carrier systems, which are easy to cause signal distortion due to the unsatisfactory channel characteristics in the broadband service of high-speed data transmission.

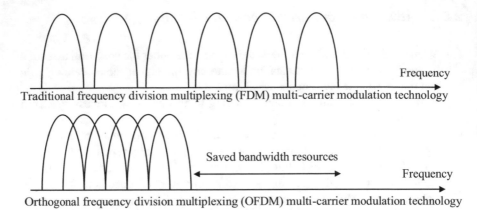

Fig. 2.15 Comparison of bandwidth utilization between FDM and OFDM

Besides using the equalizer, multi-carrier modulation technology is also one of the ways to solve these problems. The concept is to cut a large bandwidth into smaller sub-channels to transmit the signal. When using these narrower sub-channels, the channel frequency response of each sub-carrier within the sub-channel appears to be flat, which is the concept of frequency division multiplexing (FDM). Of course, FDM can reduce the multipath effect's influence, but each sub-carrier cannot overlap, and the bandwidth resources are limited. If the carrier on the spectrum can be reused, the spectral efficiency is greatly improved. So some scholars put forward orthogonal frequency division multiplexing (OFDM) technology [4]. It has become a major research hotspot in radio communication in recent years. As can be seen from Fig. 2.15, compared with single-carrier modulation, it has stronger anti-multipath propagation, anti-frequency selective fading ability, and higher spectrum utilization than FDM. In addition, OFDM can also use inverse discrete Fourier transform/discrete Fourier transform (IDFT/DFT) to replace multi-carrier modulation and demodulation. Therefore, it is widely used in high-speed radio and UWA high-rate data transmission.

2.6.1 Spectrum Features

The complex form of OFDM signals can be expressed as

$$S_{OFDM}(t) = \sum_{k=0}^{N-1} B_k e^{j2\pi f_k t + \varphi_k}. \tag{2.10}$$

In Eq. (2.10), B_k is the complex input data in channel k. In order to make the N channel signal can be completely separated, they need to meet the orthogonal

Fig. 2.16 Frequency
spectrum characteristics of
OFDM

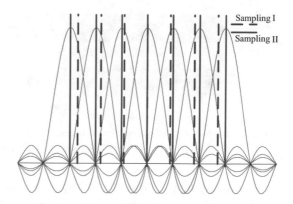

condition. Under the condition that any two sub-carriers are orthogonal within the
code duration T_B, the sub-carrier frequency is satisfied $f_k = k/2T_B$, K as an integer,
and the minimum sub-carrier frequency interval is $\Delta f = f_k - f_i = n/T_B$.

If the duration of the code element in these sub-channels is T_B, the spectral
density is shown in Fig. 2.16.

2.6.2   Fundamentals of OFDM

The fundamental principle of OFDM is to encode and distribute high-speed information data onto N mutually orthogonal carriers in parallel. The modulation rate is very
low $(1/N)$ on each carrier. The duration interval of the modulated symbols is much
larger than the time span of the UWA channel, which allows it to provide effective
protection of the transmitted digital signal in the UWA channel with large distortion
and bursty pulse interference. OFDM is insensitive to multipath time delay spread. If
the signal occupies a bandwidth larger than the UWA channel coherence bandwidth,
the multipath effect may cause some frequency components strengthen and some
frequency components to attenuate (frequency selective fading). OFDM frequency-
domain coding and interleaving establish a link between the scattered parallel data.
Hence, data corrupted by partial multipath fading or/and interference, i.e., located
in attenuated frequency components, can be recovered by the received data of the
strengthened frequency components, i.e., frequency diversity is achieved. Though
OFDM is still frequency division multiplexing, it is completely different from the
FDM scheme. It moves the different carriers to zero frequency and then integrates
them in one code element cycle. The data rate of OFDM is also associated with sub-
carrier numbers. Both the modulation and demodulation principle block diagram is
shown in Fig. 2.17.

At the transmit side, OFDM is modulated. The serial data are sequentially
encoded, modulated, converted to parallel, and sent to the computing unit, where
IFFT transformation is carried out, and then guard interval is added. After that, it

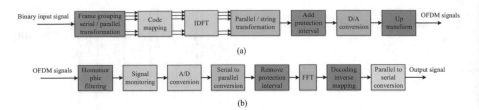

(a)

(b)

Fig. 2.17 Modulation and demodulation schematic diagram of OFDM. (**a**) Block diagram of OFDM modulation principle. (**b**) Block diagram of OFDM demodulation principle

is transformed into an analog signal by D/A conversion and sent to the channel for transmission. At the receiver part, the analog OFDM signal received via the UWA channel is converted into a serial digital signal after A/D conversion and then sent to the computing unit, where FFT operation is performed after removing the protected area. Finally, the original source signals can be restored after tandem conversion and decoding. In this schematic, the insertion of guard intervals is an essential part. The guard interval may not contain signals, but it also introduces inter-code interference (ICI). The ICI destroys the orthogonality between the sub-carriers. Suppose the guard interval is composed of the cyclic spread of the signal. The circular prefix is introduced, and the circular prefix whose length meets the requirement of eliminating ISI can eliminate ICI. Of course, the guard interval insertion will reduce the data transmission efficiency to $N/(N + L)$, where L is the length of the inserted protection gap.

In order to reduce the overall communication system BER to ensure system decoding, error correction codes can be added to the OFDM system. Marine experiments have proven that convolutional codes are effective in error correction in OFDM communication systems. Although OFDM is a useful solution to the frequency selective fading caused by multipath effects, it cannot suppress fading by itself. At the same time, there are additive noises (such as Gaussian white noise, impulse interference, etc.) in the channel, and these factors are associated with the generation of random and burst errors. This requires further protection of the transmitted data with channel coding, i.e., coded orthogonal frequency division multiplexing (coded OFDM or COFDM). Among all channel coding techniques, trellis-coded modulation used in conjunction with frequency and time interleaving is an effective method to deal with the UWA channel fading.

Figure 2.18 is the block diagram of COFDM applied to the high-definition television (HDTV) transmission system. The system uses cascaded error correction codes to eliminate the error codes. The outer code is RS code, and the inner code is convolutional code. TCM integrates convolutional code and modulation in the design. TCM improves the BER performance by 8 dB in a white noise environment compared with the traditional technique, so the inner code uses TCM. The TCM code has strong resistance to Gaussian white noise but poor resistance to the impulse noise occurring at intervals. The interleaving makes the pulse signal disperse into each RS code after error correction by TCM to facilitate RS code elimination.

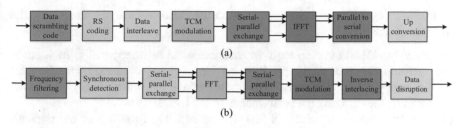

(a)

(b)

Fig. 2.18 Block diagram of COFDM applied in HDTV transmission system. (**a**) Transmitter. (**b**) Receiver

2.6.3 Advantages and Disadvantages of OFDM in UWAC

OFDM system has attracted much attention in recent years due to its advantages of being suitable for broadband transmission. In the UWA channel, OFDM systems have the following advantages over conventional single-carrier modulation or general non-overlapping multi-carrier transmission systems:

(1) OFDM divides the entire frequency band of the system into subbands whose bandwidth is smaller than the coherent bandwidth of the channel. Multiple sub-channels can make the OFDM system have a stronger ability to resist frequency attenuation, reduce the generation of burst errors, and effectively resist inter-symbol interference caused by multipath expansion.

(2) The protection interval of the circular prefix inserted between the data can effectively reduce the inter-code interference, and the circular prefix can reduce the synchronization requirements of the system.

(3) The spectrum of the sub-channels overlaps and is orthogonal, reducing the mutual influence between the receiving channels, enhancing the ability to receive signals, and utilizing the spectrum resources to a large extent. It is very critical for the UWA environment with limited spectrum resources.

(4) The modulation method on each sub-carrier of an OFDM system can be controlled flexibly. It is easy to be allocated through the dynamic modulation method to make full use of the sub-channel with small fading and avoid the adverse impact of the deep fading sub-carrier channel on system performance. This can also be effective against narrowband interference.

(5) By inserting a frequency-domain pilot, the OFDM system can easily realize frequency-domain equalization to compensate for frequency selective fading caused by the channel.

(6) OFDM system can be further evolved into the OFDMA (i.e., multiple access technologies) system, through which users can share and transmit information.

(7) The modulation and demodulation in each sub-channel of OFDM can be processed by fast Fourier transform and inverse fast Fourier transform, which reduces the system's complexity.

Due to its own technical characteristics, OFDM also needs to pay attention to the following issues:

(1) Since the OFDM system requires each carrier to be orthogonal to each other, attention should be paid to the inter-carrier interference caused by the Doppler frequency offset. This interference will damage the orthogonality of its sub-carriers, thus limiting the performance of the OFDM system.
(2) The transmit signal of an OFDM system is a superposition of the transmit signals on multiple sub-carriers. The OFDM generated by the superposition of multiple signals with different frequencies, amplitudes and phases has a large peak-to-average power ratio, so care should be taken to suppress the peak-to-average power ratio.
(3) Because the OFDM system should reduce the effect of inter-code interference, it must ensure that the oscillation frequencies of the transmitter and receiver remain the same frequency and phase, the sampling frequency of the receiver is the same as that of the transmitter, and the start and end times of modulation and demodulation are the same.

2.7 Multi-Input Multi-Output Technology

The worldwide demand for UWAC capacity is growing rapidly with the development of UWA network communications, underwater internet, and multimedia services. However, the available UWA frequency bands are very limited. If the utilization of communication frequency bands is not significantly improved, it is impossible to meet the growing demand for communication capacity. The theoretical analysis shows that the information capacity of a UWAC system can be significantly increased by increasing the number of transmitting and receiving antennas. For information systems with multiple transmit and receive antennas, the receiver has a state with known independent flat fading channels. The information capacity increases linearly with a small value between the number of transmitting and receiving antennas. Such a communication system with multiple transmitting and receiving antennas is called a multiple-input multiple-output (MIMO) communication system [5, 6]. The system can gain diversity gain and multiplexing gain by using multiple transmit and multiple receive antennas.

? Questions

1. Assume that the binary information sequence sent is 1 0 1 1 0 0 1, the symbol rate is 2000 Baud, and the carrier signal is $\sin(8\pi \times 10^3 t)$. Try to determine the following:

(1) How many carrier cycles does each symbol contain?

(2) Draw the waveforms of OOK, 2PSK, and 2DPSK signals, and observe the characteristics of each waveform.

(3) Calculate the first spectral zero bandwidth of 2ASK, 2PSK, and 2DPSK signals.

2. Assume that the binary information is 0 1 0 1, the code rate transmitted by 2FSK system is 1200 Baud, and the carrier frequency of the modulated signal is, respectively, 4800 Hz (corresponding to "1" code) and 2400 Hz (corresponding to "0" code).

(1) If the envelope detection method is adopted for demodulation, try to draw the time waveform at each point.

(2) If coherent demodulation is adopted, try to draw the demodulation principle block diagram and draw the time waveform of each point.

3. The spread spectrum gain of a certain system is 40 dB, and the internal loss of the system is $L_s = 2$ dB. The output signal-to-noise ratio of the relevant decoder is $(S/N)_{out} \geq 10$ dB. In order to ensure the normal operation of the system, what is the interference tolerance of the system?

4. The pseudo-random code rate of a direct sequence spread spectrum system is 5 Mbit/s, and the signal rate is 8 kbit/s, and what are the spread spectrum bandwidth and processing gain of the signal?

5. What are the main differences between spread spectrum communication systems and traditional modulation communication systems?

6. Explain the advantages and disadvantages of OFDM technology in UWAC.

References

1. M. Stojanovic, Recent advances in high-speed underwater acoustic communications. IEEE J. Ocean. Eng. **21**(2), 125–136 1996
2. J.G. Proakis, D.G. Manolakis, *Digital Signal Processing* (PHI Publication, New Delhi, 2004)
3. M. Johnson, L. Freitag, M. Stojanovic, Improved Doppler tracking and correction for underwater acoustic communications, in *1997 IEEE International Conference on Acoustics, Speech, and Signal Processing*, vol. 1. (IEEE, Piscataway, 1997), pp. 575–578
4. B.S. Sharif, J. Neasham, O.R. Hinton et al., A computationally efficient Doppler compensation system for underwater acoustic communications. IEEE J. Ocean. Eng. **25**(1), 52–61 (2000)
5. T.H. Eggen, A.B. Baggeroer, J.C. Preisig, Communication over Doppler spread channels. Part I: channel and receiver presentation. IEEE J. Ocean. Eng. **25**(1), 62–71 (2000)
6. T.H. Eggen, J.C. Preisig, A.B. Baggeroer, Communication over Doppler spread channels. II. Receiver characterization and practical results. IEEE J. Ocean. Eng. **26**(4), 612–621 (2001)

Chapter 3
Signal Processing in UWAC System

3.1 Diversity in UWAC System

Diversity techniques investigate how multipath signals can be used to improve the performance of a system. Originally developed for the field of terrestrial communication, the technique is now also borrowed by UWACs.

Diversity techniques utilize multiple signal paths with independent fading characteristics [1–4]. These signal paths with approximately equal average signal strength transmit the same information. Diversity techniques combine these signals properly at the receiver side and greatly reduce the effects of multipath fading, thus improving transmission reliability. Independent paths can be obtained at the receiver side through the spatial domain, time domain, and frequency domain. There are several specific methods as follows:

(1) Spatial Diversity

Spatial diversity is an efficient way to communicate as it can compensate for the loss of signal caused by the underwater fading channel. When it is used together with an equalizer, it can significantly improve the communication quality of UWAC. The principle of spatial diversity in UWAC is shown in Fig. 3.1. A transmitting transducer is used at the transmitter side, and an array of hydrophones is employed at the receiver side. The distance d between the receiving hydrophones should be large enough to ensure that the individual hydrophone output signals' fading characteristics are independent of each other. In the ideal case, the distance between the receiving hydrophones should meet the half-wavelength principle, which is the same as the spatial diversity technology in wireless communications, i.e., $d \geq \frac{\lambda}{2}$, where λ is the wavelength. Under this condition, it can be ensured that the signals received by each branch do not generate resolution ambiguity. The increase in the number of branches enhances the diversity effect for spatial diversity, but the complexity of diversity

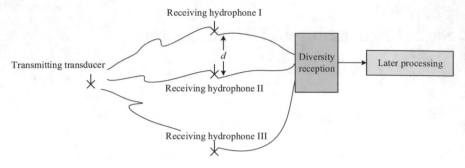

Fig. 3.1 Spatial diversity diagram

also increases. Therefore, the increase of diversity gain becomes slow with the increase of the branches' number.

(2) Angle Diversity

Due to the different environments, the signals arriving at the receiving end with different paths may come from different directions. At the receiving end, directional antennas are used to point to different signal arrival directions separately, so the decorrelation of the multipath signals received by the directional antennas is achieved.

(3) Frequency Diversity

The information is transmitted out separately at different carrier frequencies, and as long as the interval between carrier frequencies is large enough (larger than the coherence bandwidth), then signals with uncorrelated fading characteristics can be obtained at the receiving end. The advantage of this method is that the number of antennas is reduced compared with spatial diversity; however, the disadvantage is that it takes up more spectrum resources and requires a wideband transmitter or multiple transmitters at the transmitter end.

(4) Time Diversity

Time-domain diversity takes advantage of the fact that two sampling points that are far apart in time (bigger than the coherence time) are uncorrelated with each other. Since the coherence time is inversely proportional to the movement speed of the mobile station, temporal diversity does not work when the mobile user is stationary.

3.2 Equalization in UWAC System

With UWAC technology development, equalization technology is very important and can effectively deal with signal interference [5–7]. This technology is mainly used in the variable parameter UWA channel. The information in the variable parameter channel will produce a variety of propagation routes and time delays.

This book proposes an optimization method for UWA channel equalization based on the OFDM UWAC system to improve the system's communication quality and communication efficiency.

3.2.1   Development Process of Equalization

The origin of equalization technique can be traced back to the 1930s. At that time, communication technology was underdeveloped, and the pressure of competition in the communications field was relatively small. People's daily life and even the national military did not have much demand for communications systems. Besides, most communication networks and communication technology provided services for land-based wired communications, and there were many areas for improvement. Therefore, instead of sticking to the development of equalization techniques, people became more interested in coding and noise removal to improve communication systems. Therefore, the equalization technique did not receive much attention at that time. It was 1960s when the time-domain linear equalization techniques applicable to radio communication systems were proposed. Simultaneously, the zero-forcing (ZF) and minimum mean square error (MMSE) criteria have been proposed to determine the equalizer's tapping coefficients. It can be said that this is the first channel equalization technique to be widely used in communication systems. However, with the rapid development in the field of wireless communication, the environment of communication channels has become more complex. People demand higher communication quality and throughput, so simple linear equalization techniques cannot meet communication requirements, and nonlinear equalization techniques are thus born.

The feedback technique can better compensate for the loss of channel impulse response, but it also increases the computational complexity and raises the system's hardware requirements. With further exploration of nature, communication systems have also involved more complex time-varying channels, such as UWA channels. The signal transmission characteristics of these channels are not ideal, which requires further development of communication technologies. In terms of channel equalization, adaptive equalization techniques and blind equalization techniques have been studied comprehensively and extensively. They are also the most widely used equalization techniques today. The adaptive equalization technique was first investigated by Lucky in 1965. He investigated the application of this technique to quantized voice communication systems. He used a zero-forcing algorithm based on the zero-forcing criterion to select the tapping coefficients in his experiments. Lucky uses the degree of peak distortion as a measurement criterion to define the zero-forcing criterion. Moreover, the simple human-controlled time-domain equalization technique is transformed into an equalization technique automatically adjusted by the system. The above is regarded as a significant technological achievement in the field of communication. At the same time, Widrow designed the Least Mean Square (LMS) algorithm in 1966. Together with Lucky, he laid the foundation for developing digital communication systems with fast demodulation equalization techniques. After this, more and more people have started to study equalization techniques, including time-domain equalization, frequency-domain equalization, and adaptive equalization algorithms. The LMS algorithm application to complex signal adaptive equalization was described and analyzed in 1969 in the paper by Proakis and Miller, which advanced the equalization technique forward. In 1975, Proakis synthesized the relevant research results up to that year and published an influential paper. He re-states the importance of equalization techniques for communication systems and brings the research on adaptive equalization techniques to a new enthusiasm level.

In the application of adaptive equalization, "whether the receiver side of the communication system can process and compensate data information more quickly to obtain a useful signal" becomes an important consideration for the communication system. Moreover, the convergence speed of the algorithm determines how fast the adaptive equalizer can process the data. Godard proposed a faster equalization convergence algorithm in 1974. In 1978, Picinbono derived the recursive least mean square (RLS) algorithm and the Kalman filtering algorithm to improve communication systems' performance. Until 1980s, many scholars and experts in the world improved and refined the RLS algorithm from various perspectives. On the other hand, Ungerboeck proposed the grid-coded modulation technique in 1976. This technology has enabled the further development of commercial high-speed modems. This modem can transmit up to 28,800-bit rates over the telephone channel, which has led to the further development of adaptive equalization technology.

Another class of adaptive equalization algorithms, namely blind equalization techniques, has also received attention simultaneously. Blind equalization was originally proposed by Sato in 1975 and was applied to PAM communication systems. Subsequently, Godard et al. extended this algorithm to two-dimensional and multidimensional signals. The blind equalization technique based on the second-order and higher-order moments of the received signal was first proposed by Nikias in 1991 and improved by Tong in 1994. The maximum likelihood criterion for joint channel estimation and data detection was first proposed by Weber et al. The current research focuses on the convergence of stochastic gradient blind equalization algorithms and high-speed time-varying channel equalization.

The utilization of equalization techniques to improve communication quality is also one of the current research hotspots in UWAC systems. Among them, channel estimation is not only a key technology to improve communication reliability and accuracy in UWAC but a challenge in UWAC research. The usual channel estimation requires a known training sequence at the input and solves the transfer function or inverse transfer function of the channel. The equalization algorithm research mainly includes speeding up the convergence, reducing the computational complexity, and reducing the error. The performance of the maximum likelihood detector for multipath channels is optimal. However, the high computational complexity and such a long channel delay of the UWA channel make it impossible to apply the method to practical systems. Therefore this detector is only used as a reference for performance comparison. Another method to overcome inter-code interference is the Decision Feedback Equalizer (DFE). However, the computational burden of DFE is also quite enormous for a UWA channel with a time delay of several hundred symbols. Simultaneously, these algorithms are basically a direct transposition of the equalization algorithms of land-based wireless communication into UWAC. These algorithms do not exploit the UWA channel's sparse characteristics; but instead, the complexity of the algorithms is extremely high due to the severe time-varying multipath characteristics of the UWA channel. Even if channel sparsity is utilized, the DFE misrepresentation is severe due to the similar strength of each path of the UWA channel and the low signal-to-noise ratio. Therefore, the judgment feedback structure of the equalizer is not well adapted to the UWA channel. Besides, blind equalization techniques require a large amount of data to compute statistical information without using training sequences, which makes the convergence rate relatively slow. Blind equalization techniques are also usually not suitable for UWA channels with fast time-varying characteristics.

For time- and frequency selective channels, the UWA channel's time-varying features can be fitted to arbitrary accuracy using an orthogonal Basis Expansion Model (BEM). However, this builds on a large increase in the number of tasks for channel estimation. Because the amount of unknowns also increases with modeling accuracy. If the sparse nature of the UWA channel is exploited, the modeling effort can be reduced. Researchers have used subspace fitting, zero-tap detection, and Monte Carlo Markov chains to exploit channel sparsity for discrimination and equalization under the assumption of frequency selective time-invariant channels. Although there are many studies on time-domain-based UWA channel identification and equalization, few studies have been reported on designing identification and equalization algorithms for dual-selective UWA channels with respect to their sparsity. Considering the characteristics of UWA channels, orthogonal frequency division multiplexing (OFDM) technology to achieve high-speed UWAC is one of the recent research hotspots.

Based on the OFDM UWAC system, this book proposes an optimization method for UWA channel equalization to improve the communication system's quality and efficiency.

3.2.2   Classification and Characteristics of Equalization

According to the ray acoustic theory, the acoustic signal in the UWA channel generate time delays, i.e., the multipath phenomenon. The signal produce frequency selective fading in the frequency domain after passing through the multipath channel. For single-carrier communication systems such as QAM and MPSK, multipath interference causes the code elements to overlap each other and causes code element interference, especially in complex channels. Even without noise interference, the directly demodulated signal is not usable. For this single-carrier transmission phenomenon, an equalizer is invented to overcome the code interference and recover the original signal's information.

There are various ways to classify equalizers, including time-domain equalizers and frequency-domain equalizers for different operating domains, and training sequence equalizers and blind equalizers for the presence or absence of a priori sequences. For single-carrier signals, time-domain equalizers are the more commonly used equalization; training sequence equalizers and blind equalizers have their advantages and disadvantages but are of high research value.

Equalizer systems with training sequences usually use adaptive means to adjust the coefficients of the channel equalizer. The linear structured equalizer is the original adaptive equalizer system. This algorithm has a simple structure with only one tap used to weight the input signal and adjusts the taps' weight coefficients by comparing them with the training sequence. A major drawback of the linear equalizer is that only the input signal is used to adjust the coefficients, and usually, the system error of the equalizer converges slowly, which seriously limits linear equalization in high-speed real-time UWACs and complex UWA channels. To overcome all the defects of a linear equalizer, people invented the judgment feedback equalizer (DFE). DFE is based on the linear equalizer. The previous judgment result feedback is weighted and fed back to the decision device. The feedforward and feedback end jointly determine the output of the equalizer. The advantage of the judgment feedback equalizer is that it speeds up the convergence of the judgment error. Simultaneously, its compensation performance is better due to the feedback end. However, compared to the linear equalizer, the judgment feedback equalizer also has some drawbacks. When the judgment error occurs, the error will be spread to the subsequent judgment through the feedback end. Therefore, under certain conditions, the performance of the judgment feedback equalizer may not be as good as the linear equalizer with the same adaptive algorithm.

The adaptive equalizer requires an adaptive criterion algorithm to adjust the tap coefficients' weights, and the most frequently used adaptive criterion is the mean squared error (MSE) criterion. MSE is an algorithm that uses the average of the estimated and standard covariates as the reference standard. The least mean square error (LMS) algorithm adjusts the equalizer coefficients directly by comparing the mean square error. The advantage of this algorithm is that it is fast and easy to implement in hardware. The disadvantage is that the algorithm's convergence speed is slow, and the random noise has a significant impact on the algorithm. So it is not easy to obtain a shallow error convergence curve. To overcome the disadvantages of slow convergence and apparent interference by the noise in the LMS algorithm, people have proposed recursive least squares (RLS) algorithm. This algorithm is adaptive to the squared error and has a more obvious enhancement in convergence speed and stability than the LMS algorithm. However, this algorithm also has the disadvantage of being more computationally intensive, and the algorithm is more complex to implement in hardware. Therefore, the most appropriate adaptive algorithm for equalization should be selected according to the specific conditions and user requirements.

The low carrier frequency of UWAC limits the carrier bandwidth and affects the communication speed too. Adding training sequences to the transmit signal enhances the communication signal's redundancy and objectively reduces the communication system's effective transmission rate. Therefore, blind equalization algorithms that do not require training sequences began to receive attention. Japanese scholars in the mid-1970s proposed the idea of blind equalization followed by more scholars from all over the world. This theory has been supplemented and improved over the years, and the blind equalization technique has gradually matured. The Bussgang-like algorithm of blind equalization has become the most widely studied blind equalization algorithm due to its adaptive properties. Constant-mode blind equalization (CMA) is a specific form of Bussgang-like blind equalization. This algorithm's cost function depends on the magnitude of the sequence at the receiver and is not sensitive to the phase; moreover, this algorithm has a small mean square error.

3.2.3 Basic Principles of Equalization

Channel equalization is a usual tool in communication systems. According to the research perspective and field, channel equalization is divided into two categories: frequency-domain equalization and time-domain equalization. Various distortions occur when a signal passes through a channel, including frequency shifts and waveform distortions. The frequency-domain equalization method uses an equalizer to compensate for the distortion frequency-domain characteristics. And the time-domain equalization method compensates for the signal waveform characteristics from the signal waveform distortion perspective. The purpose of both equalization methods is to achieve distortion-free transmission and ensure communication quality.

3.2.3.1 Time-Domain Equalization Algorithm

According to Nyquist's first criterion, the maximum transmission code element rate of a communication system is twice the bandwidth. In a practical system, the channel characteristics make it impossible to reach the maximum transmission rate. If the signal is still transmitted at a rate of two code elements per Hertz, the intersymbol interference is generated, and the signal waveform is distorted. If an adjustable lateral filter is added between the filter and the sampling adjudicator at the receiver, the intersymbol interference can be theoretically eliminated under certain conditions. Let the total characteristics of the baseband transmission system be $H(\omega)$. The frequency characteristic of the transversal filter is $T(\omega)$, the characteristics of the communication system after adding the transversal filter are $H'(\omega) = T(\omega)H(\omega)$.

As long as satisfied

$$\sum_i H'\left(\omega + \frac{2\pi i}{T_s}\right) = T_s, \ |\omega| \leq \frac{\pi}{T_s}. \tag{3.1}$$

Then the system has no intersymbol interference in theory. If $T(\omega)$ is a periodic function with period $2\pi/T_s$, given by

$$\sum_i H\left(\omega + \frac{2\pi i}{T_s}\right) T\left(\omega + \frac{2\pi i}{T_s}\right) = T_s, \ |\omega| \leq \frac{\pi}{T_s}. \tag{3.2}$$

We have

$$T(\omega) = \frac{T_s}{\sum\limits_i H\left(\omega + \frac{2\pi i}{T_s}\right)}, \ |\omega| \leq \frac{\pi}{T_s}. \tag{3.3}$$

For a periodic function $T(\omega)$, its Fourier series has the following form:

$$T(\omega) = \sum_{-\infty}^{\infty} C_n e^{-jnT_s\omega} \tag{3.4}$$

$$= \frac{T_s}{2\pi} \sum_{-\infty}^{\infty} \int_{-\pi/T_s}^{\pi/T_s} \frac{T_s}{\sum\limits_i H\left(\omega + 2\pi i/T_s\right)} e^{jnT_s\omega} d\omega \times e^{-jnT_s\omega}. \tag{3.5}$$

The Fourier inverse transform of $T(\omega)$ can be obtained as the unit impulse response $h(t)$ in the time domain:

$$h(t) = F^{-1}[T(\omega)] = \sum_{n=-\infty}^{\infty} C_n \delta(t - nT_s). \tag{3.6}$$

From the above derivation, it is clear that C_n is completely determined by $H(\omega)$ and the transverse filter is able to make the system impulse response be $h(t)$. The network diagram is shown in Fig. 3.2.

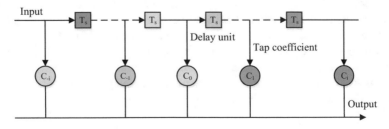

Fig. 3.2 Transverse filter structure diagram

The infinitely long transversal filter shown in Fig. 3.2 is not possible in engineering. And taking into account the possible deviation of the tap coefficients causing errors, we cannot eliminate the inter-code crosstalk in practice. From a theoretical point of view, the more taps the transversal filter has, the better the balanced signal will be; the more accurate the taps are, the closer the output will be to the original transmit signal taps. The tap coefficients' accuracy is a specific value for fixed hardware conditions, so the tap coefficients' accuracy becomes the key to equalizer performance. The usual methods for adjusting a linear equalizer's tap factor are the minimum peak method and the minimum mean square error method.

Consider a transverse filter with $2N + 1$ tapped coefficients, the unit impulse response $e(t)$ should be

$$e(t) = \sum_{i=-N}^{N} C_i \delta (t - iT_s).$$

(3.7)

And let the input signal of the equalizer be $x(t)$ and the output be $y(t)$, we can get

$$y(t) = x(t) * e(t) = \sum_{i=-N}^{N} C_i x (t - iT_s),$$

(3.8)

$$y(kT_s) = \sum_{i=-N}^{N} C_i x (kT_s - iT_s) = \sum_{i=-N}^{N} C_i x ((k - i)T_s).$$

(3.9)

The above equation can be simplified as

$$y_k = \sum_{i=-N}^{N} C_i x_{k-i}.$$

(3.10)

Equation (3.10) shows that the kth sampling moment's output value is the summation of $2N + 1$ sampling coefficients multiplied by $2N + 1$ input sample values, respectively. If the source input is a single pulse at zero moment, it is easy to know that y_0 is the useful signal, and all the others y_k are considered as inter-code crosstalk. Only when N is infinite it is possible to make all y_k other than y_0 be zero. When the value of N is a constant, there is a certain amount of distortion in the output signal. To make the distortion as small as possible, we generally consider two perspectives: minimum peak ratio and minimum mean square ratio.

Let D be a measure of signal distortion, where a smaller D means less distortion, and $D = 0$ when inter-code crosstalk is completely eliminated. For the peak ratio minimization algorithm, it is defined as

$$D = \frac{1}{y_0} \sum_{\substack{k=-\infty \\ k \neq 0}}^{\infty} |y_k|. \tag{3.11}$$

For the mean square ratio minimum algorithm, D is defined as

$$D = \frac{1}{y_0^2} \sum_{\substack{k=-\infty \\ k \neq 0}}^{\infty} y_k^2. \tag{3.12}$$

The minimum peak ratio algorithm, also known as the zero-forcing algorithm, determines the tap coefficients in the equalizer as follows: let $x_0 = y_0 = 1$ and perform normalization, then we have

$$y_0 = \sum_{i=-N}^{N} C_i x_{-i} = C_0 x_0 + \sum_{\substack{i=-N \\ i \neq 0}}^{N} C_i x_{-i} = 1. \tag{3.13}$$

Thus

$$C_0 = 1 - \sum_{\substack{i=-N \\ i \neq 0}}^{N} C_i x_{-i}. \tag{3.14}$$

Substituting the above equation into Eq. (3.10) yields

$$y_k = \sum_{\substack{i=-N \\ i \neq 0}}^{N} C_i (x_{k-i} - x_k x_{-i}) + x_k. \tag{3.15}$$

Substituting the above equation into Eq. (3.11) again, we have

$$D = \sum_{\substack{k=-\infty \\ k \neq 0}}^{\infty} \sum_{\substack{i=-N \\ i \neq 0}}^{N} C_i (x_{k-i} - x_k x_{-i}) + x_k. \tag{3.16}$$

It can be seen that when the channel response sequence is determined, the value of D is completely determined by C_i. Therefore, the minimum value of the function about C_i is required. Lucky has proved that under the initial distortion condition (the peak ratio of the sequence is less than 1), the minimum value of D always occur when N y_k around y_0 are zero. That is

$$y_k = \begin{cases} 0, 1 \le |k| \le N \\ 1, \quad k = 0 \end{cases} \Rightarrow \begin{cases} \sum\limits_{i=-N}^{N} C_i x_{k-i} = 0, k = \pm 1 \cdots \pm N \\ \sum\limits_{i=-N}^{N} C_i x_{-i} = 1, \quad k = 0 \end{cases} . \quad (3.17)$$

The matrix form can be expressed as follows:

$$\begin{bmatrix} x_0 & x_{-1} & \cdots & x_{-2N} \\ \vdots & \vdots & \cdots & \vdots \\ x_N & x_{N-1} & \cdots & x_{-N} \\ \vdots & \vdots & \cdots & \vdots \\ x_{2N} & x_{2N-1} & \cdots & x_0 \end{bmatrix} \begin{bmatrix} C_{-N} \\ \vdots \\ C_{-1} \\ C_0 \\ C_1 \\ \vdots \\ C_N \end{bmatrix} = \begin{bmatrix} 0 \\ \vdots \\ 0 \\ 1 \\ 0 \\ \vdots \end{bmatrix} . \quad (3.18)$$

The $2N + 1$ equations given above are combined in an equations system to solve the $2N + 1$ unknowns. Thus, the tapped coefficients can make the N sequences before and after a single pulse signal free of intersymbol interference. In engineering, these tap coefficients are found utilizing coefficient adjustment, called zero-forcing adjustment. The equalizer designed by this method is called a zero-forcing equalizer.

One of the most straightforward implementations of the zero-forcing equalizer is the preset auto-equalizer, whose block diagram is shown in Fig. 3.3. When a single pulse is input, the output captures each y_k and judges the sample value. If it is positive, the system will decrease C_k by one order of magnitude. If it is negative, the system will increase C_k by one order of magnitude until y_k is zero.

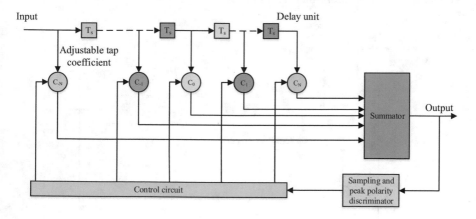

Fig. 3.3 Block diagram of the preset equalizer

A zero-forcing equalizer designed according to the principle of minimum peak distortion can equalize the signal. However, the disadvantage is that it must be used when the original distortion is less than 1. In contrast, an adaptive equalizer consisting of a minimum mean squared error criterion does not need to consider the original distortion size. The so-called adaptive means that the system can adjust its tap coefficient according to the change of channel characteristics so that the system can always be in a better equalization effect. Unlike the previous case, the source sends a sequence of signals instead of a single pulse signal in the adaptive case.

Let the mean square error of the signal be

$$\overline{e^2} = E(y_k - a_k)^2, \text{ where } e_k = y_k - a_k. \tag{3.19}$$

Substituting Eq. (3.10) into the above equation yields

$$\overline{e^2} = E\left(\sum_{i=-N}^{N} C_i x_{k-i} - a_k\right)^2. \tag{3.20}$$

Similarly, $\overline{e^2}$ is a function of C_i. Therefore, the partial derivative of C_i is obtained by making each partial derivative being zero, and the corresponding smallest e^2 can be obtained by

$$\frac{\partial \overline{e^2}}{\partial C_i} = 2E\left[e_k x_{k-i}\right] = 0. \tag{3.21}$$

The above equation holds provided that e_k and x_{k-i} are uncorrelated. In a real system, the statistical average of the product of e_k and x_{k-i} can be used to reflect whether C_i is taken to the optimal value. The block diagram of the adaptive equalizer using this algorithm is as shown in Fig. 3.4.

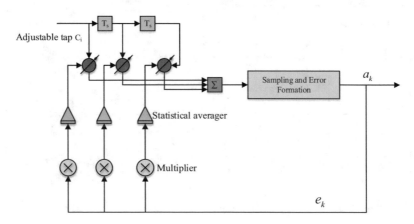

Fig. 3.4 Adaptive equalizer structure diagram

3.2.3.2 Frequency-Domain Equalization

In addition to the previously mentioned zero-forcing equalization method and the minimum mean square error method, the diagonalized equalization algorithm is a simpler frequency-domain equalization algorithm with some reference value. The following is a theoretical analysis of these common equalization algorithms from the perspective of frequency-domain equalization.

(1) Diagonalized Frequency-Domain Equalization Algorithm
In the UWAC system, the UWA channel's time-varying nature will inevitably lead to mutual interference between sub-carrier signals. In addition, in the frequency-domain channel matrix, this mutual interference between sub-carriers is manifested by non-zero matrix elements beyond the main diagonal. The main diagonal entries in the channel matrix indicate the fading of each sub-carrier, while the other elements show the characteristics of the inter-channel interference. The frequency-domain channel matrix is divided into two matrices: a diagonal matrix and an interference matrix after removing the diagonal matrix. This can be expressed as

$$H = D + G, \tag{3.22}$$

where D denotes the matrix in the absence of inter-subcarrier interference, as in the following equation:

$$D = \text{diag}\,(H_{1.1}, H_{2.2}, \ldots, H_{N.N})\,. \tag{3.23}$$

Matrix G is the diagonal matrix with sub-carrier interference only.

$$G = H - D, \tag{3.24}$$

$$Y = DX + GX + W. \tag{3.25}$$

The simplest equalization matrix simplification in the frequency-domain equalization technique, i.e., the diagonalization of the channel matrix, results in a diagonalized equalization judgment output of

$$Z = D'Y = \left(D^H D\right)^{-1} D^H Y, \tag{3.26}$$

where D' is the pseudo-inverse matrix of D.

The diagonalized approximation is the crudest and simplest compared to other kinds of balanced matrix simplifications. It directly ignores the effect of inter-subcarrier interference, and therefore its performance is also the least desirable. According to different communication performance requirements, such as BER performance, algorithm complexity, convergence speed, tracking capability, fault tolerance, different kinds of equalization matrix simplification approaches can be used.

(2) Zero-Forcing Equalization Algorithm

The zero-forcing equalization algorithm is a simple equalization algorithm that was first proposed by Lucky in 1965 and became a significant research result for improving communication systems. The zero-forcing equalization adjusts each sub-carrier in the channel so that the signal on the current sub-carrier is disturbed by the other sub-carriers to zero. The zero-forcing equalization can effectively avoid inter-subcarrier interference and improve the communication quality, but there are also shortcomings. Its neglect of channel noise (i.e., environmental noise) in the communication channel leads to its role in the communication system is limited to the weakening of multipath effects and cannot eliminate the impact of environmental noise on the system. In the UWA communication environment, the ambient noise is larger. The Gaussian white noise power of the channel is also larger, which makes the gain of zero-forcing equalization in this environment smaller and the equalization effect worse. The time-domain form of the received signal under L multiple paths after zero-forcing equalization is

$$y(n) = \sum_{l=0}^{L-1} h(n, l) x(n - l) + w(n). \tag{3.27}$$

The parameters and symbols in the above equation are consistent with the previous text. A discrete Fourier transform of Eq. (3.27) yields that

$$Y = HX + W. \tag{3.28}$$

In Eq. (3.28), X and Y are the input and output signals in the frequency domain, respectively. H is the channel matrix in the frequency domain, and the representation of the ambient noise in the frequency domain is denoted as W. Assuming that the estimated value of the original transmits signal obtained by equalization at the receiver is $\hat{X}$. The sum of squared system errors, ignoring the system noise, can be expressed as

$$U(\hat{X}) = (Y - H\hat{X})^H (Y - H\hat{X}). \tag{3.29}$$

To find the minimum value of this error in the equalization algorithm, we can first find the partial derivative of the left-hand side of Equation (3.29) to $\hat{X}$ and make it zero as follows:

$$\frac{\partial U(\hat{X})}{\partial (\hat{X})} = 2H^H (Y - H\hat{X}) = 0. \tag{3.30}$$

We can obtain

$$Z = H^{-1}Y + H^{-1}W = X + H^{-1}W. \tag{3.31}$$

Z in Eq. (3.31) represents the output signal of the system after equalization. The elements in the frequency-domain channel matrix are specific representations of the channel characteristics. The elements on the diagonal of the frequency-domain channel matrix indicate the gain of each sub-carrier. A smaller value indicates serious cannel fading. Equation (3.31) shows that when the frequency-domain channel matrix is more faded, the diagonal elements in H^{-1} are relatively larger, and thus the final noise signal is also larger. Therefore, for a communication system with a severe fading channel, the system's noise neglect characteristic with zero-forcing equalization seriously affects its final communication quality.

When the channel has deep zero points, the frequency-domain channel matrix H may not have full rank resulting in the non-existence of an inverse matrix. Therefore, the frequency-domain channel matrix's pseudo-inverse matrix is usually used to represent its inverse matrix. Then Eq. (3.31) can be converted into the following equation:

$$Z = H'Y = \left(H^H H\right)^{-1} H^H Y = X + \left(H^H H\right)^{-1} H^H W. \tag{3.32}$$

The frequency-domain channel is represented by a pseudo-inverse matrix H', which is calculated as follows:

$$H' = \left(H^H H\right)^{-1} H^H. \tag{3.33}$$

(3) Minimum Mean Square Error Equalization Algorithm

The minimum mean square error (MMSE) equalization uses the MMSE criterion as the condition for equalization judgments. The MMSE criterion determines whether equalization is successful by determining whether MSE between the output signal and the original transmit signal reaches a minimum value. The demodulated signal before equalization is denoted as:

$$X = GY = GHX + GW. \tag{3.34}$$

Then we need to find the optimal matrix G that minimizes the MSE of the following equation:

$$E\left[(X - \hat{X})^H (X - \hat{X})\right] = E\left[(X - GY)^H (X - GY)\right]. \tag{3.35}$$

We can solve for G as

$$G = \left(HH^H + \Gamma^{-1}\right) H^H, \tag{3.36}$$

$$Z = GY = \left[1 - \left(1 + \Gamma H^H H\right)^{-1}\right] X + GW, \tag{3.37}$$

where $\Gamma = C_x/\sigma_W^2$, C_x denotes the autocorrelation matrix of the input signal. Because the input signal is random, so only the elements on the diagonal of the input sequence autocorrelation matrix C_x are not zero. σ_w^2 denotes the Gaussian white noise variance in the frequency-domain sub-carrier of the received signal.

Compared with the zero-forcing equalization algorithm, the MMSE equalization algorithm takes the system noise power into account, which can better improve the communication system's performance. From Eq. (3.36), we can see that the MMSE equalization algorithm, which includes the system noise power in operation, does not cause noise amplification due to the large signal attenuation and can better resist the system noise. However, the MMSE equalization algorithm has a higher computational complexity and is more computationally intensive compared with the ZF equalization algorithm.

3.2.3.3 Performance Analysis of Frequency-Domain Equalization

Let the energy of the ith sequence of the time-domain signal sequences be ε_i, then the power of the time-domain signal ε can be found as

$$\varepsilon = \frac{1}{N} \sum_{i=-K}^{K} \varepsilon_i. \tag{3.38}$$

Thus

$$\sum_{l=0}^{L-1} E\left\{\left|h_{n,l}\right|^2\right\} = 1. \tag{3.39}$$

We use the orthogonal Fourier inverse transform, and the time-domain signal with the same power is obtained. SNR at the receiver is

$$\gamma = \frac{\varepsilon}{\sigma_w^2} = \frac{1}{N} \sum_{i=-K}^{K} \frac{\varepsilon_i}{\sigma_W^2}, \tag{3.40}$$

where $\sigma_w^2 = \sigma_W^2$ is the time-domain noise variance. In a particularly known equalization algorithm, SNR of the received signal is known, i.e., the theoretical BER or Symbol Error Ratio (SER) of the system can be calculated. This section focuses on the theoretical performance analysis of the diagonal simplified equalization algorithm, the zero-forcing equalization algorithm, and the minimum means square error equalization algorithm.

(1) Performance analysis of the zero-forcing equalization algorithm

Analyzing Eq. (3.30), the system noise power contained in the kth sub-carrier before equalization under the ZF equalization algorithm is

$$\sigma_{wk}^2 = E\left\{\left|H'_{(k,:)}W\right|^2\right\} \tag{3.41}$$

$$= \sum_{k=0}^{N-1}\sum_{k=0}^{N-1}[H']_{kk}[H]_{kk}^* E\left\{W_k W_k^*\right\} \tag{3.42}$$

$$= \sum_{k=0}^{N-1}\left|[H']_{kk}\right|^2\sigma_w^2, \quad k = -K, \ldots, K. \tag{3.43}$$

Therefore, SNR of the kth sub-carrier has the following form:

$$\Gamma = \frac{\varepsilon_k}{\sigma_{wk}^2} = \frac{\varepsilon_k}{\sigma_w^2|[H']_{kk}|^2}, \quad k = -K, \ldots, K. \tag{3.44}$$

Once SNR of each active sub-carrier is known, for a given modulation type, the BER of each active sub-carrier can be obtained, which is then used to calculate the BER of the whole system.

(2) Performance analysis of the minimum mean square error equalization algorithm

The minimum mean square error equalization is obtained from Eq. (3.35)

$$Z = GY = GHX + GW. \tag{3.45}$$

The matrix GH in the above equation is split diagonally into the matrices Q and Φ, then Eq. (3.45) is equivalent to the following equation:

$$Z = QX + \Phi X + GW, \tag{3.46}$$

where ΦX represents the residual inter-subcarrier interference. So with reference to the above equation, SNR of the kth sub-carrier can be expressed as

$$\Gamma = \frac{|Q_{kk}|^2\varepsilon_k}{\sum_{k=-K}^{K}\varepsilon_k|\Phi_{kk}|^2 + \sigma_w^2\sum_{k=0}^{N-1}|G_{kk}|^2}, \quad k = -K, \ldots, K. \tag{3.47}$$

(3) Performance analysis of diagonalized frequency-domain equalization algorithm

Applying the above splitting method, we have

$$Z = X + D'\Phi X + D'W. \tag{3.48}$$

Let $\Phi^d = D'\Phi$, and the SNR of the kth sub-carrier can be expressed as

$$\Gamma = \frac{\left|Q_{K,K}\right|^2 \varepsilon_k}{\sum\limits_{k=-K}^{K} \left|\left[\Phi^d\right]_{k,k}\right|^2 + \sigma_w^2 \sum\limits_{k=0}^{N-1} |[D']_{kk}|^2}, \quad k = -K, \ldots, K. \quad (3.49)$$

3.3 Doppler Frequency-Shift Compensation for UWAC Systems

When the transmitter and receiver have relative motion in UWAC, the frequency of the received signal will change. This phenomenon is called the Doppler effect, and the change in frequency is called the Doppler shift. In addition to the relative motion of communication equipment, the reflection of sound waves from the undulating sea surface, the refraction of sound waves by turbulence in the water, and other phenomena will also introduce Doppler shift. This can make the change in Doppler deviation at the receiving end inconsistent but a discontinuous distribution, which is called Doppler shift diffusion. Because of the low propagation speed of acoustic waves, making the same movement speed in the UWAC Doppler effect is 100,000 times more serious than in terrestrial communication.

An increase in the relative velocity of motion increases the Doppler shift. The Doppler shift has a spreading or compressing effect on the time-domain data frames, which seriously deteriorate the receiver's carrier frequency tracking and synchronization of phase symbols at the receiver end and eventually increase in the BER. Small Doppler shifts may be within the tolerance range of the coherent receiver algorithm and have little impact on performance. However, when the Doppler shift is severe, the coherent receiver algorithm performance degrades or even fails to work properly. For example, in practical UWAC systems such as submarines and cableless underwater robots, the relative motion between the transmitter and receiver is inevitable. The impact of the resulting Doppler shift on the receiver is far beyond the tolerance of the adaptive equalizer and phase tracking system at the receiver end to work properly. Therefore, eliminating or reducing the impact of the Doppler shift is an urgent problem in UWAC [8, 9].

3.3.1 Principle of Doppler Shift Compensation

The expression for the Doppler shift is

$$\Delta f = f_c \frac{v_s - v_r}{c - v_s}, \quad (3.50)$$

where f_c is the transmit signal (carrier frequency), v_s is the transmitter (source) speed, v_r is the receiver speed, and c is the speed of sound. In the presence of relative motion, the data frame is compressed if $\Delta f > 0$, or spread if $\Delta f < 0$. From Doppler shift's effect on the spreading (or compression) of the received data frames, the received signal can be described in the time domain by the following expression:

$$r(t) = s\big((1 + \Delta)t\big), \tag{3.51}$$

where $s(t)$ is the transmitted signal, $r(t)$ is the signal received by the receiver with an additional Doppler shift, and Δ is the data frame spread (or compression) ratio due to the Doppler shift. Considering the discrete signal case, let the sampling period of the receiver be T_s, then at the nth sampling point, we have

$$r[nT_s] = s[n(1 + \Delta)T_s]. \tag{3.52}$$

If Δ is known, the received signal can be resampled by interpolation and extraction to recover the original signal:

$$s[nT_s] = r\left[\left(\frac{n}{1 + \Delta}\right)T_s\right]. \tag{3.53}$$

Let $T_s' = \frac{T_s}{1+\Delta}$, we have

$$s[nT_s] = r[nT_s']. \tag{3.54}$$

Therefore, the frequency at which the received signal is resampled as

$$f_s' = (1 + \Delta)f_s, \tag{3.55}$$

where f_s is the original sampling frequency and f_s' is the resampling frequency.

The above method is the basic idea of Doppler shift compensation. Figure 3.5 shows the structure of this Doppler shift compensation principle. There are two core points: first, the determination of the data frame spreading (or compression) ratio Δ caused by the Doppler shift. Second, the resampling of the received signal, i.e., the compression (or spreading) of the received signal using interpolation and extraction to recover the signal to eliminate the carrier and symbol shifts.

3.3.2 Doppler Shift Detection and Compensation

The Doppler shift can be estimated by ambiguity function. The ambiguity function describes the signal variation when the matched filter is mismatched due to the

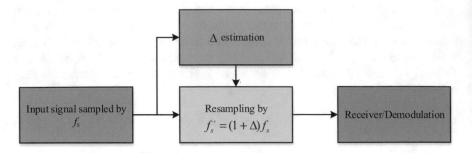

Fig. 3.5 Structure diagram of Doppler shift compensation principle

distance delay and Doppler shift under the condition of no distortion of the signal waveform. For a continuous broadband signal, the ambiguity function is defined as

$$\chi_s(\tau, \Delta) = (1 + \Delta) \int_{-\infty}^{\infty} s((1 + \Delta)t)s(t - \tau)dt. \tag{3.56}$$

Considering the received signal $r(t)$, the mutual ambiguity function can be defined as follows:

$$\chi_{sr}(\tau, \Delta) = (1 + \Delta) \int_{-\infty}^{\infty} s((1 + \Delta)t)r(t - \tau)dt. \tag{3.57}$$

It can be seen that the mutual ambiguity function of the above equation is equivalent to a two-dimensional correlation function. To estimate the Doppler shift in $r(t)$, we can make $\tau = 0$ and then search in Δ to find the Δ value corresponding to the maximum amplitude in the mutual ambiguity function $\chi_{sr}(0, \Delta)$, i.e.,

$$\chi_{sr}(0, \Delta) = (1 + \Delta) \int_{-\infty}^{\infty} s((1 + \Delta)t)r(t)dt. \tag{3.58}$$

As already described, the Doppler shift widens or compresses the data frame, and Δ reflects the size of the shift. Let the length of the received data frame be T_{rp}, and the length of the transmitted data frame be T_{tp}. Since the transmit data frame's length is known, Δ can be estimated when the received data frame length T_{rp} is measured. Let $\hat{\Delta}$ be the estimated value of Δ, then we have

$$\hat{\Delta} = \frac{T_{rp}}{T_{tp}} - 1. \tag{3.59}$$

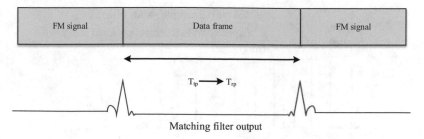

Fig. 3.6 Measuring data frame length using matched filters

The T_{rp} is measured by inserting a signal of a known waveform (the chirp signal) before and after the data frame so that a correlator (i.e., a matched filter) is used to calculate the correlation function between the chirp signal and the received signal (i.e., the fuzziness function defined above). Moreover, the length of the received data frame can be measured by detecting the time interval between the two peaks of the correlator output. In turn, $\hat{\Delta}$ is estimated. Figure 3.6 shows a schematic diagram of this T_{rp} measurement method.

The selection of the signal for Doppler shift estimation (the signal inserted before and after the data frame) should be considered in terms of both the Doppler shift's resolution and the receiver's tolerance for the Doppler shift. Linear FM signals, also known as chirp signals, have a sharp ambiguity function with both a certain resolution and a high tolerance, making them ideal for use in Doppler shift estimation signals. It is characterized by a linear time-dependent variation of the instantaneous frequency, and the mathematical expression is

$$s(t) = \cos\left(2\pi f_0 t + \pi k t^2\right). \tag{3.60}$$

In particular, it is noted that in order for the copy correlator to have a sufficiently high detection performance and accuracy of the time delay estimation, i.e., for the main ridge of the ambiguity function of the chirp signal to be very narrow, the bandwidth product BT of the chirp signal must be sufficiently large, which in general should satisfy $BT \geq 100$.

Once the Doppler shift is estimated, the received data can be resampled to compensate for the Doppler shift. The specific process is as follows: the original signal sequence is first sampled by an integer factor of I, which is achieved by inserting $I - 1$ zeros into two adjacent points of the original signal sequence and then interpolating low-pass filtering. Then, each value of the output sequence is found by linear interpolation, as shown in Fig. 3.7.

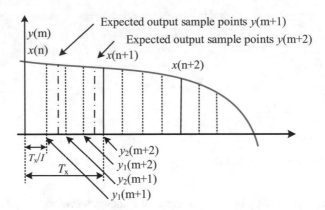

Fig. 3.7 Sample rate conversion of the original signal

where $x(n)$, $x(n+1)$, and $x(n+2)$ are the three adjacent sample points of the original input signal sequence, and the sampling period is T_x. $y_1(m+1)$, $y_2(m+1)$, $y_1(m+2)$, and $y_2(m+2)$ are the adjacent sample points obtained by increasing the sampling rate of the original signal sequence by integer factor I, and the sampling period is T_x/I. $y(m)$, $y(m+1)$, and $y(m+2)$ are the desired final output sample points. Let r be any sampling rate conversion factor (r can be a decimal), the sampling moment corresponding to the desired output sample point $y(m+1)$ can be found by r. Assume that this sample point is in the sampling time interval corresponding to $y_1(m+1)$ and $y_2(m+1)$, and normalize the sampling interval of these two points to $1/I$, and then denote by t_{m+1} the position of the sampling moment of $y(m+1)$ in the normalized sampling interval. From the linear interpolation algorithm, we get

$$y(m+1) = (1 - It_{m+1}) y_1(m+1) + It_{m+1} I y_2(m+1). \tag{3.61}$$

3.4   UWAC System Based on Frequency-Shift Keying

Based on extensive research on the characteristics of UWA channels and their corresponding modulation and coding, many researchers at home and abroad have proposed individual UWAC system design models for certain UWA system characteristics, respectively. A design example, a digital UWAC system based on frequency-shift keying, is presented below.

The UWA channel in the ocean is an enormously complex random time–space–frequency variable reference channel, which is subjected to many factors such as narrowband, high noise, strong multipath interference, etc. Therefore, it is of key significance to select the appropriate modulation method and transmission mode to guarantee the high speed and low BER of UWAC. The system is a prototype design scheme of UWAC based on a parallel transmission system, which adopts

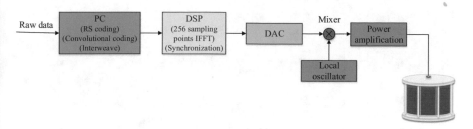

Fig. 3.8 Working principle of UWAC transmitter

cascade code with strong error correction capability and MFSK modulation. And the diversity technology adopts double-sideband modulation with suppressed carriers. The received signal is modulated by fast spectrum analysis at the receiver side, and hard-judgment Viterbi decoding and BM iterative decoding are performed. The experiments show that the transmission rate of this UWAC system is 200 bps and the bit error rate reaches below 10^{-5}–10^{-6}. The operating principle of the transmitter is shown in Fig. 3.8.

The raw data is sent to the PC for channel encoding. RS encoding uses a (15, 9) system code, which is equivalent to the binary (60, 36) system code. A code group has a code length of 60 bits, and the input information is 36 bits. The original information (or compressed original information) is grouped in 4×36 bits as a group, containing 18 bytes of the original information. The information is then convolutionally encoded with (2, 1, 7), which has a code rate of 1/2 and a constraint length of 7240 bits. 480 bits of non-systematic code are formed after the information is convolutionally encoded. Finally, the tail bits formed by the shift register in the 14 bits convolutional code encoder are also added. The interleaving method uses a 32-row × 15-column network, using the row-write, column-readout method, and the last 14 bits of tail bits in the last row are not involved in interleaving. The whole channel can be divided into a total of 8 sub-channels. Each channel is assigned four frequencies. For the 8 subchannels, there are 32 frequency resources with a frequency interval of 80 Hz. Four frequencies for each channel are allocated as follows: from 80 Hz, let $f_n = 80n$, where $n = 1, 2, \ldots, 32$, with $n = 1, 2, \ldots, 8$ as the first group, $n = 25, 26, \ldots, 32$ as the fourth group. Then the four frequencies assigned to each channel are the four frequencies used to form a sub-channel, i.e., for the first sub-channel, its four frequencies are f_1, f_9, f_{17}, f_{25}, the four frequencies of the eighth sub-channel are $f_8, f_{16}, f_{24}, f_{32}$, and so on. The modulation of each sub-channel is 4FSK mode. First, make 2bits correspond to a frequency of a channel. DSP in the frequency domain forms eight frequencies at a time. The initial phase of these eight frequencies is random to ensure that the impact of power dispersion is minimized. The sampling frequency is set to 20,480 Hz, and a 256-point IFFT is performed to form a time-domain waveform with eight different frequencies superimposed with the random initial phase. Then a double-band modulation capable of suppressing the carrier is performed, resulting in a composite waveform with 16 carrier frequencies superimposed. The double-

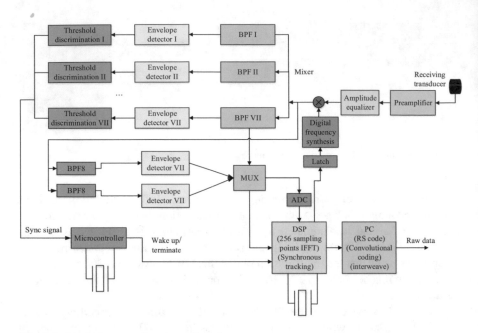

Fig. 3.9 Block diagram of receiver principle

band modulation can have the effect of frequency dichroism at the cost of doubling the bandwidth occupied by the signal and a decrease in the SNR of more than 3 dB. The operating principle of the receiver is shown in Fig. 3.9.

On the receive side the process is divided into four parts: synchronous capture → Doppler shift compensation → signal demodulation → channel decoding. Before the synchronous capture, the DSP is in the "hibernation" state. The microcontroller is in the search state. The latch latches a constant code to the digital frequency synthesizer. The resulting local oscillation frequency is f_1. When the seven synchronous sequences $f_{sym1}, \ldots, f_{sym7}$ enter the mixer in order to mix with the local oscillator frequency in the search state, resulting in seven differential frequencies $f_1 \sim f_7$. They are exactly the center frequency of the 7 narrowband filters behind the mixer. Each narrowband filter has a 3 dB bandwidth of ± 150 Hz. That is, as long as the Doppler shift is not greater than 150 Hz, the synchronous sequence can be detected completely. The rules of synchronous sequence detection in the microcontroller are: as long as three or more frequencies are detected in the order of the transmitter, and the time interval between them is less than a certain error value, then the microcontroller outputs the synchronous indication signal to "wake up" the DSP and enter the synchronous tracking state. After the DSP enters the synchronous tracking state, it generates four frequency hopping sequences at 10 ms interval: $f_{1.1}$, $f_{1.2}$, $f_{1.3}$, $f_{1.4}$, they are mixed with the received fine synchronous sequences: f_{sym3}, f_{sym1}, f_{sym4}, f_{sym2}, and the difference frequency sequence is f_8, f_9, f_8, f_9. The sampling value of the waveform information after using the envelope detector for synchronous tracking is 2, 1, 0.5, 0.5 ms. The timing error of channel code element reception after synchronous locking is ±0.5 ms or less.

After completing the synchronous tracking, the DSP switches the multiswitch MUX to the mixer output and samples the mixer output directly, followed by a 256-point FFT spectrum analysis. The signal used for Doppler shift correction, MF_{CD4}, consists of four single-frequency signals superimposed: f_{D1}, f_{D2}, f_{D3}, f_{D4}. These four frequency values are known for both the transmitter and the receiver. And the DSP can correct the fundamental frequency values based on the results of the spectral analysis compared to the known standard frequency values to eliminate the effect of the Doppler shift.

The received signal is first transformed to the fundamental frequency by a mixer and then sampled with 256 sampling points, where $f_s = \frac{256}{12.5\,\text{ms}} = 20,480\,\text{Hz}$. The FFT spectrum analysis is performed on the sampled values of multiple sinusoidal signals superimposed on each other. Since the interval of dividing the baseband frequency is only 80 Hz, the biggest difficulty is spectrum leakage, which is caused by the phase discontinuity at the truncation according to the period extension of the signal truncation. There are three reasons for the phase discontinuity at the truncation:

(1) The sampling frequency is not exactly an integer multiple of the fundamental frequency.
(2) The frequency of the received signal has a small frequency shift.
(3) The code element of the received signal is timed ahead or has a serious lag of more than 7.5 ms.

Although the effect of spectral leakage can be reduced by adding a window, the resolution of the spectral line is also reduced. The system adopts the following measures to solve the problem of spectral leakage:

(1) The IFFT synthesis time-domain waveform is used at the transmitter side to ensure that the period delay at the truncation of the transmit signal at its integer multiple periods is continuous in phase.
(2) The Doppler compensation at the receiver side can eliminate the effect of the Doppler shift so that the sampling frequency is an integer multiple of the fundamental frequency.
(3) The code element timing at the receiver side may have errors, and the synchronous tracking technique can be used to reduce the errors.

With the use of synchronous tracking technology, the error can be controlled within 0.5 ms. Simultaneously, in the spectrum analysis, deliberately take the lagging channel code element arrival time 1 ms as the starting point. The subsequent 12.5 ms sampling points to do FFT spectrum analysis due to each code element's modulation in the anti-code interference protection interval of 7.5 ms. When the signal passes through the underwater acoustic channel, there must be a self-multipath signal extended to this protection interval, and FFT spectrum analysis is completed by using the extended self multipath signal and the first 11.5 ms signal.

Without any window function, the spectral resolution is the highest. Without adding any window, the function makes the highest spectral resolution.

(4) Under the premise of having certain a priori knowledge, adaptive spectral enhancement technology can also improve spectral resolution. Because each code element is represented by two frequencies that mirror each other, only the one with the larger power spectrum during the spectral analysis is taken as the basis for the verdict and converted into the corresponding binary code element.

In the last step, channel decoding is performed. The decoding process of the channel consists of deinterleaving, hard-judgment Viterbi decoding, and RS iterative decoding. Since deinterleaving must wait until all the codes of a frame are received, the decoding delay is at least the time of one frame of data.

? Questions

1. What kinds of signal processing techniques can be used to process signals in order to improve the reliability of communication signal reception? And explain their characteristics.
2. What are the basic principles of time diversity, frequency diversity and spatial diversity?
3. What is the impact of the multipath transmission characteristics of the UWA channel on the performance of the communication system?
4. What are the commonly used equalization methods?
5. What are the main factors causing the Doppler shift and Doppler expansion of the UWA channel? What are the effects of Doppler shift and Doppler expansion on the communication system?
6. What is the basic principle of Doppler estimation based on fuzzy function?
7. If the transmitter speed $v_s = 10$ m/s, the receiver speed $v_r = 5$ m/s, calculate the Doppler expansion of a signal with the carrier frequency $f_c = 10$ kHz (The sound speed $c = 1500$ m/s).
8. Consider a three-tap transverse filter, where $C_{-1} = -1/4$, $C_0 = 1$, $C_1 = -1/2$. The values of the equalizer input $x(t)$ at each sampling point are $x_{-1} = 1/4$, $x_0 = 1$, $x_{+1} = 1/2$, and the others are 0. Calculate the equalizer's output $y(t)$ at each sampling point.
9. Design a zero-forcing equalizer with 3 taps to reduce inter-code crosstalk. Given that $x_{-2} = 0$, $x_{-1} = 0.1$, $x_0 = 1$, $x_1 = -0.2$, $x_2 = 0.1$, calculate the coefficient of 3 taps.

References

1. G.B. Henderson, A. Tweedy, G.S. Howe et al., Investigation of adaptive beamformer performance and experimental verification of applications in high data rate digital underwater communications, in *Proceedings of OCEANS'94*, vol. 1 (IEEE, Piscataway, 1994), pp. I/296–I/301
2. Q. Wen, J.A. Ritcey, Spatial diversity equalization applied to underwater communications. IEEE J. Ocean. Eng. **19**(2), 227–241 (1994)
3. B.G. Song, J.A. Ritcey, Spatial diversity equalization for MIMO ocean acoustic communication channels. IEEE J. Ocean. Eng. **21**(4), 505–512 (1996)
4. Y. Gong, K.B. Letaief, Performance evaluation and analysis of space-time coding in unequalized multipath fading links. IEEE Trans. Commun. **48**(11), 1778–1782 (2000)
5. J.G. Proakis, Adaptive equalization techniques for acoustic telemetry channels. IEEE J. Ocean. Eng. **16**(1), 21–31 (1991)
6. M. Stojanovic, High-speed underwater acoustic communications, in *Underwater Acoustic Digital Signal Processing and Communication Systems* (Springer, Boston, MA, 2002), pp. 1–35
7. F. Pancaldi, G.M. Vitetta, R. Kalbasi et al., Single-carrier frequency domain equalization. IEEE Signal Process. Mag. **25**(5), 37–56 (2008)
8. M. Johnson, L. Freitag, M. Stojanovic, Improved Doppler tracking and correction for underwater acoustic communications, in *1997 IEEE International Conference on Acoustics, Speech, and Signal Processing*, vol. 1 (IEEE, Piscataway, 1997), pp. 575–578
9. T.H. Eggen, A.B. Baggeroer, J.C. Preisig, Communication over Doppler spread channels. Part I: channel and receiver presentation. IEEE J. Ocean. Eng. **25**(1), 62–71 (2000)

Chapter 4
UWA Network Technology

4.1 Introduction

Like other communication technologies, UWAC technology has also undergone significant development from analog to digital. The early analog UWAC technology represented by UWA telephone generally used SSB modulation technology. The operating frequency is generally from 8 to 11 kHz. As it is analog technology, it requires a considerable amount of transmitting power, about in the order of a hundred watts. Underwater phones can provide end-to-end analog voice communication at several kilometers apart without forming a network.

Advancements in semiconductor technology, represented by ultra-large-scale integrated circuits, have dramatically changed the face of communications technology, including UWAC technology. Many algorithms that were considered too complex and unrealistic in the past have been gradually applied with the advancement of semiconductor technology, and digital UWAC technology has gradually replaced analog UWAC technology.

Previously, oceanographic instruments placed on the seafloor were recovered by releasing the instruments to obtain the collected data, which was inefficient and poor in real time. All data would lose if they failed. If the data is transmitted to the surface by UWAC equipment, the data can be obtained in real time, greatly improving efficiency. If several such underwater communication nodes are placed in a given region, underwater networks are an effective way to organize them and improve the whole system's performance. According to the function of underwater networks, there are currently sensor networks for undersea observation, channel detection, hydrological data collection, underwater combat networks for network-centric warfare, alert detection networks for underwater target reconnaissance and detection in coastal waters, etc. In the underwater network, various types of nodes are linked together through the UWAC network to collect, process, analyze information, and transmit control commands through the UWA network. A UWAC network is a wireless network established with the help of UWAC [1].

© The Author(s), under exclusive license to Springer Nature Switzerland AG 2022
Y. Lou, N. Ahmed, *Underwater Communications and Networks*, Textbooks in Telecommunication Engineering, https://doi.org/10.1007/978-3-030-86649-5_4

Fig. 4.1 UWAC network diagram

Generally speaking, a UWAC network includes not only underwater fixed nodes but also underwater mobile nodes and surface gateway nodes (Fig. 4.1). Underwater fixed nodes are generally used for marine environmental monitoring, marine information acquisition, and seabed-based equipment placement. Underwater mobile nodes are usually underwater vessels such as underwater cableless robots and submarines. The information collected by the surface gateway is usually transmitted by radio to the control center on land. Under the remote control of the controlling center, different nodes of the UWA network can collaborate to accomplish tasks such as information acquisition, monitoring, and operations. In the UWA network, the information transmitted between the nodes may include information data collected by the node sensors or the control commands and navigation information issued by the central control node or gateway. With the help of the UWA network, the performance of the original system can be greatly improved.

The UWAC network is a kind of radio communication network by nature, which has many similarities with the radio communication network. However, there are significant differences between UWA propagation characteristics and radio propagation characteristics in space, making UWAC networks significantly different from radio communication networks in terms of technology. It therefore is not feasible to directly transplant technologies that have proven to be successful in radio communication networks to UWAC networks. So we have to study the communication network technology adapted to the underwater environment according to the UWAC network's characteristics.

4.2   Characteristics of UWA Network

With respect to functionality, UWA networks are similar to wireless communication networks, but due to the idiosyncratic complexity of the propagation characteristics of UWA channels and the special characteristics of the structure and deployment of underwater nodes, UWA networks have some characteristics as follows [2, 3]:

(1) Dynamic topology

The topology of the UWA network is dynamically adjusted due to the addition of new nodes, the failure of underwater nodes due to seawater erosion or power depletion, the interruption of the UWAC link due to changes in channel transmission conditions or the movement of underwater nodes, etc.

(2) Self-organization

Many underwater nodes are deployed in the sea via ships. Therefore, the location of the nodes cannot be predetermined, and the mutual neighboring relationship between the nodes cannot be known in advance. This requires the sensor nodes to have self-organization capability, which can be automatically configured and managed to form multi-hop UWA networks through topology control mechanisms and networking protocols.

(3) Limited node energy

Many underwater nodes rely on battery power and have limited energy. Because of the extensive deployment of underwater nodes in the sea and the long recovery period, it is not practical to replenish the energy by replacing the batteries. Therefore, the efficient use of energy to maximize the network life cycle is the primary problem facing UWA networks.

(4) High reliability requirements

The underwater environment is harsh, underwater nodes are easily damaged or failed, and nodes are often deployed randomly. This requires underwater nodes to be more robust against the challenging environment. The maintenance of UWA networks is quite difficult, and therefore the software and hardware of UWA networks must have strong robustness and fault tolerance.

4.3   Topology of UWA Network

In order to reduce the complexity of network protocol design, the underwater acoustic communications network (UACN) hierarchy is divided into the physical layer, data link layer, and network layer. The Open Systems Interconnection (OSI) model and transmission control protocol/internet protocol (TCP/IP) are proposed from a practical application point of view, with reference to the International Organization for Standardization. The structure of UACN is hierarchical, with each layer building on the next and providing certain services in a covert form to the upper layer, as shown in Fig. 4.2 [4, 5].

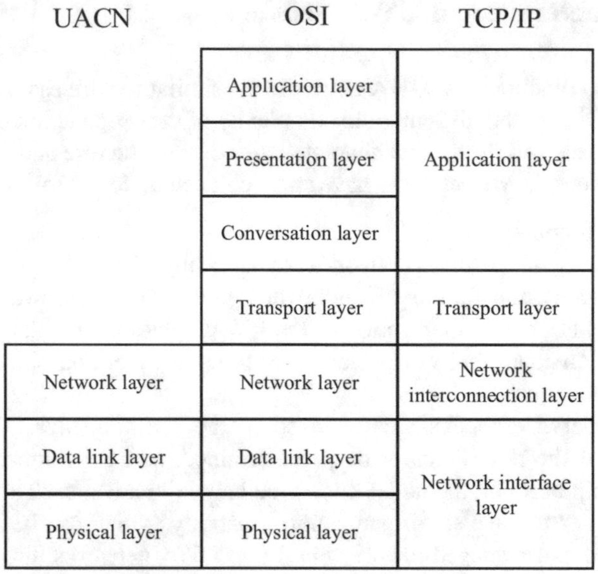

| UACN | OSI | TCP/IP |

	Application layer	
	Presentation layer	Application layer
	Conversation layer	
	Transport layer	Transport layer
Network layer	Network layer	Network interconnection layer
Data link layer	Data link layer	Network interface layer
Physical layer	Physical layer	

Fig. 4.2 UWAC network, open system interconnection model and transmission control protocol/internet interconnection protocol architecture

The main function of the physical layer is to accomplish the modulation and demodulation of the information flow data and to shield the transmission medium and physical devices. At the transmitter side, modulation is performed to modulate the digital information represented by 0 and 1 into acoustic signals for propagation in the aquatic channel. At the receiving end, the received acoustic signals are demodulated to recover the original bitstream information. The physical layer transmits only the digitally modulated information consisting of 0 and 1, ignoring the exact meaning of each bit.

The goal of the physical layer communication algorithm is to improve the communication system's reliability and effectiveness as much as possible. The channel's bandwidth is limited due to signal amplitude fading and inter-code interference caused by the UWA channel's multipath effect. The main technologies currently used in the physical layer of UACN are single-carrier PSK, MFSK, OFDM, and other algorithms. To effectively achieve robust UWAC, the key techniques usually used in UWAC are judgment feedback equalization technique, time-reversal mirror technique, sparse channel estimation, equalization technique, broadband Doppler compensation technique, etc. These methods are used in different network protocols for control frames or data frames due to differences in communication rates, robustness, etc.

The data link layer is the intermediate layer of UACN, and its ultimate goal is to enable each end node to share bandwidth resources fairly and efficiently. The data link layer of UACN focuses on media access control (MAC) protocols.

The MAC protocols in UWAC are usually divided into two categories, which are fixed-assignment and random-competition categories. Fixed allocation MAC protocols mainly include frequency division multiple access (FDMA), time division multiple access (TDMA), and code division multiple access (CDMA) protocols. Random contention protocols include ALOHA protocol, multiple access with collision avoidance (MACA), and carrier sense multiple access (CSMA). Due to the different requirements of UACN applications, different single multiple access techniques or mixed multiple access methods should be used.

The network layer's main role is to solve the path planning of how the packets reach the receiver from the sender, traffic control, and congestion control. The various nodes in the network must work together to complete the routing. Its functions include neighbor discovery, packet routing, congestion control, etc. The neighbor discovery function collects information about the topology in the network. The routing protocol can discover and maintain the routes to the destination nodes for the packets.

The designed routing protocol needs to be fast, accurate, efficient, and have good scalability. When the topological structure of the communication network changes, routes can be found quickly, and accurate routing information can still be provided. The protocol should be able to choose the best path based on network congestion. In addition, the control information for maintaining routing should be minimized to reduce the overhead of routing protocols. Since the size of UACN is variable, the designed routing protocol should be able to adapt to it.

For the UACN, the UWAC nodes and the links between the nodes form the network topology. The design of the physical layer, data link layer, and network layer protocols all relies heavily on the network topology design, which affects the reliability of the network. The topological structure of UACN mainly includes three types: centralized, distributed, and multi-hop, as shown in Fig. 4.3.

A centralized network, also known as a star network, has a central node that communicates with each child node. Because all nodes need to communicate with the central node, the central node is very dependent on the central node. If the central node has problems, the whole network will be paralyzed. The advantage of this type of network is simple structure, easy to maintain, but this structure is only suitable for small-scale UWAC networks, not for large-scale shallow underwater communication networks with requirements for security and stability.

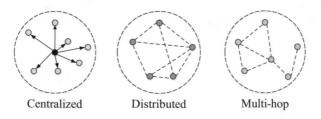

Centralized   Distributed   Multi-hop

Fig. 4.3 Topology of UACN

In distributed networks, since all nodes are connected to each other, they can communicate directly. Therefore, there is no need for routing, which is suitable for small-scale communication networks. When the nodes are far away from each other, the transmitting power needs to be increased. So the energy consumption is high, and it is not suitable for large-scale UACN.

Multi-hop networks, also known as ad hoc networks, are connected between adjacent nodes for direct communication, and the transmission of information is achieved by multiple forwarding between adjacent nodes. By increasing the number of nodes, a large range of communication can be achieved, but higher requirements are placed on the routing algorithm. Although multi-hop networks prolong the end-to-end delay of signals due to forwarding, they reduce power consumption while ensuring reliable operation.

4.4   Network Protocol Architecture of UWA Network

4.4.1   Applications of UWA Network

4.4.1.1   Introduction to Mobile Autonomous Network (MANET)

Mobile ad hoc network (MANET) technology has been increasingly used in UWAC networks in recent years. A self-organizing network (ad hoc) is a collection of wireless nodes (mobile or fixed nodes). This network does not need the existing network infrastructure or temporary networks' dynamic formation compared with the cellular network. The nodes collaborate with each other wirelessly to communicate and complete the information interaction. Since this network does not require any predefined infrastructure, it has gained wide interest in military and commercial applications. Mobile self-assembly networks can extend the network coverage and provide wireless connectivity to areas with poor or no coverage (the edge of the cellular network). The main differences between cellular networks and ad hoc wireless networks are shown in Fig. 4.4 [6–8].

Ad Hoc Wireless Network	Cellular Network
Stable connection and good communication environment	Irregular connection and poor communication environment
No base station	Fixed cell sites and base stations
Multi-hop dynamic network topology	Dynamic backbone topology
High cost and long installation time	Cost-effective, short installation time

Fig. 4.4 Cellular and ad hoc wireless networks

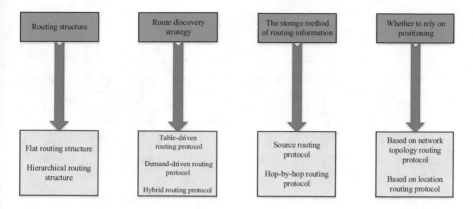

Fig. 4.5 Classification of common routing protocols

Routing can be classified in a variety of ways. We often classify routing protocols based on characteristics such as the routing structure and the application environment of the protocol. The classification of common routing protocols is shown in Fig. 4.5.

4.4.1.2   Routing Protocol for UWA Mobile Self-organizing Networks

Due to the complex and variable characteristics of the UWA channel as well as the network topology, the protocols need to be highly adaptable and perform well in terms of energy consumption and network performance when we choose UWA routing protocols for the variable and uncertain network topology. The study of routing protocols for UWAC networks focuses on table-driven routing protocols, on-demand routing protocols, and hybrid routing protocols.

(1) Table-Driven Routing Protocols
Each node periodically exchanges information to maintain a routing table, through which information is passed to establish a route from the source node to the destination node. When the source node needs to send a packet to a node in the network, it only needs to look up the stored routing table to get the destination node's routing information and send the packet without waiting for the path to be established with a small transmission delay. When there is a link error or a topology change, the node updates the routing table and broadcasts the message throughout the network. The disadvantage of table-driven routing protocols is that the maintenance of routes between nodes without passing information consumes a large amount of network bandwidth. Table-driven routing protocols commonly used in MANET networks are shown in Fig. 4.6.

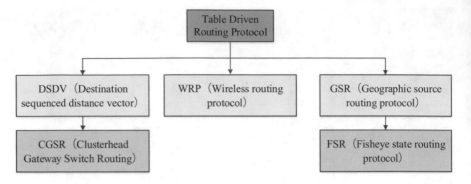

Fig. 4.6 Table-driven routing protocols commonly used in MANET networks

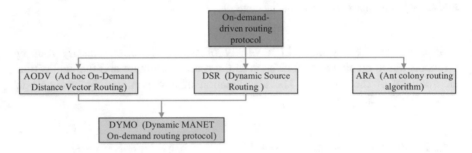

Fig. 4.7 Common on-demand driven routing protocols for MANET networks

(2) On-Demand Driven Routing Protocol

This protocol consists of two processes, route discovery and route maintenance, where nodes do not need to periodically exchange information for maintaining the routing table. When the source node sends a packet, it turns on the route discovery function and waits for the path to be established to send the packet if there is no local route information to the destination node. On-demand routing protocols do not require periodic exchange of routing information as table-driven routing protocols do, saving many bandwidth resources, but require route establishment before sending packets. So the latency is high. The on-demand routing protocols commonly used in MANET networks are shown in Fig. 4.7.

(3) Hybrid Routing Protocols

Hybrid routing protocols are a new generation of routing protocols with both the advantages of table-driven routing protocols and on-demand driven routing protocols at the same time. In a hybrid routing protocol, nearby nodes' routes are maintained using a table-driven approach, and the route discovery approach is used for distant nodes. Hybrid routing protocols reduce the overhead of route discovery and increase the scalability of the network. Hybrid routing protocols commonly used in MANET networks are shown in Fig. 4.8.

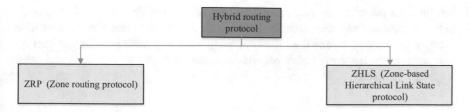

Fig. 4.8 Hybrid routing protocols commonly used in MANET networks

4.4.2 Routing Protocol Design Issues

A lot of high-performance routing protocols have recently developed, but these routing protocols are for land-based wireless networks. Due to the UWAC channel's inherent characteristics, these protocols cannot be directly applied to underwater networks. According to the UWAC network's characteristics and limitations, designing a high-performance routing protocol needs to consider the following issues.

4.4.2.1 Energy Consumption and Time Delay

As the propagation of UWAC has a narrow bandwidth and high delay, data transmission is prone to data congestion and high delay, which can cause information transmission failure and collapse of the entire network. Underwater nodes must work with electrical energy supply, so energy control is an important factor in designing routing protocol for UWAC networks. The efficiency of the protocol can be improved by controlling the energy balance of the network nodes.

4.4.2.2 Scalability

The expansion of the UWAC network is reflected in the number of network nodes' distribution, the coverage area of the network, the transmission delay of the network, and the network survival cycle. The problems arising from these expansions need to provide solutions and measures.

4.4.2.3 Security

For any network, the security of the network is critical. The UWA channel is an open channel. Anyone can go through the underwater phone to get the information in the channel. If there are no corresponding encryption means, our information transmitted through the underwater phone channel will be easily accessed by others and threaten our information's security. Therefore, the design of routing protocols needs to consider network security issues. The common acoustic communication security technology means the following categories: key management and encryption technology, intrusion detection, routing security, trust management, and other security technology.

4.4.2.4   Self-Adaptive

UWAC network consists of multiple network nodes, where each node contains a large number of electronic components, while underwater environment is complex and changing. When one of the multiple nodes fails due to its own or environmental reasons, the network topology changes, thereby affecting the network communication capability. Replacing the failed nodes is time-consuming and difficult to achieve. Each node in the UWA network needs to have strong self-adaptability and react quickly when some nodes have problems maintaining normal communication in the network.

4.4.3   Routing Protocol Optimization Techniques

MANET network is a collection of wireless mobile nodes. The network's wireless topology changes rapidly and unpredictably, which can degrade the network's performance. And it is important to optimize the protocol in different application environments. The routing protocols are improved in network quality, energy balance, topology, and security issues. Three technical approaches are summarized to optimize the network for the underwater environment's characteristics: energy balancing techniques, cross-layer design, and adaptive techniques.

4.4.3.1   Energy Balancing Technique

Since the underwater network nodes have limited energy and are not easily replaced, energy control is needed for the network nodes. Traditional self-organizing network protocols do not have energy control capability. To maximize the network's survival cycle and avoid dead nodes that degrade the network performance, energy balancing techniques are used to improve the energy balance of nodes. An ad hoc on-demand distance vector (AODV) routing protocol based on an ant colony algorithm is proposed to address high network energy problems and large delays. It changes the traditional wireless self-assembled network on-demand plane distance vector routing method of relying on transmission time and hop count to find transmission paths and improves the network connectivity and node energy balance.

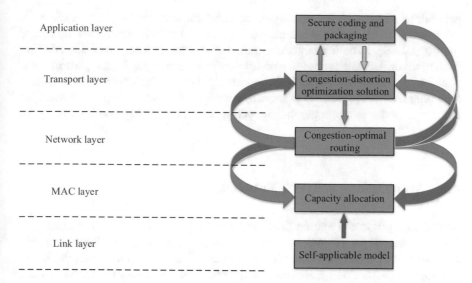

Fig. 4.9 Cross-layer design framework

4.4.3.2 Cross-Layer Design

Cross-layer design enables information exchange and intermodulation optimization between different layers in the network and improves the network's overall performance. Figure 4.9 shows the cross-layer design framework. In the data link layer, the adaptive modulation can improve the link rate when the channel changes and expand the network usage. The MAC layer and the network layer jointly manage the network traffic and reduce the network's congestion. An energy-efficient cross-layer protocol is proposed, which considers node energy and data retransmission based on Minimum Battery Cost Routing (MBCR). The cross-layer design method uses energy consumption as the routing rules to avoid using energy drained nodes. It collects the node's remaining energy, transmits power information at the physical layer, and adjusts the data link layer's transmission power and the control information of the network layer according to the actual situation. At the same time, it also calculates the energy consumption of the destination node reaching the network layer and selects a route based on the calculation result. The experimental results show that the improved protocol reduces network delay and energy consumption.

4.4.3.3 Adaptive Technology

In multi-hop mobile networks, as nodes can join and leave at any time, it will cause constant changes in the network topology, which affects the packet rate, throughput, and end-to-end delay of the network. The use of adaptive technology allows the network to select the appropriate path mechanism according to its own situation, enhancing the network's performance and reliability. For optimized link state routing (OLSR), a routing protocol parameter adjustment mechanism and adaptive routing algorithm are proposed and implemented using the Linux platform. An adaptive algorithm AODV-AOW that combines the adaptive on-demand weighting algorithm (AOW) is proposed to solve the problem of network congestion when the number of nodes is too large, thus reducing the network delay and improving the network efficiency.

4.5   Applications of UWA Network

UWAC network in the marine field has important civil and military applications. China's research for UWAC networks compared with the United States, Europe, and some other developed countries started late. But after more than a decade of development, China has achieved fruitful results in the fields of UWAC coding technology, modulation and demodulation technology, and UWA channel simulation research (Fig. 4.10).

(1) The key part of information transmission in the UWAC network is the UWA modem. The UWAC technology is the basis for the research of UWAC network technology. The UWAC network system has a high communication rate and has important significance in improving the network's overall performance.

(2) The UWA network has the characteristics of prolonged propagation, low propagation rate, and error-prone propagation process, which leads to wireless communication protocols that cannot be used in the UWAC network. So it is important to increase the research on the structure of the UWA network and network protocols to improve the network's efficiency and stability.

(3) UWAC network technology has wide application in marine information collection, marine environment monitoring, disaster forecasting, distributed tactical surveillance, and other fields. In order to guarantee the territorial security and maintain maritime rights and interests of individual countries, it is urgent to establish a perfect underwater information network.

With the advent of information age, the human demand for information and energy is increasing, and the research hotspots are gradually expanding from land to ocean. UACN has a wide range of applications in both military and civil fields. In the civil field, UACN can be widely used in marine scientific research, meteorological and underwater logical data acquisition, environmental pollution monitoring, marine geological exploration, ocean current movement data acquisition, marine disaster warning and rescue, and marine rare organism protection. In the military field, in order to build the integrated information network of sea, land, air, and space, and provide the information transmission and interconnection functions required by maritime battlefield communication, detection, positioning, and navigation, UWA network is an important part of future information warfare.

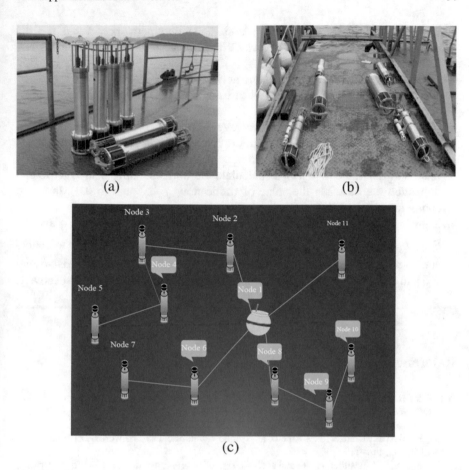

Fig. 4.10 Some research results of underwater communication network

It can display the underwater and surface combat situation information, realize regional sea area monitoring, threat target warning identification, submarine long-distance positioning, navigation, etc. Therefore, it is necessary to provide a unified information transmission in the military and civilian maritime information system. Hence, UACN research has great potential for development in both military and civilian maritime information development fields.

? Questions

1. What is the main function of the UWA network? What are its main components?
2. What are the characteristics of a UWA network?

3. What are the main topologies in communication networks? What are the advantages and disadvantages of each? What topologies are mainly used in UWA networks?
4. What are the characteristics of ad hoc networks?
5. What are the problems that arise when using the ALOHA protocol for multiple access in UWAC networks?
6. Compare the advantages and disadvantages of using the MACA protocol to control multiple access and polling to control multiple access in UWAC networks.
7. Assume that the capacity C between node A and node B in a UWAC network is fixed and the value is 1 Mbps. Calculate the corresponding delay metric and utilization metric when the value of traffic flow f is equal to 0.8 Mbps and approximates to 1 Mbps.
8. Knowing that the normalized Shannon capacity expression is $C_B = \frac{C}{B}$ and the Shannon capacity formula is $C = B\log_2\left(1 + \frac{P}{N_0 B}\right)$, try to use C_B to denote energy per bit required to transmit at rates approaching the normalized capacity E_b, and calculate the value of E_b when C_B equals 1 and tends to 0, respectively.

References

1. J. Rice, B. Creber, C. Fletcher et al., Evolution of Seaweb underwater acoustic networking, in *OCEANS 2000 MTS/IEEE Conference and Exhibition. Conference Proceedings (Cat. No. 00CH37158)*, vol. 3 (IEEE, Piscataway, 2000), pp. 2007–2017
2. J.G. Proakis, E.M. Sozer, J.A. Rice et al., Shallow water acoustic networks. IEEE Commun. Mag. **39**(11), 114–119 (2001)
3. G. Lapierre, L. Chevallier, F. Gallaud et al., Design of a communication protocol for underwater acoustic modems and networks, in *MTS/IEEE Oceans 2001. An Ocean Odyssey. Conference Proceedings (IEEE Cat. No. 01CH37295)*, vol. 4 (IEEE, Piscataway, 2001), pp. 2220–2226
4. Y. Wang, Y. Jiao, S. Zeng, A MAC protocol for underwater acoustic networks. J. Xiamen Univ. (Nat. Sci.) **2007**, S2 (2007)
5. A. Kebkal, K. Kebkal, M. Komar, Data-link protocol for underwater acoustic networks, in *Europe Oceans 2005*, vol. 2 (IEEE, Piscataway, 2005), pp. 1174–1180
6. M. Chitre, L. Freitag, E. Sozer et al., *An Architecture for Underwater Networks* (IEEE, Piscataway, 2006)
7. J.M. Jornet, M. Stojanovic, Distributed power control for underwater acoustic networks, in *OCEANS 2008* (IEEE, Piscataway, 2008), pp. 1–7
8. M. Stojanovic, Underwater acoustic communications: design considerations on the physical layer, in *2008 Fifth Annual Conference on Wireless on Demand Network Systems and Services* (IEEE, Piscataway, 2008), pp. 1–10

Chapter 5
UWAC Challenges and Research Trends

5.1 Challenges of UWAC Technology

At present, the practical performance of UWAC and UWA networks has deviated significantly from the theoretical analysis expected to achieve the ideal results, mainly because the actual channel conditions faced by acoustic propagation underwater are very different from the channel assumptions commonly used in theoretical analysis or laboratory simulations.

The UWA channel is one of the most difficult communication channels to date. Due to the slow propagation speed of underwater sound, the absorption of sound by seawater and the complexity and variability of the seawater medium will have serious disturbances to the UWA signal, especially in shallow channels, there are strong multipath and large signal fluctuations to the design of the UWAC system. How to overcome multipath interference, time–frequency and Doppler expansion, and time-varying fading is the problem of UWAC technology that needs to pay attention to solve

1. UWA channel complexity. The complexity of UWA channel is mainly shown in the following aspects:

   (1) the random undulation of the sea surface;
   (2) random, non-uniform, non-stationary seawater medium and its distribution of bubble layer;
   (3) different scales of cold and warm water mass;
   (4) laminar flow, turbulence, internal waves, and vortex;
   (5) dispersed inhomogeneous body, such as fish, plankton;
   (6) random spatial and temporal changes in the speed of sound;
   (7) random uneven seafloor; and
   (8) a variety of ocean noise, such as waves, wind, and rain, ship noise, and biological noise.

© The Author(s), under exclusive license to Springer Nature Switzerland AG 2022
Y. Lou, N. Ahmed, *Underwater Communications and Networks*, Textbooks in
Telecommunication Engineering, https://doi.org/10.1007/978-3-030-86649-5_5

2. UWA channel variability. Variability is manifested in the following aspects:

   (1) Time variability
       There are multiple types of time variation, such as surface waves: weak wind
       (1–3 s), medium wind (8–10 s), and strong wind (20–30 s); internal waves:
       shallow sea (minutes to tens of minutes) and deep sea (tens of minutes to
       hours); turbulence (hours to days); vortexes (days to months); and sound
       velocity profiles and mixed-layer thickness (seasonal variations).
   (2) Spatial variability
       Spatial variability is related to water depth,· sound velocity distribution,
       bottom characteristics, and non-uniform bodies (organisms, bubbles). There-
       fore, the sound field structure varies from place to place, such as the spatial
       variability of a multiplex structure.
   (3) Frequency variability
       For different frequencies, acoustic propagation has different attenuation,
       producing serious ups and downs of the spectral characteristics.

3. Strong multipath and finite frequency-domain bandwidth.

   (1) Strong multipath
       Two types of multipath: "macro" multipath reflected from the sea surface
       and seafloor (more intense in shallow waters) and "micro" multipath formed
       by refraction (more obvious in deep waters).
   (2) Finite frequency band
       The presence of absorption effect results in different bandwidths for different
       transmission distances.

UWAC is a generally accepted method of communication in the water. However,
due to the complex, random, and variable characteristics of UWA channels, there is
no commercial UWAC system that can meet consumers' needs in terms of communication distance, communication rate, communication reliability, communication
direction (horizontal or vertical channel), and information transmission form at the
same time.

On the other side, limited by one of the most complex channels, i.e., the
UWA channel, the UWA network faces great challenges in terms of design, signal
processing, etc. There are several main disadvantages that limit the performance of
UWA network[1]:

1. Limited communication capability
   Compared with terrestrial networks, the design and construction of UWA networks are more difficult, largely due to the complex time, space, and frequency
   variation of the UWA physical layer, strong multipath, high noise, and Doppler
   effects.
2. Low communication efficiency and unstable topology
   Because the network nodes work underwater for a long time, the energy supply
   is severely limited. And the life of the battery usually directly determines the life

of the network node, so the problem of resource saving needs to be considered as much as possible when designing the system.

The above two points imply that the scale of UWA network is severely limited, resulting in the current UWA network not being able to scale and be long-lasting. The large-scale network makes the research more complex. Moreover, the larger the scale, the larger the coverage area, and the more nodes, the greater the communication delay. This will reduce the network's overall performance, increase energy consumption, and further shorten the network life. Therefore, current research is still limited to small networks for short periods.

5.2   Research Trends in UWAC Technology

5.2.1   In-band Full-Duplex (IBFD) UWAC

5.2.1.1   Performance Benefits of IBFD Technology

Currently, there are many studies and results about communication network protocols. However, technically mature terrestrial wireless network protocols fail in underwater communication networks due to long transmission delays, limited bandwidth, and high channel spatial and temporal variability of UWAC. At the same time, the shortage of available spectrum resources for UWA channels, severe Doppler expansion, and higher environmental noise also cause disadvantages such as inefficiency and limited throughput of underwater communication networks. Therefore, to improve the spectral efficiency and system throughput of UWAC networks under the severe bandwidth limitation of UWAC is the core problem faced by UWAC network technology.

To solve the contradiction between the increasing demand for wireless services and the scarce spectrum resources, the radio communication community has proposed the co-channel simultaneous full-duplex technology, which can transmit and receive electromagnetic wave signals simultaneously at the same frequency resources and at the same moment and therefore theoretically can double the existing spectrum efficiency. This provides a new way of thinking to improve the performance of UWAC networks by applying the in-band full-duplex (IBFD) technology to the relaying node of UWAC networks, which can transmit and receive signals simultaneously and reduce the delay time of end-to-end communication between the nodes of the passing network without increasing the passband range [2–5].

Compared with the traditional half-duplex UWAC system, IBFD UWAC technology allows the communication system to add a new optional mode in the uplink and downlink according to the network throughput requirements. It can switch between full-duplex mode and half-duplex mode flexibly. In particular, when two nodes transmit the same frequency band information at the same time, the eavesdropper receives the superposition of the signals sent by the two nodes. This makes the eavesdropper unable to demodulate the signal sent by any node correctly, thus ensuring the security of the communication network to a certain extent.

According to Shannon's theorem, the channel capacity of a half-duplex communication system with fixed bandwidth and after a channel with the Gaussian white noise interference can be expressed as

$$C_{\mathrm{HD}} = B\log_2\left(1 + \frac{S_e}{N}\right), \tag{5.1}$$

where C_{HD} is the channel capacity of the half-duplex communication system (bit/s), B is the bandwidth of the half-duplex communication system (Hz), S_e is the average power of the desired signal (W), and N is the received noise power (W).

For an IBFD communication system, which allows simultaneous bi-directional communication within the same bandwidth, the channel capacity can be expressed as

$$C_{\mathrm{IBFD}} = 2B\log_2\left(1 + \frac{S_e}{I_{\mathrm{IBFD}} + N}\right), \tag{5.2}$$

where C_{IBFD} is the channel capacity of IBFD communication system (bit/s), and I_{IBFD} is the residual interference after self-interference cancellation. By transforming Eqs. (5.1) and (5.2), we can obtain the spectral efficiency SE_{HD} and SE_{IBFD} (bit/s/Hz) of half-duplex and IBFD communication systems as

$$SE_{\mathrm{HD}} = \frac{C_{\mathrm{HD}}}{B} = \log_2\left(1 + \frac{S_e}{N}\right), \tag{5.3}$$

$$SE_{\mathrm{IBFD}} = \frac{C_{\mathrm{IBFD}}}{B} = 2\log_2\left(1 + \frac{S_e}{I_{\mathrm{IBFD}} + N}\right). \tag{5.4}$$

The spectral efficiency gain (SEG) for IBFD communication systems is defined as

$$SEG = \frac{\frac{C_{\mathrm{IBFD}}}{B}}{\frac{C_{\mathrm{HD}}}{B}} = \frac{2\log_2(1 + \frac{S_e}{I_{\mathrm{IBFD}} + N})}{\log_2(1 + \frac{S_e}{N})}. \tag{5.5}$$

The IBFD communication system SEG variation curve with SNR and interference noise ratio (INR) can be obtained by simulating Eq. (5.5) as shown in Fig. 5.1. This demonstrates that the IBFD spectral efficiency gain increases with the decrease of the residual interference signal energy when the expected SNR is constant. When the self-interference signal is completely eliminated, the spectrum efficiency of an IBFD communication system can be doubled theoretically. Therefore, to effectively eliminate the strong self-interference signal is the key problem to be solved in the IBFD communication system.

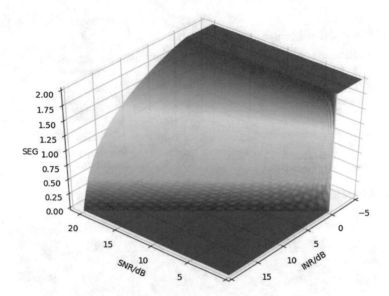

Fig. 5.1 Spectrum efficiency gain curve of IBFD communication system

A typical IBFD-UWA modem structure is shown in Fig. 5.2. A prototype IBFD-UWA modem produced according to the structure in Fig. 5.2 is shown in Fig. 5.3. The working scenario and self-interference propagation schematic of the IBFD-UWA modem in a typical shallow sea environment is shown in Fig. 5.4. The transmit signal $x[k]$, when transmitted through the transmit link to the transmit transducer, is represented by the corresponding equivalent baseband as $x_{TX}[k]$. The effect of the nonlinear gain of the amplifier leads to the nonlinear distortion contained in $x_{TX}[k]$. Subsequently, $x_{TX}[k]$ at the transmitting transducer propagates through the UWA channel and reaches the receiving hydrophone, forming the self-interference. Assuming that the UWA channel has linear properties, the effect of the channel on the channel can be described by the channel impulse response $h_u[m, k]$, i.e.,

$$x_{SI}[k] = \sum_{m=0}^{L_u-1} h_u[m, k]x_{TX}[n - k]. \tag{5.6}$$

Considering that the hydrophone receives both the desired signal $d[k]$ and the observed noise $w[k]$, the received signal can be expressed as

$$x_{RX}[k] = x_{SI}[k] + x_d[k] + n[k]. \tag{5.7}$$

The purpose of self-interference suppression is to reconstruct the estimated value $\hat{x}_{SI}[k]$ of $x_{SI}[k]$ using the estimated channel $\hat{h}_u[m, k]$, to obtain the suppressed signal $x_c[k]$ by superimposing the received signal $x_{RX}[k]$ with opposite phase copy

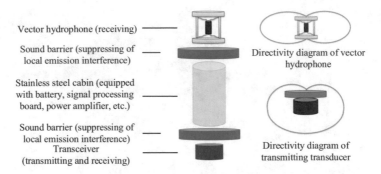

Vector hydrophone (receiving)

Sound barrier (suppressing of local emission interference)

Stainless steel cabin (equipped with battery, signal processing board, power amplifier, etc.)

Sound barrier (suppressing of local emission interference)
Transceiver
(transmitting and receiving)

Directivity diagram of vector hydrophone

Directivity diagram of transmitting transducer

Fig. 5.2 A typical IBFD-UWA modem structure

Fig. 5.3 A prototype IBFD-UWA modem

$\hat{x}_{SI}[k]$ in the analog or digital domain, and to minimize the value of residual self-interference $x_{SI}[k] - \hat{x}_{SI}[k]$ in $x_c[k]$. In IBFD UWA systems, due to the effect of simultaneous transceiver operation mechanism, the local transducer transmit signal forms a serious self-interference signal, which is received by the local hydrophone along with the distal desired signal.

Therefore, in order to reduce the self-interference energy of available data, effective analog and digital self-interference suppression are implemented in the analog domain and digital domain, respectively. This can improve the probability of correct demodulation of the signal required by the far end and is the key technical way to improve the performance of IBFD-UWA communication. It can be seen that a full understanding of the shallow sea self-interference signal transmission mechanism, effective observation of the self-interference channel space–time–frequency characteristics, and improvement of the estimation accuracy are the basis for achieving high suppression performance.

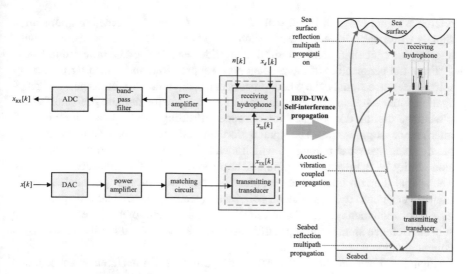

Fig. 5.4 Schematic diagram of IBFD-UWA modem

The self-interference signal in IBFD communication systems can be divided into three main categories in terms of generation mechanism, i.e., linear component, nonlinear component, and transmitter noise. The linear component is the largest energy component of the self-interference signal, which is composed of the interference signal propagated directly from the transmitter to the near-end receiver and after several environmental reflections to the near-end receiver. And it can be expressed as a linear combination of different time-delayed copies of the emitted local reference signal. The nonlinear component is generally caused by the circuit nonlinear devices in the IBFD communication system, mainly introduced by the power amplifier. If this component is not offset, it will affect the overall system self-interference offset effect. The transmitter noise generated by the transmitter analog circuit part of the high-power components is generally lower than the transmit signal energy with about 60 dB. From the energy point of view, the component is still much larger than the desired signal energy.

5.2.1.2   Self-interference Cancellation Technology

The self-interference signal strength needs to be reduced to the background noise level at the receiver to achieve the desired effect of IBFD communication. So, in addition to the removal of the linear and nonlinear components, the cancellation of the transmitter noise needs to be accomplished. Due to the random nature of the transmitter noise, which cannot be predicted by modeling, for example, it needs to be captured during the cancellation of the transmitter noise. This means that any full-duplex system needs to have an interference cancellation process on the analog domain or an interference cancellation process for the transmitter noise with the reference signal range in the digital domain. In addition, for some communication signals with peak-to-average ratios such as OFDM, further improvements in self-interference cancellation are required to provide peak-to-average ratio redundancy.

From the perspective of the propagation, suppression, and offsetting process of self-interference signals, the whole process can be divided into four types, i.e., propagation domain self-interference channel modeling, analog domain self-interference offsetting, digital domain self-interference offsetting, and spatial domain self-interference suppression. The spatial domain self-interference suppression is a passive interference suppression, which means it can further improve the suppression ability of a full-duplex communication system for self-interference signals. The spatial domain interference suppression is mainly through physical isolation, transceiver directivity, and antenna polarization techniques to obtain the gain effect by spatial redundancy. The above domains gradually complete the elimination of self-interference signals through different technical means and finally realize IBFD communication.

Figure 5.5 shows the performance of various self-interference cancellation schemes, including the amplifier-assisted scheme using the amplifier output signal as the reference signal and the direct self-interference cancellation scheme. Figure 5.5 compares the effect of different effective quantization bits in terms of the self-interference cancellation algorithm's performance.

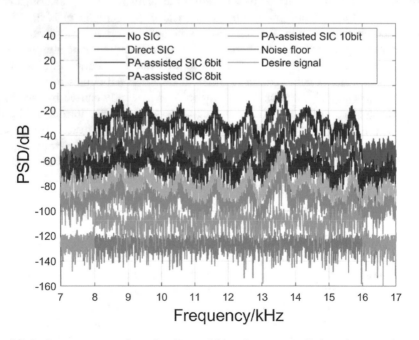

Fig. 5.5 Performance comparison of various serial interference cancellation schemes

Compared to the radio frequency (RF) counterpart, full-duplex UWA communication technology is more complex due to the specific complexity of the UWA channel, which exists mainly in the following areas:

1. Due to the slower speed of sound, the delay of the self-interference signal and the corresponding number of self-interference channel taps of full-duplex UWA communication are much larger than those of the RF counterpart. Therefore, the analog self-interference canceling circuit scheme with delay line structure commonly used in full-duplex RF communications cannot be implemented, so a new analog self-interference canceling method needs to be developed.
2. Due to the effects of scattering, diffraction, and bypassing of sound field propagation, the passive isolation scheme is not significant, thus requiring a higher canceling performance of the self-interference canceling scheme. In addition, the ocean noise has non-Gaussian characteristics and the presence of impulse noise, thus putting higher demands on the robustness of the adaptive algorithm. Finally, the mechanical structure of the acoustic field and the UWA communicator will generate coupled vibrations, thus making the actual self-interference channel structure more complex, which will further increase the difficulty of self-interference cancellation.
3. Since UWA communication is broadband by nature, the narrowband self-interference canceling scheme commonly used in full-duplex wireless communication cannot be used. In addition, because of the complex time-varying multipath effects in the actual marine environment, especially in the shallow sea environment, an effective time-varying broadband self-interference canceling scheme needs to be developed for full-duplex UWAC technology.

According to the current status of research, IBFD communication is mainly based on land-based wireless communication research. And the research on IBFD UWAC technology is in the initial stage, mainly focusing on propagation channel estimation, measurement, and digital dry self-turbulence offset technology, which has not yet formed a perfect theoretical framework.

5.2.2 Advanced Modulation Schemes

The very significant practical importance of OFDM technology for wireless communications is attributed to the computational advantages brought by the FFT algorithm, such as the use of orthogonal frequency division multiplexing multiple access (OFDMA) for LTE systems. The main advantage of orthogonal multiple access (OMA) is that inter-user interference can be avoided under ideal conditions, and this can largely simplify system and protocol design, such as detection, channel estimation, and resource allocation. However, on the other hand, the dimensionality of the orthogonal decomposition limits the number of users that an OMA system can support. In fact, orthogonality is often lost because of the effects of frequency selectivity, phase noise, and frequency fading of the channel. In addition, because other users cannot share the bandwidth resources occupied by the users with poor channel conditions, the spectral efficiency of these OMA schemes is low. In case of fifth generation (5G) mobile internet with the explosive growth of data traffic, these OMA schemes are not applicable. Furthermore, OMA is not optimal from an information theory perspective.

OFDM systems have the following drawbacks:

1. OFDM-based systems have a high peak-to-average power ratio. This is due to the statistical probability of many independent sub-channels superimposed together in an unknown way in OFDM. Therefore, it is necessary to use a low-power linear power amplifier at the transmitter.
2. The rectangular-wave pulse waveform of OFDM signals leads to large spectral partials and very high out-of-band radiation.
3. Carrier orthogonality is sensitive to frequency fading and phase noise of the carrier.

Therefore, OFDM systems are not suitable for data transmission in many 5G applications. As a result, some non-orthogonal waveforms have been proposed and regarded as an alternative to the 5G physical layer. For example, generalized frequency division multiplexing (GFDM) [6], based on independent block modulation where each block consists of several sub-signals and sub-carriers, employs cyclic prefixes and cyclic filtering in GFDM. And in particular, GFDM employs tailing techniques to reduce the length of the signal pulse tail by cyclic filtering. The cyclic signal framework of GFDM also enables the utilization of one CP for the whole data block that contains multiple GFDM signals, which offers greater spectral efficiency than conventional OFDM. In fact, GFDM is an adaptable physical layer scheme because it includes both CP-OFDM and single-carrier frequency domain equalization (SC-DFE) as special cases. In addition, GFDM allows adjusting the time and frequency interval of each data symbol with respect to the channel properties and the application type.

OMA's shortcomings can be avoided by non-orthogonal multiple access (NOMA) techniques. Specifically, non-orthogonal resource assignment implies that the number of available orthogonal resources does not severely restrict the number of supported users or devices. Thus, by utilizing non-orthogonal resource allocation, NOMA can support more users than OMA. Multiple users in NOMA are scheduled on the same resource (e.g., on the same time, spectral, and spatial dimensions), and in NOMA, each user can utilize the whole bandwidth for the entire communication time, which may increase the user's data rate and reduce latency. In other words, since NOMA uses the power domain for multiple access, different power levels can serve different users. Therefore, a certain amount of inter-user interference introduced by non-orthogonal transmissions is acceptable and is removed by the receiver through serial interference cancellation (SIC). Due to its efficient resource utilization, both academia and industry consider NOMA as a core technology for 5G systems.

OFDM can efficiently deal with the multipath fading of the UWA channel and its large delay expansion. In fact, its symbols are so long that it can overcome

ISI (inter-code crosstalk). However, OFDM also has the following drawbacks. The relative motion of the receiver and transmitter in the UWA channel is prone to time-varying frequency shifts. The addition of a cyclic prefix (CP) at the beginning of each symbol as a protective interval to handle ISI problems makes OFDM-based communication systems inefficient in terms of spectrum utilization. High out-of-band (OOB) leakage exists due to the use of rectangular pulse shaping filters at the transmitter end. OFDM also has a high mean-to-peak ratio due to the random summation of in-phase sub-carriers. OFDM systems use part of the symbol length for CP, which is greatly increased in UWAC due to longer delay propagation since it must be equal to the length of the channel impulse response. Because the UWA channel introduces dispersion, long symbols will lead to increased degradation based on ICI with lower channel utilization.

For GFDM UWA systems, the transmitted data is first classified into sub-signals and sub-carriers in terms of two-dimensional blocks(frequency and time), which are then filtered by a pulse-shaping prototype, as illustrated in Fig. 5.6. The time and frequency response of the filter prototype may change accordingly if the requirements are different. The better spectral efficiency and resistance to ICI and ISI of GFDM in practical applications make it ideal for underwater communications. By making a slight sacrifice in computational complexity, GFDM provides a more robust and applicable solution for underwater medium transmission. Available simulations have demonstrated that GFDM systems exhibit better SER performance than OFDM systems in UWA channels. In addition, the gain in spectral efficiency of GFDM systems is an outstanding advantage over other candidate systems, as GFDM is able to exploit fragmented spectrum and different spectral efficiencies

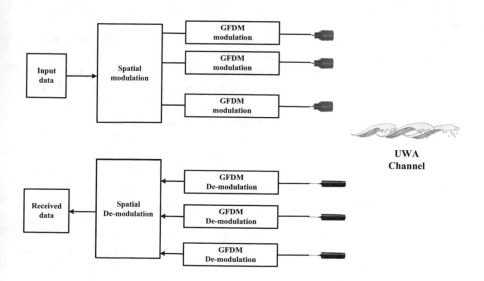

Fig. 5.6 Schematic diagram of GFDM-based UWAC system

through different pulse shapes and their roll factors. For these reasons, GFDM is an ideal choice in spectrum-constrained UWA networks.

Compared to cyclic prefix OFDM (CP-OFDM) systems, GFDM reduces OOB leakage. GFDM offers high flexibility in frequency-domain localization, MIMO integration, and scalability for UWA fields with limited power and bandwidth. Due to GFDM's higher spectral efficiency, flexibility, and better BER performance, GFDM is still a popular technology today, despite its higher computational demand than OFDM.

5.2.3   Massive MIMO

Multi-antenna technology is a critical technology in wireless communication systems and has a broad application prospect. Traditional point-to-point MIMO systems suffer from several shortcomings in practice. Firstly, the number of antennas is limited by the size, power consumption, and price of the mobile terminal (e.g., smartphone), which negatively affects the multiplexing gain. Secondly, the multiplexing gain can be weakened or even lost due to strong interference, unfavorable channel conditions, and narrow spacing between antennas due to the size of the mobile terminal. In traditional multiple-input multiple-output designs, the number of antennas is limited to 20 or less. However, today it has been proven that when the number of antennas in a base station is increased to several hundreds or thousands, MIMO communication systems between multiple users show some excellent features, resulting in the so-called massive MIMO systems. Since 2010, a lot of theoretical and practical research on massive MIMO technology has been carried out.

Unlike traditional multi-user MIMO systems that use relatively few base station antennas, the scale of antenna used for large-scale MIMO systems will reach hundreds or even thousands. Although the large increase of antennas is a new challenge for the design and implementation of transceiver signals, massive MIMO systems still have their unique advantages in signal processing and communication. For instance, the base station has far more antennas than the number of users in the system. Simply matched filter precoding (downlink) and simply matched filter detection (uplink) provide the best performance and facilitate better processing of low-complexity signals between the base station and user terminals. In addition, some random impairments such as small-scale fading and noise will be averaged out due to the increased number of base station antennas.

Improving energy efficiency is also a potential for massive MIMO. It has been verified that if the number of transmitting antennas N_t becomes larger and assuming the other parameters of the system remain constant, the transmitting energy per user in a multi-user massive MIMO system is reduced to the original $1/N_t$ and $1/\sqrt{N_t}$ with perfect and imperfect CSI information without affecting the bootstrap throughput and stability. Therefore, large-scale MIMO systems are a more energy-efficient and green communication technology. In addition, the main concerns of security and privacy for future wireless communication systems are well addressed in massive MIMO systems. In fact, due to a large number of free space degrees, massive MIMO can be used to protect cellular network systems from active and passive eavesdropping.

Because of the good performance, massive MIMO is used as a core technology in 5G systems. However, there are still many research challenges and issues with this technology. For example, in a massive MIMO system, the hardware overhead will increase significantly. Although there is an option to use cheaper hardware components, but this increases the hardware corruption rate, bringing phase noise, in-phase/positive phase imbalance and non-linear amplifiers. These problems must be addressed in an appropriate way, or performance is degraded. New resource allocation and user-related algorithms are also needed to address the channel hardening effect due to a large number of base station antennas.

The potential benefits of Massive MIMO provide a viable scheme for achieving high-quality UWACs. However, because of the bandwidth nature of UWA channels, existing large-scale wireless MIMO technologies cannot be directly transposed to the UWAC field. UWAC offers significant performance superiority in the underwater environment compared with RF and optical waves, but the underwater channel still faces the problems of frequency selective fading, low sound speed, and very strong multipath interference. The UWAC system can be considered broadband because of the large ratio of the signal bandwidth to the center frequency. Therefore, the OFDM technique is widely used in UWAC for low complexity dispersive channels.

To further improve the throughput and reliability of UWAC systems, research on various aspects of MIMO technology in UWAC has been carried out. Many sea trials have demonstrated that MIMO-OFDM is an attractive choice for UWAC, and data rates of 125.7 kb/s can be achieved at a bandwidth of 62.5 kHz using iterative detection and Doppler compensation techniques. In addition, another popular scheme is the DFT precoded MIMO-OFDM UWA transmission.

However, the data rate of existing UWA systems is limited to a few hundred kilobits per second. With the demand for large data volumes (e.g., real-time video), higher demands are placed on UWAC systems. Military applications make it necessary to take into account the security in the underwater channel as well. As mentioned above, massive MIMO has the advantages of background noise averaging, security, and fast transmission rates, so massive MIMO for underwater environments is starting to receive attention and research. Large-scale use of receiving hydrophones has been used to establish highly stable underwater sensing networks. Real-time video and multi-user UWA transmissions based on filtered multi-carrier modulation by using a large number of receive hydrophones at the receiver side have been investigated separately. Moreover, with multiple transducers deployed at the transmitter and multiple hydrophones deployed at the receiver simultaneously, the obtained high beamforming gain is very helpful in combating serious propagation path loss in the UWA channel while greatly improving the transmission capacity.

Regard a shallow water large-scale MIMO UWA simulation, where transmitter and receiver equipped with many transducers and hydrophones, respectively.

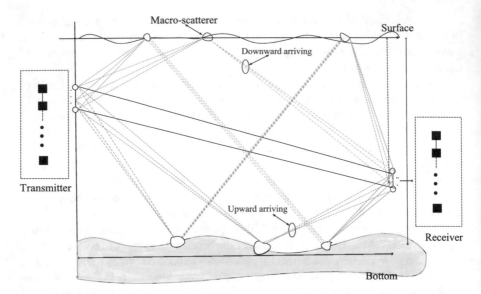

Fig. 5.7 Illustration of a UWA point-to-point massive MIMO channel

Considering the macro scattering effect, a large-scale MIMO channel model based on geometry is established, as shown in Fig. 5.7 [7]. In the shallow water channel model, it can be seen that each element of the channel in beam domain tends to be independent, depending on the joint variance of angle of departure and angle of arrival pairs, which is independent of time and frequency. The input covariance matrix can be optimally built from the reduced dimension matrix. When the number of sensors in the transmitter is infinite, the beam space transmission can be realized to maximize the rate. If the amount of receiving hydrophones tends to infinity, the water-filling algorithm can be used. Thus, the optimal power allocation of beam space can be achieved, and the modeling and analysis of large-scale underwater MIMO can be completed. Due to the complexity of underwater channels and the advantages of large-scale MIMO, the application of large-scale MIMO in UWA system (i.e., large aperture array in both transmitter and receiver) is worthy of serious study, which can serve the future underwater wireless communication.

5.2.4   Machine Learning-Based UWAC Technology

In 1950, Alan Turing discussed the possibility of creating "intelligent machines" in computing machinery and intelligence. The introduction of the Turing test led to the development of artificial intelligence in a variety of research areas. In the development process of artificial intelligence, it was hoped that computers could automatically analyze data and discover patterns in it to help people improve

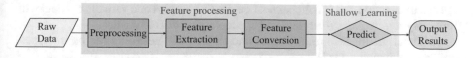

Fig. 5.8 Flow chart of machine learning data processing

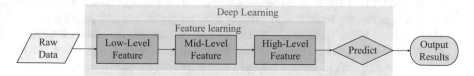

Fig. 5.9 Flow chart of deep learning data processing

productivity, and machine learning began to emerge and gradually became an important part of the artificial intelligence field. Machine learning data processing flow chart is shown in Fig. 5.8. According to different learning methods, machine learning can be divided into unsupervised learning, semi-supervised learning, supervised learning, and reinforcement learning.

As the data to be processed becomes more complicated, models with a certain "depth" need to be constructed in order for them to make a better feature representation of the data information. Therefore, deep learning began to develop rapidly as a subproblem of machine learning. The data processing flow chart of deep learning is shown in Fig. 5.9. Unlike shallow learning in machine learning, deep learning requires multi-step feature transformation between the data features of the original data from the bottom features to the top features. Meanwhile, the model used in deep learning is mainly a neural network model, which can solve the contribution assignment problem (Credit Assignment Problem) by the error backpropagation algorithm. In recent years, with the increase of computer computing power and the widespread popularity of high-performance GPUs, deep learning based on neural networks has become the most mainstream artificial intelligence technology at present. Deep learning is mainly divided into three types: feedforward neural networks, convolutional neural networks, and recurrent neural networks depending on the neural network models used.

Thanks to the rapidly developing information technology and big data, artificial intelligence-based technologies have been widely used in many fields such as computer vision, natural language processing, and speech recognition. In order to break through the difficult problems of traditional UWAC technology, the existing artificial intelligence technology will provide new ideas for UWAC technology.

As the information carrier for long-distance transmission in the marine environment, most of the underwater communication technology research is based on acoustic signals. The difference between UWAC and terrestrial wireless communication lies in the special nature of the UWA channel. As the marine environment is affected by various factors, such as depth, salinity, temperature, refraction, and interface reflection, the UWA channel becomes a complex and variable channel with time-varying, frequency-varying, and space-varying characteristics.

Artificial intelligence technologies and UWAC and networks can be integrated with the following ways [8]:

1. Artificial intelligence-based UWA channel estimation and equalization techniques. In OFDM UWA communication systems, obtaining channel state information is necessary to deal with various channel distortions and interference in channel estimation. Traditional channel estimation methods such as least squares (LS) and MMSE can lead to severe degradation in performance estimation due to complex multipath channels. Channel estimators based on deep neural network (DNN) models can improve the performance of the communication system by training the deep neural network model and then estimate the channel impulse response. For fast fading UWA channels, the support vector machine (SVM)-based blind equalization algorithm for UWA channels can complete convergence by secondary planning iterations. In addition, SVM can also be used for blind equalizer weight coefficient initialization.

2. UWA adaptive modulation communication technology based on artificial intelligence. In UWAC, significant propagation delay is the main challenge to realize an adaptive modulation scheme. Due to the long propagation delay, when the transmitter receives the channel state information, it is often out of date, limiting the gain of feedback and even leading to communication failure. The problem of adaptive modulation can be solved by reinforcement learning. Dyna-Q algorithm in reinforcement learning, on the one hand, predicts the channel state and calculates the communication throughput according to the actual and simulated data communication experience of each modulation order in different channel states. On the other hand, in order to maximize the communication throughput, the Dyna-Q algorithm selects the modulation order according to the predicted channel state, the calculated throughput, and the channel state returned by the receiver.

3. Artificial intelligence-based quality prediction technology for UWA communication. A machine learning model based on logistic regression (LogR) is used to capture the effects of environmental factors such as wind speed, tide, current speed, and some specific factors on the performance of the UWA network link. In order to adapt the single-carrier UWAC system to the time-varying nature of the UWA channel, a DNN model is used in the receiver section. An online training mode and an online testing mode are used alternately for the model. This improves the system performance of the UWA communication system under time-varying UWA channel conditions.

4. Artificial intelligence-based security technology for UWA communication. Jamming attacks can further lead to denial-of-service attacks and pose a serious threat to underwater sensor networks (UWSNs). The interaction between underwater sensors and jammers in UWSNs can be considered as an underwater jamming game in which the participants choose their transmit power based on SNR and the transmission cost of legitimate signals. A sensor network power control strategy can be implemented based on the Q-learning algorithm in reinforcement learning.

1. What are the main challenges of UWA communication?
2. What are the characteristics of UWA network?
3. What performance gains can be obtained with full duplex technology? What are the technical difficulties in realizing full-duplex UWA communication?
4. Briefly describe the advantages of GFDM over OFDM.
5. Briefly describe the definition of massive MIMO and its advantages.
6. Briefly describe how artificial intelligence and UWAC communication and networks can be combined.
7. Given that the average receive power of the desired signal is 40 dBW and the noise power is 1 dBW, calculate the spectral efficiency of the IBFD-UWAC when the energy of the self-interference signal is 10 dBW and 0.001 dBW, respectively.
8. In a UWAC network with a single base station, the transmit power of the base station is 100 W with single transducer. If the number of base station transmit transducers is changed to 100, under perfect and imperfect CSI information, what is the maximum amount the transmit power of the base station can be reduced to, respectively?

References

1. A. Song, M. Stojanovic, M. Chitre, Editorial underwater acoustic communications: Where we stand and what is next? IEEE J. Ocean. Eng. **44**(1), 1–6 (2019)
2. Y. Zhao, G. Qiao, S. Liu et al., Self-interference channel modeling for in-band full-duplex underwater acoustic modem. Appl. Acoust. **175**, 107687 (2021)
3. B.A. Jebur, C.T. Healy, C.C. Tsimenidis et al., In-band full-duplex interference for underwater acoustic communication systems, in *OCEANS 2019-Marseille* (IEEE, Piscataway, 2019), pp. 1–6
4. Y. Wang, Y. Li, L. Shen et al., Acoustic-domain self-interference cancellation for full-duplex underwater acoustic communication systems, in *2019 Asia-Pacific Signal and Information Processing Association Annual Summit and Conference (APSIPA ASC)* (IEEE, Piscataway, 2019), pp. 1112–1116
5. L. Shen, B. Henson, Y. Zakharov et al., Digital self-interference cancellation for full-duplex underwater acoustic systems. IEEE Trans. Circuits Syst. II Express Briefs **67**(1), 192–196 (2019)
6. M. Murad, I.A. Tasadduq, P. Otero, Towards multicarrier waveforms beyond OFDM: performance analysis of GFDM modulation for underwater acoustic channels. IEEE Access **8**, 222782–222799 (2020)
7. W. Wu, X. Gao, C. Sun et al., Shallow underwater acoustic massive MIMO communications. IEEE Trans. Signal Process. **69**, 1124–1139 (2021)
8. R. Wang, A. Yadav, E.A. Makled et al., Optimal power allocation for full-duplex underwater relay networks with energy harvesting: a reinforcement learning approach. IEEE Wirel. Commun. Lett. **9**(2), 223–227 (2019)

Part II
Underwater Optical Wireless Communication and Networks

Chapter 6
Basic Principles of Underwater Optical Communication

6.1 Introduction

In recent years, with the intensification of global warming and the gradual depletion of terrestrial resources, the study of ocean exploration systems has received more and more attention. A typical air–sea integrated communication network is shown in Fig. 6.1. Underwater wireless communication (UWC) technology is a crucial technology to realize ocean exploration. UWC refers to data transmission through wireless carriers in unguided water environments. UWAC has become one of the most commonly used UWC technologies because of the considerable communication distance it can provide. However, UWAC systems have disadvantages such as low bandwidth, high latency, multipath effect, and Doppler shift. These disadvantages lead to spatial and temporal variations in the acoustic channel, limiting the system's available bandwidth. Besides, the RF signal is subject to a high degree of attenuation in seawater due to the need for high power transmitters and antennas of enormous size. Optical communication technology has a much higher bandwidth and communication rate than RF and acoustic technologies. With the increasing human demand for underwater high-speed data transmission, UWOC technology has become a key solution to meet the demand for underwater high-speed communication [1].

Compared to UWA and RF technologies, UWOC technology offers ultra-high data rates, very low link latency, and very low energy consumption. Specifically, UWOC technology enables UWC at Gbps over distances of tens of meters. The advantages of high speed make UWOC suitable for many underwater high-speed communication applications, such as underwater video transmission and large-scale underwater data sensing. Because light travels much faster than sound waves in water, UWOC links have extremely low latency. In addition, UWOC offers a higher level of communication security than acoustic and RF communications.

© The Author(s), under exclusive license to Springer Nature Switzerland AG 2022   119
Y. Lou, N. Ahmed, *Underwater Communications and Networks*, Textbooks in Telecommunication Engineering, https://doi.org/10.1007/978-3-030-86649-5_6

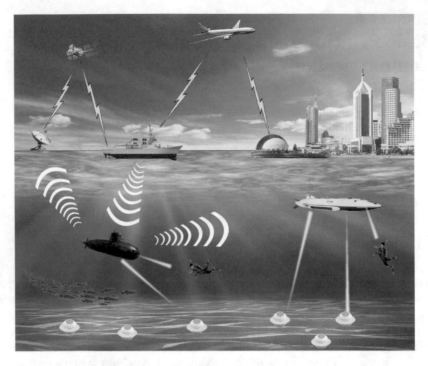

Fig. 6.1 Underwater wireless communication network

Figure 6.1 shows an underwater wireless network incorporating multiple communication technologies. In contrast to UWA and RF communication systems that transmit information via diffuse waves, most UWOC systems are implemented via line of sight (LOS) transmission and are almost impossible to eavesdrop on. Besides, optical components' energy consumption and cost are much lower than those of acoustic and RF components. Compared to acoustic transducers and RF transceivers, which consume huge amounts of energy, the key optical components of underwater transceivers such as laser diodes (LD) and light-emitting diode (LED) in UWOC systems have the advantage of small size and low cost. These advantages facilitate the large-scale commercialization of UWOC while accelerating the realization of underwater wireless communication networks.

Despite the incredible advances in UWOC technology, there are still various obstacles in the UWOC field that are yet to overcome. First, in seawater, optical transceivers' links may be disconnected for short periods due to sea surface fluctuations, ocean turbulence, and subsea turbulence. Second, scattering and turbulence can have a serious negative impact on the UWOC, leading to multipath fading. Although blue-green light has a high transmission in seawater compared to other wavelengths of the beam, however, the above factors still lead to severe degradation of UWOC system performance and limit the communication range of UWOC.

There is still a plenty of room for improvement in the communication performance of UWOC systems. For developers of underwater wireless optical communication technology, further research and analysis of new theoretical models remain the focus of making UWOC a reality soon. In order to improve the robustness of UWOC systems and increase the system operating life, adaptive techniques to save energy and ensure link connectivity also need to be investigated. Besides, rational coding and modulation techniques can improve wireless optical communication systems' reliability in marine channels with spatial and temporal variations. Finally, routing protocols suitable for underwater wireless optical communication networks still need further research.

This book focuses on the subsystem technologies of UWOC systems such as coding and modulation, analyzes the key technologies to improve the performance of UWOC systems, and discusses the feasibility of solutions to improve the performance of underwater wireless communication network.

6.1.1 History

In 1917, Einstein pointed out that in addition to spontaneous radiation, particles at higher energy levels could also leap to lower energy levels in another way. With this as a basis, the laser was introduced in 1960. The invention of the laser gave an impetus to the research of wireless optical communication technology. Since 1960, the technology of terrestrial free space optical communication (FSO) has been developed at a high speed. However, because the exploration of underwater wireless communication was still in its infancy and the light beam had severe attenuation during underwater propagation, UWOC was not considered as an effective UWC method and its development almost came to a standstill from 1960s to 1980s. It was in 1990s, after nearly 20 years of experimental and theoretical research on underwater light propagation, that UWOC was considered as a promising UWC solution. The discovery of the underwater blue-green light transmission window provided the key theoretical support for the UWOC technology [2]. Duntley proposed in 1963 that, influenced by algal photosynthesis, blue-green visible light with wavelengths from 450 to 550 nm has a higher transmittance and lower attenuation in seawater compared to other wavelengths of light. Therefore, underwater optical communication sensor system can work in the blue and green spectral range. Figure 6.2 depicts the underwater attenuation of visible light at different wavelengths. Subsequently, the existence of a blue-green transmittance window in water was experimentally confirmed by Gilbert et al.

Early UWOC research was mainly for military needs. In 1976, Karp proposed an optical communication system to be deployed between underwater sensors and ground-based stations and evaluated the feasibility of the communication system. In early 1990s, the U.S. Navy conducted a large number of UWOC experiments. In 1992, researchers experimentally demonstrated that UWOC with 50 Mbps communication rate could be achieved in a 9 m long underwater link using

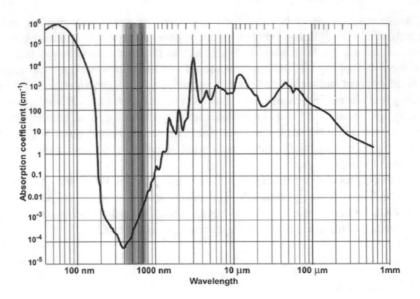

Fig. 6.2 Attenuation of the ocean to light of different wavelengths

a 514 nm ion laser. In 1995, based on theoretical analysis, a UWOC system using LED as transmitters was capable of 10 Mbps wireless optical communication at distances of 20 and 30 m. In 2005, Vasilescu et al. designed a point-to-point wireless optical communication link that is capable of optical communication at a data rate of 320 Kbps over a range of 2.2 m. This technique provided a theoretical basis for use in the deployment of underwater optical sensor networks.

Since the twenty-first century, reducing the transmitter–receiver pointing requirements and improving the performance of communication systems have become the focus of research in UWOC technology. To reduce the pointing requirements of UWOC system, in 2006, researchers tested an LED-based omnidirectional optical communication link in a submarine observatory. In addition to enhance the reliability of UWOC, in 2005, using frequency modulation technique, Chancey et al. established a communication link with 5 m distance and a communication rate of 10 Mbps. Simpson et al. extended Chancey's work by studying modulation and error correction techniques dedicated to UWOC systems. In 2010, a high power LED array bi-directional communication system using a 470 nm light source was proposed. The system used discrete pulse interval modulation technique to achieve wireless optical communication up to 50 m. It can be seen that almost all UWOC systems based on LED transmitters can achieve wireless communication at Mbps order of magnitude data rate in pure seawater. However, their transmission range is limited by the diffusion angle, transmitting power, and water environment.

In the last decade, laser diode (LD) has been widely used as a light source in UWOC systems in order to obtain higher data rates and communication atmosphere. In 2008, Hanson and Radic conducted experiments with LD at a wavelength of

532 nm as a light source, and the UWOC system covered a distance of 2 m with data rate of up to 1 Gbps. In 2015, Oubei et al. conducted experiments with a laser at a wavelength of 405 nm, and at a transmission range of 4.8 m, the UWOC link was able to provide a high communication rate of 1.45 Gbps. To further improve the transmission distance and data rate, Oubei and Duran et al. conducted experiments. The experimentally tested UWOC link was able to achieve wireless communication at a data rate of 2.3 Gbps over a transmission distance of 7 m, respectively, and the communication rate could be increased to 2.488 Gbps in a 1 m link. Based on this, using OFDM, Oubei et al. further improved the data rate of the UWOC system. The experimental results show that the data rate of UWOC system with QAM-OFDM technique in clear water medium is up to 4.8 Gbps. In addition, for high attenuation underwater channel, Wu et al. conducted experiments on UWOC system in turbid water. It was tested that the UWOC system applying 32 QAM-OFDM was able to achieve a communication rate of 4.88 Gbps over a transmission distance of 6 m.

UWOC technology is still heavily used in military applications, and UWOC has not been successfully mass-marketed. Only a very few UWOC products have been commercialized. These products include BlueComm (produced by Sonardyne), which enables 20 Mbps underwater data transmission at a distance of 200 m, and LC50CCA-50/40WD dual-source underwater wireless optical communication device (produced by SDFSO), which achieves 50 Mbps at a distance of 20 m. In addition, multimodal UWC systems consisting of UWAC and UWOC technologies have received a lot of attention in recent years, where acoustic waves are used for underwater telecommunication and downlink data transmission, and optical waves are mostly used for uplink data transmission. However, the research on multimodal UWC systems is still in the initial stage and more theoretical studies are needed.

6.1.2 Classification of Underwater Wireless Optical Communication Links

UWOC links have two key parameters, namely transmitter beamwidth and field of view (FOV) of the detector. Based on this, there are four typical configurations of UWOC links according to the different deployment methods between UWOC transceivers [3]:

(1) Point-to-point line of sight (LOS) configuration
(2) Diffused LOS configuration
(3) Retroreflector-based LOS configuration
(4) Non-line of sight (NLOS) configuration

Fig. 6.3 The point-to-point LOS configuration

6.1.2.1   LOS Configuration

The point-to-point LOS configuration is the simplest and most common UWOC configuration in terms of structure and is shown in Fig. 6.3. In a point-to-point LOS configuration, the transmitter is perfectly aligned with the receiver. At the receiving end, the beam coverage is usually smaller or close to the receiving lens area. In the LOS configuration, the energy loss caused by the beam diffusion is usually not considered. Therefore, the optical signal power arriving at the receiver is obtained by multiplying the transmit power, the path loss factor, and the gain of the focusing mirror. Point-to-point LOS UWOC systems typically use light sources with low divergence angles (e.g., LD) to achieve higher communication rates and distances, so the transmitter and receiver must be closely aligned with each other. Achieving this requirement is challenging in turbulent water environments. In addition, when the sender and receiver are mobile nodes (frogmen or underwater robots), the alignment of the transmitter and receiver is even more difficult.

Advantages
(1) Long communication distance
(2) Very high data rates
(3) Increased energy efficiency with minimal path loss
(4) Reduced multipath distortion
(5) Minimal impact from background light
(6) Simple construction
(7) Low cost

Disadvantages
(1) Point-to-point links are very susceptible to blocking, leading to communication interruptions.
(2) Lower flexibility, as the transmitter and receiver must be closely aligned with each other, which is more difficult to achieve in channels with temporal variations.

Application
The point-to-point LOS link performs well in an ideal underwater environment. Since the transmitter emits the beam with a small diffusion angle directly to the receiver, the link enables wireless communication with low scattering effect and therefore reduces the energy loss. Due to the high concentration of beam energy, the background light negatively impacts the link. In addition, optimal

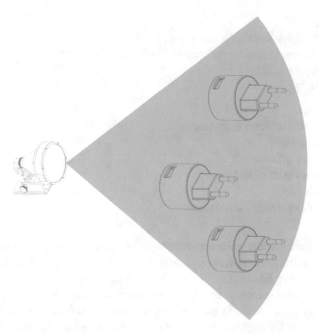

Fig. 6.4 Diffused LOS configuration

performance of the point-to-point LOS link is achieved when the transceiver is stationary, such as when two sensor nodes are fixed at seafloor. Conversely, for mobile platforms such as underwater robots or combat divers, effective targeting and tracking mechanisms must keep the transmitter and receiver aligned. In addition to alignment issues, marine organisms' activity may cause blurring or interruption of the beam. Therefore, in order to establish LOS links with robustness, it is necessary to exclude the interference of marine organisms to the links.

6.1.2.2 Diffused LOS Configuration

Diffused LOS configuration, shown in Fig. 6.4, uses diffuse light sources, such as LED with large divergence angles, so the use of diffused LOS configuration relaxes the requirement for precise pointing of the transmitter–receiver. In addition, the beam is more susceptible to the negative effects of the marine environment because the beam has a large diffusion angle. Scattering and absorption effects, geometric diffusion of the beam, and background light radiation can all have serious adverse effects on the link communication performance. Therefore, the shorter link distance and lower communication rate are the main factors limiting the performance of diffused LOS configuration compared to the LOS configuration.

Advantages
(1) Single to multi-point broadcasting is possible
(2) Transmitter and receiver do not need to be strictly aligned
(3) Simple construction
(4) Low cost

Disadvantages
(1) Shorter communication distance
(2) Lower data rates
(3) Low energy efficiency
(4) Severe multipath distortion
(5) High path loss
(6) Poor confidentiality performance

Application
For single-point-to-multi-point broadcast applications, diffusion LOS links perform
well. Although the links possess a shorter range of action, optical communications
can provide higher bandwidth (up to Gbps), making short-range communication
from a single-point source to multiple nodes feasible through a diffuse LOS architecture. These nodes include submarines, unmanned underwater vehicle (UUV),
frogmen, undersea sensors, surface base stations, etc. For example, the link can
accommodate surface stations to communicate with a group of small UUVs
operating together to control and coordinate the actions of a group of UUVs,
allowing them to play an important role in operations such as mine countermeasures
(MCM), underwater search and rescue, and subsea exploration.

6.1.2.3   Retroreflector-Based LOS Configuration

Retroreflector-based LOS configurations can be seen as special point-to-point LOS
configurations and are shown in Fig. 6.5. This configuration is generally used to
implement underwater full-duplex systems with limited power and space. This
configuration is usually divided into two parts: a transmitter–transceiver and a
retroreflector–transceiver. The former uses a high power light source with a small
diffusion angle. The beam is emitted from the transmitter–transceiver, passes
through the clear water, reaches the modulated retroreflector, and is reflected from
the retroreflector. During the reflection, the receiving end can perform additional

Fig. 6.5 Retroreflector-based LOS configuration

encoding of the reflected light to make it accompanied by response information. Since no additional light source is usually required at the receiver side in retroreflector-based LOS configuration, the power consumption and device size at the receiver side are low, resulting in a longer receiver operating life. However, one of the disadvantages of this configuration is that the backscattering of the transmitted light signal may interfere with the reflected light signal, thus introducing additional noise and reducing the reliability of the UWOC. In addition, as the light signal is reflected by the retroreflector, it passes through the channel twice before returning to the transmitter, and the beam receives more attenuation and is subject to higher background noise, resulting in reduced system reliability.

Advantages
(1) Full-duplex communication is achieved
(2) Low power consumption and small size of the receiving end device

Disadvantages
(1) Higher transmitter-side signal power.
(2) Background light radiation has a greater impact.
(3) The transmitter–receiver needs to be tightly aligned.
(4) High path loss.
(5) Low reliability.

Application
Consider two nodes communicating using full-duplex UWOC technology, where one has more energy storage (e.g., submarine, surface station) and the other (e.g., diver and UUV) has less. In this scenario, the submarine has a higher energy available and can carry larger equipment. Therefore, the communication system can place more complex equipment with higher power consumption on a platform with more energy storage (submarine) to reduce the energy-constrained platform's burden (frogman).

6.1.2.4 NLOS Configuration

The NLOS configuration enables UWOC by reflecting light signals at the sea surface. As shown in Fig. 6.6, this configuration does not require strict pointing and tracking requirements and , therefore, overcomes the alignment limitations of LOS UWOC. In this configuration, the transmitter emits the beam to the sea surface at an angle of incidence greater than the critical angle of total internal reflection. Subsequently, the sea surface reflects the beam in the direction of the receiver. The receiver lens is approximately perpendicular to the reflected light. Random fluctuations in the sea surface caused by wind and waves are the main factors that negatively affect the NLOS configuration performance. The random fluctuations of the sea surface can intensify the beam scattering effect and make the energy more dispersed, causing severe signal dispersion.

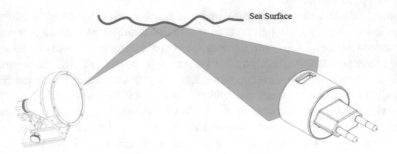

Fig. 6.6 NLOS configuration

Advantages

(1) Low pointing and tracking requirements

Disadvantages

(1) Signal dispersion
(2) High path loss
(3) High requirements for sea surface environment
(4) Vulnerable to background radiation

Application

NLOS structures perform well in scenarios where obstacles are present. Currently, underwater robots are limited to use in open water with little to no obstructions present. In complex underwater environments, cable connections between underwater robots and underwater users such as submarines are not possible. Cables can become entangled in objects such as corals and reefs, reducing the range of underwater robots activity and potentially causing damage to the environment and equipment. In addition, UWA technology has the disadvantage of high latency, making it difficult to control the underwater robots for real-time obstacle avoidance. To address this problem, NLOS enables real-time, high bandwidth communication between the host computer and the underwater robot, allowing the user to provide precise wireless real-time control of the underwater robots, enabling the underwater robots to perform obstacle avoidance in complex marine environments (e.g., underwater reef environments).

Several other geometries apply multi-beam techniques to form multidirectional transmitters and omnidirectional receivers, or both transmitters and receivers are omnidirectional. These two cases simplify the pointing and tracking requirements and are the most straightforward solutions to solve transmitters and receivers' alignment problems mechanically. However, current studies on omnidirectional transmitters and omnidirectional receivers only consider the non-visual transmission interference caused by the attenuation coefficient and do not consider the multipath, scattering (spatial and temporal), and multiple scattering effects caused by turbid seawater. Therefore, the current omnidirectional optical transmitter and omnidirectional optical receiver are only applicable to clear seawater, and theoretical study is still incomplete.

6.2 Comparison Between UWOC and Other UWC Technologies

UWOC system has a broad application prospect and can complement other UWC technologies in the field of underwater communication to effectively increase the performance of underwater wireless network (UWN). There are three feasible UWC technologies at this stage, namely: UWAC technology, underwater radio frequency (RF) technology, and UWOC technology. In order to highlight the unique advantages of UWOC technology, this subsection compares UWOC technology with underwater RF and UWAC technologies. The advantages and limitations of the three UWC technologies are summarized in Table 6.1 [1].

(1) UWAC Technology
Currently, UWAC is the most common and widespread UWC technology. The application of UWAC can be traced back to as early as the late nineteenth century. After two centuries of development, UWAC technology has advanced tremendously and can be used for almost all underwater activities. Low noise hydrophone with preamplifier produced by B&K is shown in Fig. 6.7. Acoustic waves have the advantage of relatively low underwater absorption and long coverage distances, making them the primary carrier wave for underwater wireless communications.

Table 6.1 The advantages and limitations of the three UWC techniques

UWC technologies	Benefits	Limitations
Acoustic	Most widely used UWC technology	Low data rate
		Large communication latency
	Long communication range up to 20 km	Bulky, costly, and energy consuming transceivers
		Harmful to some marine life
RF	Relatively smooth transition to cross sea surface	
	More tolerant to water turbulence	Short link range
	Loose pointing requirements	Bulky, costly, and energy consuming transceivers
	Moderate data transmission rate at close distance	
Optical	Ultra-high data rate	Sophisticated processing required to cross sea surface
	Ultra-low latency	Suffers from severe absorption and scattering
	Low-cost and compact transceivers	Moderate link range

Fig. 6.7 Low noise
hydrophone with preamplifier
˙ from B&K

However, despite the incredible advances that have been made in UWAC tech-nology, it still has some limitations. First, UWAC performance is limited by low bandwidth, high transmission loss, high latency, multipath propagation, Doppler expansion, and time-varying channels. These factors limit the system's available bandwidth (typical UWAC communication frequencies range between tens and hundreds of kHz). Although UWAC communications can support ultra-long-range communications, acoustic waves' physical properties limit UWAC communica-tions' data rate. At a range of several kilometers, the data rate can only reach tens of kbps. In addition, there are severe communication delays in long-range UWAC links due to the slow propagation speed of acoustic waves in water. Therefore, the channel capacity of UWAC channels is limited to meet the demand of underwater high-speed communication. Second, the sound velocity gradients are very different in different marine environments, making the UWAC system designed for one marine environment may fail in other environments. In addition, underwater acoustic equipment may not work efficiently during hot afternoons when subject to seawater temperature variations. Third, long-range UWAC often uses low-frequency transducers as acoustic wave transmitters. This low-frequency transducer is usually bulky, expensive, and energy consuming. Considering the cost of maintenance and engineering difficulties, purely acoustic communication devices are not economical for large-scale underwater communication network implementation. Finally, acoustic technology may also affect marine organisms that use sound waves to accomplish communication and navigation. In terms of protecting endangered marine life, acoustic waves are not the most environmentally friendly way to communicate underwater.

(2) Underwater RF Communication Technology

Underwater RF communication technology is another way of underwater wireless communication. Underwater RF communication has two significant advantages. First, underwater RF communication has higher bandwidth and faster speed compared to acoustic waves. Depending on the system's design structure, the frequency of RF waves can fluctuate between a few tens of hertz to several gigahertz. Among them, ultra-low-frequency (30–300 Hz) electromagnetic waves have been extensively used in military applications. Second, RF waves can pass more smoothly through the air/water interface compared to sound and light waves. At low frequencies, the energy loss caused by refraction at the air–water interface is close to 60 dB, and the refraction loss decreases as the electromagnetic wave frequency increases. Figure 6.8 shows Seatooth portfolio from WFS. Taking advantage of this, electromagnetic waves can be transmitted directly from the underwater node to the land base station, enabling cross-border communication. Finally, RF methods often propagate RF signals as diffuse waves, so RF communication systems are more tolerant of turbulence and seawater turbidity.

Significant constraints that hinder underwater RF technology development include the shorter link range, the absorption effect of seawater on RF signals, and seawater's electrical conductivity. Due to the high conductivity of seawater (an average conductivity of 4 Mhos/m), the absorption loss of RF signal high frequency in seawater is high, so ultra-low-frequency (30–300 Hz) electromagnetic waves can also travel only a few meters distance underwater. In addition, to achieve low-frequency RF communication, the transceiver needs to install a huge

Fig. 6.8 Seatooth portfolio from WFS

antenna. Finally, from the cost point of view, the low-frequency RF communication equipment with antennas attached is expensive.

(3) UWOC Technology

After it was first proposed in the 1990s, UWOC technology has made significant progress in three decades. Currently, UWOC technology has become the most important complementary technology to UWAC technology. First, compared to acoustic and RF systems, UWOC has ultra-high data rates, up to Gbps in dark clear water, due to the physical characteristics of optical carriers. In addition, the UWOC system has the advantage of extremely low latency due to the extremely fast speed of light propagation in water (about three quarters of the speed of light in vacuum). This high-speed and low latency advantage allows UWOC to support many real-time high data volume applications, such as large underwater wireless sensor networks, underwater robots for video surveillance, underwater robot control in real time, etc. Second, UWOC offers higher security compared to UWAC and RF communications. Because most UWOC systems use LOS configuration, the optical energy is highly concentrated and the transceiver is completely symmetrical. Instead of a diffuse broadcast scenario as in the case of acoustic and RF waves, this improves the secrecy performance of the communication system. Finally, the cost and size of optical components are much lower than their acoustic and RF counterparts. In UWOC systems, optical underwater transceivers often use laser diodes or photodiodes. Figure 6.9 shows BlueComm 100 from Sonardyne using UWOC technology. These components have the advantage of being smaller in size and lower in cost. These advantages can facilitate the large-scale commercialization of UWOC.

However, there are still several obstacles to be broken in UWOC technology. First, optical carrier wave propagation in seawater medium is negatively affected by transmission effects. These include temperature fluctuations, scattering, absorption, dispersion, and other factors; therefore, point-to-point UWOC is currently limited to short distances (less than 100 m). Second, similar to acoustic carriers, UWOC systems designed for one environment (e.g., ports) may fail in another environment (e.g., oceanic). This is because the essential characteristics of different water bodies are entirely different. As the latitude and longitude change, the composition and temperature of seawater change, and physical properties such as absorption scattering change, which may lead to model failure. Third, the directional characteristics of UWOC have high requirements for accurate pointing, reducing the connectivity

Fig. 6.9 BlueComm 100
from Sonardyne

of the network. Influenced by surface wind and waves, deep-sea currents, and ocean turbulence, the optical transceiver may be misaligned, leading to communication interruptions. Finally, the beam carrier wavelength is chosen to be blue or green to reduce the water medium's absorption effect. However, blue and green light can attract phototropic marine organisms. The activity of marine organisms may obscure transceivers' link, causing interruptions or blurring of the received signal.

6.3 UWOC System

Figure 6.10 shows a schematic design of a typical point-to-point LOS structured UWOC system. First, at the transmit side, an electrical signal is generated by the signal, which is then modulated to the optical carrier by the modulator. An optical source (LED or LD) converts the modulated electrical signal into an optical signal, which is then amplified by an amplifier. The optical signal is sent exactly to the receiving end in a precisely aligned link. At the receiving end, the light signal is passed through a filter, a focusing lens, and then converted into a current signal by a signal detector. The current signal received by the detector is converted into a voltage signal by a transimpedance amplifier and further signal processing steps are performed on the signal processor. Finally, the electrical signal is demodulated at the demodulator. Data recovery and signal analysis are performed by a terminal such as a PC or an oscilloscope.

6.3.1 Modulation

The UWOC can modulate the signal by direct modulation methods or by attaching an external modulator. Direct modulation achieves signal modulation by controlling

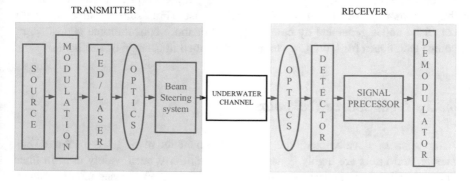

Fig. 6.10 A typical UWOC system structure

the current of driving laser. Unlike direct modulation techniques, external modulators achieve signal modulation by controlling a constant-power optical signal from the laser. Two types of modulation techniques are primarily used in UWOC system optical terminals: intensity modulation or coherent modulation. Because the optical signal detector often responds directly according to the received signal's light intensity, the intensity modulation technique has become the most widely used UWOC modulation technique under its simple form and high energy efficiency.

The purpose of applying modulation techniques is to enhance the performance of UWOC systems. The most appropriate modulation method should be selected for the performance requirements of UWOC for different application scenarios. The application scenarios include power-constrained (e.g., long-distance communication, underwater sensors with limited battery energy), bandwidth-constrained (e.g., photodetectors in multipath channels), and media turbidity. To meet the performance requirements of the above scenarios, a series of modulation techniques have been proposed, such as On/Off Keying (OOK) and pulse position modulation (PPM), as well as 2PSK, QPSK, and QAM.

6.3.2 Coding

To mitigate the effects of underwater fading, a series of UWOC coding techniques have been proposed by researchers. These include turbo, low-density parity check (LDPC), Reed–Solomon (RS), convolutional, and other channel coding schemes. The forward error correction (FEC) coding technique sacrifices the bandwidth of the UWOC system in exchange for high energy efficiency, high reliability, and long distance. FEC codes contain two types of encoding: packet codes and convolutional codes. Among them, packet codes have the advantages of simple structure and easy implementation, so they are widely used in UWOC systems. Common grouping codes include RS codes, Bose-Chaudhuri–Hocquenghem (BCH) codes, and cyclic redundancy check (CRC) codes. Packet codes are simple but do not provide optimal performance for UWOC systems, especially when the water is highly turbid. For this case, more powerful and complex codes are used. For example, turbo codes are combined codes generated by combining two or more convolutional codes. More coding techniques for UWOC systems are described in Chap. 4 of this book.

6.3.3 Light Source Technology

After years of development, the optical components market has become very mature, and lasers are highly commercialized and have been widely used in fiber optic and FSO communication systems. In a typical UWOC system, two sources commonly used are: LD and LED. Among them, LD is a diode material with single conductivity characteristics. According to its PN junction composition, the material can be divided into homojunction, single heterojunction, double heterojunction, and quantum well laser diode, where the quantum well types laser diode, because of its high output power, low threshold current, and other advantages become the mainstream of UWOC communication system. A LED is essentially the same as LD, a semiconductor light-emitting material made of PN structure. LED structure light-emitting principle and LD light-emitting principle have the following differences:

(1) LEDs are mostly double heterogeneous PN junction structure, and LDs are mostly quantum well type structure.
(2) LEDs emit light from spontaneous radiation, rather than excited radiation.
(3) LEDs do not have a minimum threshold and consume less energy.
(4) LEDs do not have an internal optical resonant cavity and cannot emit coherent light.

Compared to LEDs, LDs have higher transmit power, lower diffusion angle, longer modulation bandwidth, and shorter response time. Therefore, LDs are more suitable as a transmitter unit for high-speed UWOC applications that can meet strict alignment requirements. On the other hand, LDs have a much faster response rate with a modulation bandwidth of up to GHz, which is more than ten times that of LEDs. The advantages of small diffusion angle and collimation allow UWOC systems using LD as the light source to communicate at ranges of up to 100 m underwater. Obviously, LDs are more in line with the underwater wireless optical communication system for longer distance, large data capacity, and high-speed communication.

Compared with LDs, the advantage of LED is that no minimum threshold current is required, and it can work continuously and steadily under lower power consumption conditions: and LED is a cold light source, so there is no need to consider heat dissipation, and it can work steadily for a long time with a longer service life than LDs. LEDs have lower output power intensity, larger dispersion angle, and lower modulation bandwidth. It can be installed in short-range and low-speed UWOC systems. Because LEDs have a large beam dispersion angle, they are often used in diffused LOS UWOC systems. In addition, LEDs are inexpensive, simple in construction, and as a cold light source, their performance is not easily affected by temperature. The LED-based UWOC system has a larger beam spread angle, which reduces the alignment requirements of the transceiver. LEDs are more compatible with underwater wireless optical communication systems, closer range, stable, and low-cost requirements. Table 6.2 shows the performance of some of the optical components.

6.3.3.1 Common Lasers in UWOC Systems

Common lasers used in UWOC systems are shown in Table 6.3. Argon-ion lasers are very high power gas lasers that can usually produce several watts of green or blue light with a very high beam quality. The core component of an argon-ion

Table 6.2 Comparison of LD and LED parameters

Parameter comparison	LD	LED
Responsiveness	Fast	Slow
Light dispersion angle	6° 10°	20° 120°
Energy consumption	High	Low
Threshold current	High	None
Operating current	High	Relatively low
Working stability	General	Strong
Transmission distance	Long	Short

Table 6.3 Comparison of common lasers in UWOC systems

Types	Wavelength	Advantages	Shortcoming
Argon-ion laser	455–529 nm	High output	Less efficiency
		Less lifetime	High input power required
			Cooling required
DPSSL	532 nm(green)	High output power	Less efficiency
	493 nm(blue)	Long lifetime	Costly
		Compact size	
Ti: Sapphire	455 nm	Fast output pulses	Costly
		Adjustable frequency	Sensitive to vibrations
FPFD	532 nm	High power	Difficult to modulate
Semiconductor laser	450–470 nm (InGaN)	Highly efficient	Costly
	375–473 nm (GaN)	Compact	Prevent overcurrent damage
Fiber laser	518 nm	Compact size	Costly
		High efficiency	External modulator required
		High output power	

laser is the argon gas filled tube, made of beryllium oxide ceramic, which produces a very strong gas discharge between two of the hollow electrodes, resulting in a very high density of argon-ion plasma. Diode pumped solid-state lasers (DPSSL) have an electric-to-optical conversion efficiency of 19%, continuous operation for hundreds of thousands of hours, heat load and thermal effect about three times smaller than lamp-pumped solid-state lasers, and good beam quality and small size for full curing. Ti: Sapphire is widely used in wavelength-tunable lasers as a light-extracting solid-state laser crystal with a tunable range of 650–1100 nm and a peak of 800 nm, which is one of the widest wavelength-tunable laser crystals. The upper state lifetime of titanium gemstone is as short as 3.2 ms, and due to its high saturation power, it is commonly used as a lamp, argon-ion laser or frequency double pumped neodymium yttrium aluminum garnet laser, etc. Using self-locking technology, titanium gem lasers can directly output laser pulses with pulse widths less than 6.5 fs, which is the narrowest laser pulse of all lasers that output directly from the resonant cavity. The dual frequency technology allows the laser beam to cover a wide wavelength range from blue to deep ultraviolet, producing 193 nm lasers for use in lithography. Semiconductor and flash lamp-pumped, frequency-doubled (FPFD) both use ND:YAG (neodymium-doped yttrium aluminum garnet) crystal as the material for laser generation, which converts 808 nm visible light into 1064 nm invisible laser, but another key factor in the output laser is to make the crystal bar output laser pump source, semiconductor pumping is the use of semiconductor diodes to emit 808 nm light waves; and lamp pumping is the use of krypton lamp to pump the light, but the krypton lamp emits a broader spectrum of light, only a slightly larger peak at 808 nm, the other wavelengths of light eventually become useless when heat is dissipated. Therefore, the conversion efficiency of semiconductor-pumped lasers is much higher than lamp-pumped. Fiber laser is mainly composed of pump source, coupler, rare earth element-doped fiber, resonant cavity, and other components. The pump source is composed of one or more arrays of high power LDs, whose pump light is coupled to the rare earth element-doped fiber as the gain medium through a special pump structure. The photons on the pump wavelength get absorbed by the doped fiber medium, resulting in particle number inversion, and the excited emitted light wave is fed back and oscillated by the resonant cavity mirror to form the laser output.

6.3.4 Signal Detector

The photoelectric sensor, as the receiver of the UWOC system, is capable of converting optical signals into electrical signals. High-performance photoelectric sensors should have the advantages of wide field of view, high gain, and high signal-to-noise ratio. For blue-green light sources, there are three most common types of photosensors, including the photo multiplier tube (PMT), semiconductor photosensor, and bionic quantum photosensors. Among them, semiconductor photosensor includes positive intrinsic negative (PIN) and avalanche photodiode (APD).

6.3.4.1 PMT

Leonid Kuznetsk invented the PMT. It is a vacuum tube device that converts a weak light signal into a current signal and is composed of a cathode stage material and a multiplier stage material. Unlike other photodiodes, the PMT uses the external photoelectric effect to generate photocurrent, so it can obtain a very high signal gain and is extremely sensitive to light, and weak signals single-photon level can be detected by the photomultiplier tube. Figure 6.11 shows PMTs manufactured by Hamamatsu.

Advantages
(1) The photosensitive area is large and uniform, and the diameter of up to 22 mm is about 20 times that of the same type of photodiode.
(2) Fast response speed.

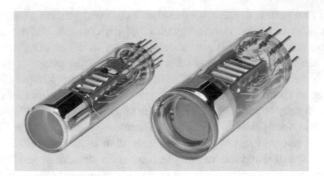

Fig. 6.11 PMTs manufactured by Hamamatsu

(3) The equivalent noise power is very low, and the detection signal strength can be as weak as a single-photon level.
(4) Signal amplification gain is huge.

Disadvantages
(1) Large size, fragile, and easily damaged tube structure, high cost in some complex environment applications.
(2) Poor stability, although the photomultiplier tube has a large gain coefficient, but the anode output current is too large, poor continuity, the signal can fluctuate easily.
(3) Higher noise, similar to avalanche photodiodes, photodiodes are extremely sensitive, and in practical application scenarios, they are susceptible to interference from avoiding background light radiation, resulting in distortion of the received signal.

Overall, the photomultiplier tube has similar characteristics to the avalanche photodiode. The incident weak light signal has high sensitivity and strong amplification effect, so the PMT is often used in applications requiring high sensitivity, such as long-range wireless optical communications.

6.3.4.2 PIN

PIN is an optical component made from a silicon-based semiconductor with a photoelectric conversion function. The PIN photodiodes are different from conventional PN photodiodes in that the depletion layer in the middle of the PN junction incorporates a layer of N-type semiconductors that is almost equivalent to silicon with a low doping concentration. According to PIN-type photodiodes, PIN-type photodiodes can be divided into two types of flat structure and table top structure. The former production process is relatively simple, low cost, the latter needs to be in the plane based on corrosion digging groove, and the production process is more complex, relatively high cost, but the advantage of a stable structure, high

Fig. 6.12 PINs manufactured by Hamamatsu

operating frequency, compared to the former more widely used. Figure 6.12 shows PINs manufactured by Hamamatsu.

Advantages

(1) Sensing wavelength range is large, 110–1900 nm, including ultraviolet, infrared light band.
(2) Fast response speed.
(3) High quantum efficiency.
(4) Strong working stability, low noise introduction, and small error.
(5) Simple structure and low cost

Disadvantages

(1) No internal gain, the output current signal is weak, usually requires additional signal amplification processing equipment
(2) Higher equivalent noise power, i.e., lower detection sensitivity and poorer ability to detect weak signals

PIN-type photodiodes have the advantages of fast response and high detection bandwidth and are widely used in wireless optical communication monitoring systems.

6.3.4.3 APD

APD is a photoconversion diode element commonly used in underwater wireless optical communications. It is based on a PN junction and differs from a PIN-type photodiode. The avalanche photodiode uses the reverse bias voltage's avalanche effect to continuously amplify the generated photocurrent-type signal to achieve higher signal gain. Figure 6.13 shows APDs manufactured by Hamamatsu.

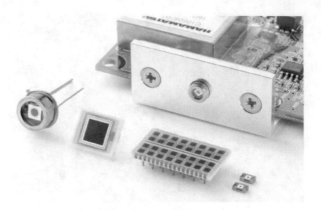

Fig. 6.13 APDs manufactured by Hamamatsu

Advantages
(1) Higher sensitivity and faster response speed, modulation bandwidth up to GHz level
(2) Low equivalent noise power

Disadvantages
(1) Higher voltage required to work, higher energy consumption.
(2) Introduction of large noise. Diode in the photoelectric conversion process will produce scattered particle noise, due to the avalanche photodiode amplification effect leads to scattered particle noise is amplified, easy to trigger the photoelectric signal distortion.

Avalanche photodiode as a new type of photoelectric conversion device gradually by the industry attention, because of its fast response, high sensitivity, and other advantages to becoming a replacement for the traditional PN-type photodiode alternative materials, in the long-distance wireless optical communication and other need to detect weak signal in the scene of application space is very broad.

6.3.5   Alignment and Compensation

In clean seawater, the UWOC has extremely high requirements for transceiver alignment to ensure the robustness of the UWOC, as the beam propagates collimated underwater. However, in the turbid marine environment, scattering causes the beam to propagate in different directions, thus slightly reducing the transceiver alignment requirements.

From the designing point of view, in order to reduce the pointing requirements of UWOC systems, some researchers use PMTs as UWOC receivers because they have large lenses (lens range from 10 to 500 mm) and also have large FOV. Thus, PMTs are able to receive the maximum amount of light energy reaching the receiving surface. However, PMTs are not only expensive, bulky, and easily damaged, but also unsuitable for multi-user environments. In addition to PMT, to improve the robustness of the UWOC system, researchers have used electronic switch aiming and tracking to enable the optical front end of the receiver to adaptively change the FOV depending on the angle of beam arrival. Finally, smart transmitters and receivers have been proposed in recent years. The smart transmitter can estimate the environmental characteristics based on the backscattered light collected by the receiver. Based on the collected information smart transmitters and receivers are able to change the beam spread angle as well as control the FOV to align the field of view with the optical signal, respectively, thus improving pointing accuracy. This approach is very beneficial in a multi-platform environment.

6.3.6   Spatial Diversity Technology

UWOC provides high-capacity, low latency links. However, due to various characteristics of underwater channels, UWOC cannot achieve long-distance transmission. Therefore, it is necessary to extend the near-range coverage of optical carriers to fully exploit their advantages in the underwater environment. Several methods to overcome the coverage distance limitation using spatial and multipath diversity (e.g., MIMO, spatial modulation techniques) in UWOC systems have been investigated.

Multiple-input multiple-output (MIMO) techniques are an effective method to improve system throughput and robustness. Compared with conventional UWOC SISO systems, UWOC MIMO systems can provide higher data rates and broad communication ranges. In addition, MIMO technology can increase the UWOC system's reliability and make the UWOC system work well in turbid waters.

In addition, cooperative diversity (relay-assisted diversity) is an effective technique to extend the communication range of UWOC. Although this scheme has been extensively studied in underwater acoustic communications, cooperative diversity is just emerging in the UWOC field. In a UWOC system using cooperative diversity scheme, the source nodes propagate optical signals to the adjacent nodes, which subsequently propagate the information to the next node again and finally to the base stations such as ships and buoys. The nodes that are responsible for relaying the information are called relay nodes. For example, in an underwater wireless sensor network, the information collected by the source node needs to be forwarded to a remote control station. In this case, the data is collected using relay nodes and forwarded to the next relay node, gradually spreading the information. This scheme not only extends the distance of UWOC communication and increases the coverage distance of underwater wireless sensor network, but also is more energy efficient.

6.4   UWC Network

UWOC has the advantages of high bit rate, low power consumption, and low latency, but its communication distance is extremely short due to the effect of scattering and absorption. In contrast, UWAC has a smaller data rate, greater power consumption, significant latency, and a broad communication range. UWAC technology can communicate at speeds of 100–5000 bps over distances of up to thousands of meters, or at higher data rates over shorter distances. Therefore, a hybrid system is necessary to take full advantage of both technologies. In hybrid systems, UWOC technology can complement UWAC technology. It enables optical communication at high data rates and low latency in short-range applications and extremely robustness in long-range applications with acoustic technology [4, 5].

A typical acoustic-optical hybrid UWC network is shown in Fig. 6.14 where underwater robots, or underwater sensors, transmit large amounts of data to a central base station via relay nodes. These underwater robots are equipped with acoustic and optical transceivers. The acoustic devices are used to enable applications such as

Fig. 6.14   A typical acoustic-optical hybrid UWC network

underwater positioning and long-range underwater wireless communication. After using UWAC for assisted alignment, the optical devices can be used for short-range wireless communication. In hybrid systems, UWOC technology is commonly used to send uplink signals due to its high bandwidth and high directionality characteristics. And downlink signals from ships, floats, and other users to control underwater robots are realized by UWAC technology. In addition, long-range low bandwidth applications such as familiar love locating or tracking can also be achieved by UWAC. The hybrid acoustic-optical system is free to choose the best transmission mode in the shortest transmission time depending on the load and type of water. Thus the hybrid system has significant advantages in terms of throughput and energy efficiency.

6.5 Summary

Since the twentieth century, underwater wireless optical communication technology has made remarkable development but still faces many challenges in increasing communication distance, suppressing attenuation, and increasing system reliability, as follows:

(1) Severe attenuation. Although absorption and scattering effects of seawater are relatively low for blue-green light beams, however, due to the interaction of photons with water molecules and dissolved particles, absorption and scattering still lead to the attenuation of the optical signal, causing multipath fading and limiting the effective communication distance of UWOC.

(2) Transmitter–receiver location uncertainty. Affected by gravity, wind friction, and non-uniform water density generated by ocean currents, seawater often behaves as a non-stationary medium, resulting in transmitter–receiver position uncertainty. Besides, some receivers, such as underwater vehicles, are located in random positions and frequently move, which requires consideration of methods to compensate for positional uncertainty to ensure proper communication.

(3) Short life span of underwater equipment. The underwater environment is very complex. It is difficult for optical devices in the deep sea to replenish their energy by absorbing solar energy. Seawater turbulence, pressure, temperature, and salinity changes can significantly affect the performance and reduce the lifetime of UWOC equipment. Extending the subsea operation time, increasing equipment reliability, and reducing equipment power consumption is are urgent problem.

1. What are the advantages and disadvantages of UWOC technology? What is the significance of studying UWOC technology?
2. How to increase the communication distance of underwater wireless optical communication technology?
3. Why do phase-modulated UWOC-based systems often use LD as the light source?
4. Check the product information of PIM, PIN, APD and compare their response curves to understand the advantages and disadvantages of different photodetectors.
5. UWOC system has the problems of severe attenuation, uncertain position, and limited working life. Which solutions have been proposed for the above problems?
6. The absorption coefficient of RF signal in seawater is approximated as $\alpha_{RF} \approx \sqrt{\pi f \mu_{RF} \sigma_{RF}}$, where f in the above equation is the signal frequency, μ_{RF} is the magnetic permeability, and σ_{RF} is the conductivity of seawater. Suppose the magnetic permeability of seawater is $\mu_{RF} = 4\pi \times 10^{-7}$ H/m and the conductivity of seawater is 4.54 S/m. Try to calculate the absorption coefficient of seawater at RF signal frequencies of 10, 100, and 1000 Hz. And try to find the absorption coefficients of optical carrier waves with a wavelength of 532 nm in coastal, harbor, and pure seawater, and compare them with the absorption coefficients of RF signals.
7. Suppose there exists a light beam with a wavelength of 0.8 μm incident on the photodetector, and the number of incident photons per second is 10^{10}, try to calculate the incident light power on the detector, if the efficiency of the photodetector to convert light energy into current is 0.65 mA/mW, try to calculate the current generated by the photodetector.
8. Suppose there exists an LD source with a wavelength of 500 nm. the beam obeys a Gaussian distribution and the beam waist spot radius $\omega_0 = 0.5$ mm. Try to find the size of the radius of the time plate and the radius of curvature of the wavefront surface at 10, 20, and 100 cm from the beam waist.

References

1. Z. Zeng, S. Fu, H. Zhang et al., A survey of underwater optical wireless communications. IEEE Commun. Surv. Tutorials **19**(1), 204–238 (2016)
2. M.Z. Chowdhury, M.K. Hasan, M. Shahjalal et al., Optical wireless hybrid networks: trends, opportunities, challenges, and research directions. IEEE Commun. Surv. Tutorials **22**(2), 930–966 (2020)

3. N. Saeed, A. Celik, T.Y. Al-Naffouri et al., Underwater optical wireless communications, networking, and localization: a survey. Ad Hoc Netw. **94**, 101935 (2019)
4. A. Al-Kinani, C.X. Wang, L. Zhou et al., Optical wireless communication channel measurements and models. IEEE Commun. Surv. Tutorials **20**(3), 1939–1962 (2018)
5. H. Kaushal, G. Kaddoum, Underwater optical wireless communication. IEEE Access **4**, 1518–1547 (2016)

Chapter 7
Underwater Wireless Optical Channel Model

7.1 Introduction

Due to the unique optical properties of the underwater environment, the propagation of light in water is very complex. In order to derive a channel model applicable to UWOC and achieve reliable UWOC communication, it is necessary to understand the physical properties of light propagation underwater to provide a basis for the design of UWOC systems [1].

The optical properties of water are divided into two main categories, which are intrinsic optical properties (IOP) and apparent optical properties (AOP). IOP depends only on the nature of the medium and describes the optical parameters of the medium, independent of the nature of the light source. IOP is mainly composed of the absorption coefficient, scattering coefficient, volume scattering function, refractive index, etc. On the other hand, AOP depends both on the medium's properties and on the geometry of the light field in the medium, such as pointing or diffuse, collimated light sources. Thus, the AOP parameters describe not only the physical properties of the medium but also the structure of the optical field in the medium. The AOP parameters consist mainly of the irradiance reflectance, the mean cosine, and the diffuse attenuation coefficient. The IOP and AOP parameters are defined in this chapter. Radiative transfer theory establishes the link between IOP and AOP. The properties of the water environment and the light field's geometry, such as surface waves, undulations of the seafloor, and the ambient light source's intensity, provide the initial conditions for the radiative transfer theory necessary to derive a model applicable to the UWOC channel. In UWOC systems, IOP is typically used to predict UWOC link performance, AOP for calculating the intensity of ambient light in the ocean. As IOP has more impact on link performance, this section's remainder will focus on IOP.

The purpose of this chapter is to introduce the model of the UWOC system. The quantitative description of environmental properties is intended to apply radiative transfer theory to the UWOC model. In addition to a brief introduction to the

Y. Lou, N. Ahmed, *Underwater Communications and Networks*, Textbooks in Telecommunication Engineering, https://doi.org/10.1007/978-3-030-86649-5_7

definition of IOP and AOP parameters, this chapter presents models for describing the optical properties of media parameters that predict the overall water medium's optical properties from parameters with biological or geological significance (e.g., chlorophyll concentration or particle matter concentration). The inseparable links between optics and biology, chemistry, and geography are thus made apparent. At the end of the chapter, the process of translating from different particle to the overall optical properties is described.

7.2   Absorption and Scattering Losses

Absorption and scattering are able to describe the attenuation of light beams during underwater propagation and are the most fundamental IOPs. Absorption refers to the absorption of the energy of photons by the water medium. Absorption describes the process of transferring light energy into other forms of energy such as heat and chemical energy. Scattering refers to the interaction of photons with molecules or atoms in the medium and causes a change in the state of motion of the photons [2, 3].

7.2.1 Absorption and Scattering Coefficients

When a light beam propagates underwater, part of it is absorbed by the medium, and part of it is scattered, resulting in an attenuation of the beam energy. To describe the absorption and scattering effects quantitatively, consider the following scenario: assume that a cubic medium exists in a homogeneous water environment (where the absorption and scattering effects do not vary with position). A beam of light is transmitted through the cubic medium. The effect of the medium on the energy of the beam is depicted in Fig. 7.1.

In Fig. 7.1, ΔV is the width of the medium cube (perpendicular to the direction of incidence of the light beam), and Δr is the thickness of the cube (parallel to

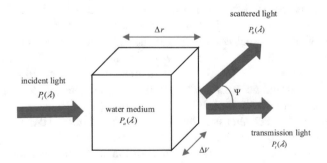

Fig. 7.1 Schematic diagram of beam scattering and absorption

the direction of incidence of the beam), P_i is the energy of the incident beam, P_a is the energy absorbed by the cube medium, and P_s is the energy of the scattered beam, where the angle between the scattered beam and the direction of the beam's incidence is ψ, called the scattering angle. With ψ as the center, the steric angle of the scattered beam irradiation range is $\Delta\Omega$. P_t is the energy of the transmitted beam which remains in the original direction. In practice, the absorption scattering effect is related to the wavelength of the beam. Assume that only Rayleigh scattering occurs during the photon's propagation, i.e., that the photon's wavelength does not change. According to the law of conservation of energy,

$$P_i(\lambda) = P_a(\lambda) + P_b(\lambda) + P_t(\lambda). \tag{7.1}$$

The relationship between scattered power, absorbed power, and incident power can be used to determine the spectral absorbance, scattering rate, and transmittance. The spectral absorptivity is

$$A(\lambda) = \frac{P_a(\lambda)}{P_i(\lambda)}. \tag{7.2}$$

The spectral scattering rate is

$$B(\lambda) = \frac{P_b(\lambda)}{P_i(\lambda)}. \tag{7.3}$$

The spectral transmittance is

$$T(\lambda) = \frac{P_t(\lambda)}{P_i(\lambda)}. \tag{7.4}$$

The spectral absorptivity, scattering rate, and transmittance add up to 1. That is, $A(\lambda) + B(\lambda) + T(\lambda) = 1$.

Spectral absorption coefficient and spectral scattering coefficient refer to the spectral absorption rate and spectral scattering rate per unit distance in water media. The spectral absorption coefficient is defined as

$$a(\lambda) = \lim_{\Delta r \to 0} \frac{\Delta A(\lambda)}{\Delta r}. \tag{7.5}$$

The spectral scattering coefficient is defined as

$$b(\lambda) = \lim_{\Delta r \to 0} \frac{\Delta B(\lambda)}{\Delta r}. \tag{7.6}$$

The spectral attenuation coefficient is defined as

$$c(\lambda) = a(\lambda) + b(\lambda). \tag{7.7}$$

The spectral absorption coefficient describes the intensity of the absorption effect of the water medium. The spectral scattering coefficient describes the intensity of the scattering effect of the beam by the water medium. The spectral attenuation coefficient is expressed as a linear sum of the absorption and scattering coefficients. Hence, the loss of light intensity per unit distance caused by scattering and absorption is expressed by $c(\lambda)$ in Eq. (7.7).

7.2.2 Scattering Phase Function

In order to describe the scattering effect of light beams as they propagate through the water under the influence of suspended particles, the concept of the bulk scattering function was introduced. In Eq. (7.3), the spectral scattering rate describes the scattered energy ratio to the total incident energy. Consider the distribution of the scattered power in different orientations. $B(\lambda, \psi)$ represents the ratio of the energy of the scattered beam to the energy of the incident beam that is scattered into the ψ-direction, covering an angle $\Delta\Omega$. Thus, the angular scattering per unit distance and per steradian angle, called the volume scattering function (VSF), can be expressed by the following equation [4]

$$\beta(\lambda, \psi) = \lim_{\Delta r \to 0, \Delta\Omega \to 0} \frac{\Delta B(\lambda, \psi)}{\Delta r \, \Delta\Omega}. \tag{7.8}$$

This function describes the variation of the scattered light intensity with the direction of scattering. If $\beta(\lambda, \psi)$ is integrated over all directions, the scattered power per unit irradiance incident on a unit volume of water, the spectral scattering coefficient, is obtained

$$b(\lambda) = \int_{\Xi} \beta(\lambda, \psi) d\Omega = 2\pi \int_0^\pi \beta(\lambda, \psi) \sin \psi d\psi. \tag{7.9}$$

In the above equation, θ is the angle between the direction of scattering and the direction of incidence of the beam. The spectral volume scattering phase function is another important marine channel optical property defined as the ratio of the volume scattering function to the scattering coefficient

$$\tilde{\beta}(\psi) = \frac{\beta(\lambda, \psi)}{b(\lambda)}. \tag{7.10}$$

Combining Eqs. (7.9) and (7.10) yields the normalization condition for the scattering phase function of the spectral body

$$2\pi \int_0^\pi \tilde{\beta}(\psi; \lambda) \sin \psi d\psi = 1. \tag{7.11}$$

Based on Eq. (7.11), the average value of the cosine of the scattering angle ψ in the direction of scattering is

$$g = 2\pi \int_0^\pi \tilde{\beta}(\psi) \cos \psi \sin \psi d\psi. \tag{7.12}$$

The mean cosine g is often called the asymmetry parameter of the phase function. In UWOC systems, the asymmetry parameter is widely used to describe a spectral volume's scattering phase function, called the Henyey-Greenstein (HG) function.

$$\tilde{\beta}(\theta) = P_{HG}(\theta, g) = \frac{1 - g^2}{4\pi \left(1 + g^2 - 2g \cos \theta\right)^{\frac{3}{2}}}. \tag{7.13}$$

The scattering phase function shows that the photon scattered is not isotropic. The scattering phase function varies with wavelength and medium composition. As early as 1972, Petzold measured the g-values of clean seawater, coastal seawater, and harbor seawater. The g-values for clean ocean, coastal and turbid harbor were 0.8708, 0.9470, and 0.9199, respectively. The g-values measured by Petzold were based on a non-coherent light source. However, light source coherence has almost no effect on the inherent optical properties of the channel. Petzold's data is therefore used to this day, and $g = 0.924$ is considered to be an approximation of the average cosine in most underwater scenarios. Despite the advantages of the HG function's simple structure for computational purposes, the formula does not account for small angular forward scattering and large angular backscattering in the underwater environment. To better match the Mie scattering theory, Bohren has made some modifications to the function, called the two-term Henyey-Greenstein (TTHG), which matches the experimental results better and is given by

$$P_{TTHG}(\theta) = \alpha P_{HG}\left(\theta, g_{fwd}\right) + (1 - \alpha)P_{HG}\left(\theta, -g_{bk}\right), \tag{7.14}$$

where P_{HG} is the HG function and α is the weight of the forward HG function describing the weight of the forward HG function concerning the inverse HG function. g_{fwd} and g_{bk} are the asymmetry factors of the forward and inverse HG phase functions, respectively.

7.2.3   Composition of the Coefficients of Absorption and Scattering

In general, absorption and scattering effects of UWOC systems cause three undesirable effects. First, the absorption effect causes light energy to be converted into thermal and chemical energy. The total energy of light is continuously reduced, limiting the UWOC communication distance. Second, the scattering effect causes the light beam to be scattered into the medium, causing the beam to spread. The limited size of the lens of the photodetector leads to the reduction of the light energy collected at the receiving end, causing the signal-to-noise ratio of the system to decrease. Third, in the underwater environment, due to the scattering effect, the photons arrive at the receiver through different paths, resulting in a more dispersed arrival time of the photons and aggravating multipath dispersion. The adverse effects of multipath phenomenon include intersymbol interference (ISI) and timing jitter.

The IOP can be modeled as a superposition of the different components of the water medium acting on the light beam. Where the absorption coefficient can be expressed as a superposition of the absorption coefficients of the four substances causing the absorption effect

$$a(\lambda) = a_w(\lambda) + a_{CDOM}(\lambda) + a_{phy}(\lambda) + a_{det}(\lambda), \qquad (7.15)$$

where $a_w(\lambda)$ is uptake by pure seawater, $a_{CDOM}(\lambda)$ is uptake by colored dissolved organic matter (CDOM), $a_{phy}(\lambda)$ is uptake by phytoplankton, and $a_{det}(\lambda)$ indicates uptake by detritus.

In addition to water molecules, pure seawater contains many dissolved salts, including NaCl, $MgCl_2$, Na_2SO_4, KCl, etc. The dissolved salts account for an average of about 35% of the total weight of pure seawater. The water molecules and the dissolved salts together determine the absorption effect of pure seawater. The corresponding absorption spectra of pure seawater are shown in Fig. 7.2. There is a visible light transmission window of 400–500 nm in pure seawater.

In addition to dissolved salts, seawater also contains varying concentrations of CDOM, where the main component of these organic substances, humic and brownish-yellow acids is produced by the decay of plants. The size of CDOM in the ocean is usually less than 0.2 mm, and when sufficient concentrations of CDOM are dissolved in seawater, the water turns slightly yellow. These organic substances

Fig. 7.2 Absorption spectra of pure seawater

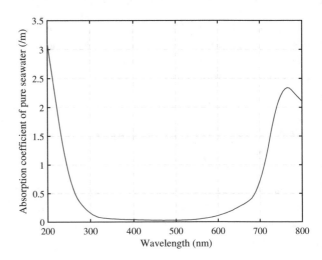

Fig. 7.3 Absorption spectra of CDOM

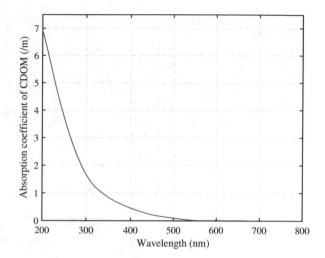

are generally known as yellow substances. Yellow substances absorb light at low wavelengths very well, but the absorption of yellow substances decreases rapidly with increasing wavelengths. The absorption is particularly pronounced in the blue and ultraviolet bands. The absorption of CDOM is relatively weak compared to the other components of seawater. However, in lakes, rivers, and coastal waters, especially at certain times (periods of extensive plant decay), the concentration of CDOM greatly increases, resulting in much higher absorption of light in the blue and violet bands of seawater. Figure 7.3 demonstrates the strong absorption of blue-violet wavelength light beams by CDOM and the weaker absorption of yellow and red light.

Phytoplankton is a strong absorber of visible light and has an important influence on the absorption characteristics of seawater. The photosynthesis of chlorophyll mainly causes the absorption of phytoplankton. Chlorophyll is a strong absorber of blue and red light wavelengths. As different phytoplankton species contain different types of photosynthetic pigments, the absorption effect varies from one phytoplankton to another. Figure 7.4 illustrates the general characteristics of phytoplankton absorption. It can be observed that there are clear absorption peaks at wavelengths of approximately 440nm and 675nm. The photosynthetic pigments absorb blue light more strongly so that the blue light absorption peak at $\lambda = 440$ nm has a higher peak value than the red light absorption peak. In contrast, phytoplankton absorption is weaker at wavelengths λ from 550 nm to 650 nm.

The detritus consists of organic and inorganic particles. Organic particles consist mainly of bacteria, tiny phytoplankton, organic debris, large particles (zooplankton and aquatic animals), etc. The suspended inorganic particles consist mainly of quartz, clay, and metal oxides. These detrital particles are classified as the same species due to their similar absorption properties. Figure 7.5 shows that the detritus' absorption properties are similar to those of CDOM, both having higher absorption properties in the lower wavelength bands.

Fig. 7.4 Absorption spectra
of phytoplankton

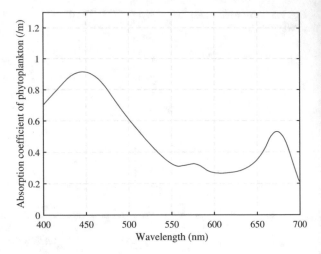

Fig. 7.5 Absorption spectra
of detritus

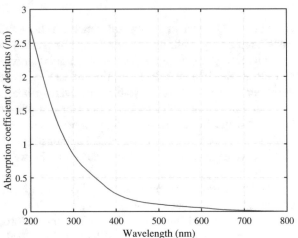

Similar to the spectral absorption coefficient, the scattering coefficient of underwater light propagation can be expressed as the sum of the scattering coefficients of different substances in the water medium.

$$b(\lambda) = b_w(\lambda) + b_{phy}(\lambda) + b_{det}(\lambda), \tag{7.16}$$

where $b_w(\lambda)$ is the scattering coefficient from pure seawater, $b_{phy}(\lambda)$ is the scattering coefficient from phytoplankton and $b_{det}(\lambda)$ indicates scattering caused by debris. The effect of wavelength on scattering is low compared to that of absorption. The main factor affecting the scattering intensity is the concentration of particulate matter in the water medium.

The scattering coefficient changes as the refractive index vary with the current, salinity, and temperature in pure seawater. Einstein-Smoluchowski proposed that

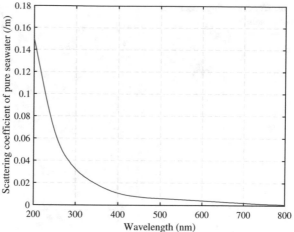

Fig. 7.6 Scattering spectra of pure seawater

the scattering effect is related to fluctuations in molecular number density with fluctuations in the refractive index and that such fluctuations give rise to scattering. In seawater, random fluctuations in the concentration of dissolved ions (e.g., Na^+, K^+, Cl^-, etc.) lead to fluctuations in the refractive index, and thus to larger scattering. Riley pointed out that microscopic spherical particles cause scattering. The wavelength of visible light is larger than the size of water molecules. Therefore, the Rayleigh scattering model can be used to approximate the light scattering caused by seawater. The corresponding scattering spectra is shown in Fig. 7.6. In pure seawater, the lower the wavelength of light, the more intense the scattering effect.

The contribution of phytoplankton and detrital material to the scattering effect exceeds 40% of the total. Since the scattered light produced by phytoplankton and detritus is mainly forward propagated, the Mie scattering model can approximately describe the scattering effect caused by phytoplankton and detritus. In addition, the value of scattering coefficient is related to the density of phytoplankton and debris in the water medium. In Figs. 7.7 and 7.8, the scattering spectra of phytoplankton and detritus with different densities are illustrated.

The effect of dissolved substances on the absorption and scattering properties of seawater has been summarized, as shown in Table 7.1. As the latitude, longitude, and depth change, the composition of the seawater medium changes dramatically. Depending on the composition of the marine medium, the intensity of the scattering and absorption phenomena varies. For example, in pure seawater, scattering is weak and absorption is the main factor limiting the performance of UWOC. In contrast, in coastal seawater, where the concentration of organic matter and suspended particles is high, scattering has a stronger effect on UWOC performance. As a result, the optical communication transmission window shifts from the blue-green band of pure seawater to the yellow-green band in coastal waters.

Fig. 7.7 Scattering spectra of phytoplankton

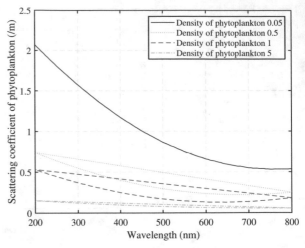

Fig. 7.8 Scattering spectra of detritus

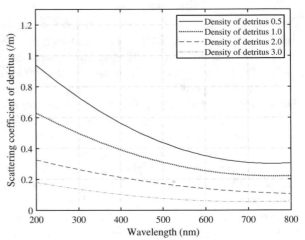

7.3   UWOC Channel Modeling

UWOC can be considered as an extension of FSO. However, the UWOC channel has unique fading characteristics compared to the FSO channel, and the FSO channel model cannot effectively describe the UWOC channel. Therefore, a reliable channel model dedicated to describe UWOC must be proposed. This chapter presents the recent advances in UWOC channel modeling techniques, including link fading modeling, geometric dissonance modeling, and turbulence modeling.

Table 7.1 Summary of absorption and scattering of seawater

Compositions	Absorption coefficient	Scattering coefficient
Pure water	–Invariant at constant temperature and pressure	–Rayleigh scattering
	–Wavelength dependent	–Wavelength dependent
Salts	–Negligible within visible spectrum	–Rayleigh scattering
	–Inversely proportional to wavelength	–Not depend on wavelength
CDOM	–Variable with the density	–Negligible
	–Inversely proportional to wavelength	
Plankton, mineral, and detritus	–Variable with the density	–Mie scattering
	–Inversely proportional to wavelength	–Variable with the density
		–Inversely proportional to wavelength

7.3.1 Bill Lambert's Law

The two main IOPs that contribute to underwater light attenuation are absorption and scattering. These two IOPs are independent of the structure of the UWOC link. Therefore, the modeling of underwater optical attenuation can be seen as an accurate description of UWOC links with different structures under the given conditions of absorption and scattering coefficients.

The attenuation of the beam in the LOS UWOC can be described using Bill Lambert's law [5]:

$$I_t = I_0 \exp(-cz), \tag{7.17}$$

where I_t is the transmitted light irradiance, and I_0 is the light source irradiance. Beer-Lambert's law has a simple exponential form and is usually used to calculate the light energy loss in UWOC. However, in practical applications, the irradiance change curves plotted by Beer-Lambert's law often do not match the actual light energy change. This is because Beer-Lambert's law assumes that all scattered photons cannot reach the receiver. In fact, some photons in water media can still be captured by the receiver after multiple scatterings. Therefore, Beer-Lambert's law severely underestimates the performance of UWOC systems. Especially in media with strong scattering effects, the light energy calculated by Beer-Lambert's law will have a larger offset from the actual situation.

7.3.2 Radiative Transfer Equation

When an incident light propagates along a given direction in the water medium and interacts with it, part of the light intensity is absorbed, and the amount of absorption is determined by the absorption coefficient; another part of the light is scattered along other directions, and the amount of scattering is determined by the scattering coefficient. The light intensity propagating along the direction will be attenuated by scattering in different directions and enhanced by radiation from different directions. Therefore, this process can be described by the radiative transfer equation (RTE) [6]:

$$
\left[\frac{1}{v}\frac{\partial}{\partial t} + \mathbf{n} \cdot \nabla\right] I(t, \mathbf{r}, \mathbf{n}) = -cI(t, \mathbf{r}, \mathbf{n}) + \int_{4\pi} \beta\left(\mathbf{r}, \mathbf{n}, \mathbf{n}'\right) I(t, \mathbf{r}, \mathbf{n})d\mathbf{n}' + E(t, \mathbf{r}, \mathbf{n}),
$$
(7.18)

where v is the speed of light, c is the beam attenuation coefficient, t is time, $\mathbf{n}$ is the direction vector, $\mathbf{r}$ is the position vector, and ∇ is the scattering operator for position $\mathbf{r}$. I is the radiance, β is the volume scattering function, and E is the source radiance.

The RTE can be reduced to Bill Lambert's Law if the following conditions are satisfied:

- The radiation source is a monochromatic direct beam.
- Refraction, reflection, or scattering in the medium can be ignored.
- There is no emission in the wavelength range of the incident beam.

The RTE was formulated in compliance with the law of conservation of energy. The left-hand side of the equation describes the variation of light intensity in different paths, and the right-hand side of the equation describes the energy loss due to absorption and scattering effects in that path and the scattered energy propagating from other locations in the medium.

7.3.2.1   Analytic Solution of the RTE Equation

Due to the complex structure of the RTE with multiple integral and differential terms at the same time, it is difficult to find an exact analytical solution. Only few analytical solutions of the RTE have been proposed in recent years. RTE can be analytically solved by using modified Stokes vector. This model takes both multiple scattering and light polarization effects into account.

7.3.2.2   Numerical Solution of the RTE Equation

The RTE is an integral-differential equation that contains several independent variables. Moreover solving RTE analytically is quite complex and therefore, RTE is solved numerically by: invariant imbedding, Monte Carlo, and discrete ordinate. Comparison of the three RTE numerical solution methods is shown in Table 7.2. Discrete ordinate attempts to solve the RTE by decomposing it into a linear combination of Fourier series and Legendre polynomials. Invariant imbedding method partitions the RTE into more manageable initial value problems and thus reducing the complexity of the problem.

Table 7.2 Comparison of typical RTE numerical solution methods

Discrete ordinate	Invariant imbedding	Stochastic model
Analytical structure	Analytical structure	Statistical structure
Analytic solution for given VSF	Applicable to one dimension	Analytic solution for a given VSF
Fast for a homogeneous channel	Fast	Fast

7.3.3 Monte Carlo Simulation for UWOC Channel Modeling

Monte Carlo (MC) method is a typical probabilistic method. The MC method is able to decompose a particular problem into a number of random events with known probability density functions to obtain an approximate answer to the exact problem. For the UWOC, each photon motion trajectory in the medium is found using a known medium characteristic and volume scattering function [7]. Finally, information on these photon trajectories is collected, and the channel characteristics, including the channel impulse response function, can be derived statistically. The MC method tracks each photon's trajectory in the medium and focuses on the photons that reach the receiving plane. In the field of computer graphics, this method is often referred to as ray tracing. Although the MC method is extremely demanding in terms of computational power, the computational effort is no longer the main limiting factor for the MC method with the rapid development of computer technology. However, it has some disadvantages: not suitable for the simulation of point sources and point detectors, cannot solve fluctuation phenomena, and computationally intensive. MC simulation is divided into three key modules: the photon's initial state, the propagation of the photon, and the reception of the photon.

The initial state of the photon is related to the structure of the transmitter. Cox has simplified the emitting part of the UWOC to a combination of a quasi-rectangular light source and a single lens so that the initial loading state of the photon can be described by known parameters such as beam waist and focal length. The propagation of a photon is shown in Fig. 7.9. The VSF function can describe each scattering of the photon in the medium; the photon's receiving process is shown in Fig. 7.10. The photon arriving at the receiver needs to satisfy two conditions: (1) the angle of incidence is less than half the FOV. (2) The incident position is within the receiver lens.

As another method for solving the RTE, deterministic numerical methods are highly mathematical. The discrete vertical coordinate method (DVCM) [6] is a typical deterministic numerical method that expands the RTE's integral terms through the sum of orthogonal bases. However, the method does not deal well with the forward peak volume scattering function (VSF). In addition to DVCM, the Invariant Embedding Method (IEM) is also a deterministic method for solving the RTE. In 2005, H. Gao proposed a direct angular discretization format based on finite elements. As the RTE's integral terms are directly discretized in the angular space,

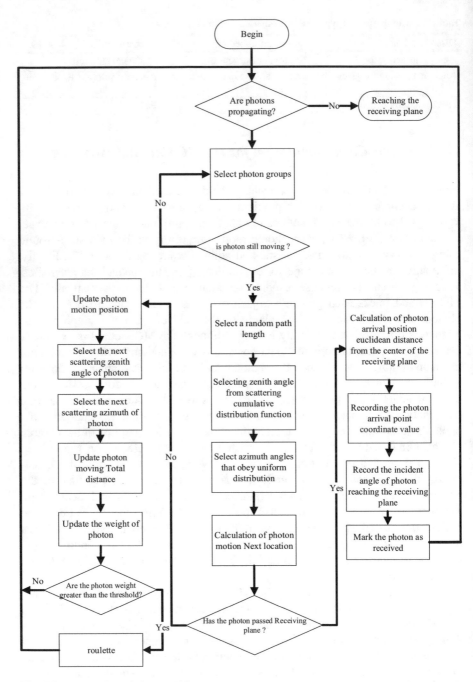

Fig. 7.9 Propagation of photons MC

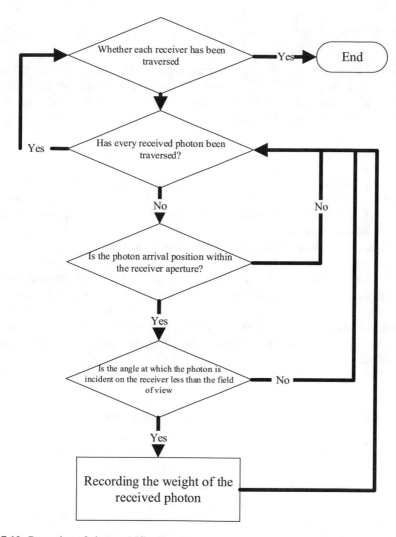

Fig. 7.10 Reception of photons MC

the scattering characteristics can be easily captured. However, few researchers have adopted the above method as an alternative to MC simulation.

To further save the computational time of the numerical method, the matrix-free Gauss-Seidel iterative method can be used to calculate the received power of the UWOC system. The Gauss-Seidel iterative method can also be used to handle elastic waves with high forward peaks. This method is able to obtain similar computational accuracy as the MC method with much shorter computation time. Specifically, a typical three-dimensional RTE can be downscaled to a two-dimensional RTE.

$$\mathbf{n} \cdot \nabla I(\mathbf{r}, \mathbf{n}) = -cI(\mathbf{r}, \mathbf{n}) + \int_{2\pi} \beta\left(\mathbf{n}, \mathbf{n}'\right) I\left(\mathbf{r}, \mathbf{n}'\right) d\mathbf{n}' + E(\mathbf{r}, \mathbf{n}), \tag{7.19}$$

where $I(\mathbf{r}, \mathbf{n})$ is the optical radiation propagating in the direction of position $\mathbf{r}$ towards $\mathbf{n}$ in units of $Wm^{-2}sr^{-1}$. $E(\mathbf{r}, \mathbf{n})$, the same as the three-dimensional RTE, and $\beta\left(\mathbf{n}, \mathbf{n}'\right)$ is the volume scattering function, which is related to the scattering phase function as shown in Eq. (7.20).

$$\beta\left(\mathbf{n}, \mathbf{n}'\right) = b\tilde{\beta}(\theta), \tag{7.20}$$

where θ is the angle between the direction vector $\mathbf{n}$ and $\mathbf{n}'$. To numerically solve the integral-differential RTE in Eq. (7.19), the angular and spatial variables are first discretized, and then the fully discretized system of large linear equations is solved by Gauss-Seidel iteration.

Compared to LOS UWOC channel modeling, non-LOS UWOC channel modeling has received less attention. The scattering and attenuation effects are independent of the structure of the UWOC system. Therefore, the difference between LOS channels and NLOS channels is mainly whether the beam is fully internally reflected at the sea surface or not. Similar to the LOS channel modeling, the NLOS channel model can be obtained by analytical and numerical methods. However, in recent years, most NLOS channel models have been derived by numerical methods such as MC.

7.4   Modeling of UWOC Geometric Misalignments

Similar to the work on modeling UWOC optical attenuation, the alignment error of UWOC links can also be modeled by analytical and numerical methods. In the analytical approach, Brandon first proposed modeling the irradiance distribution of a light source in a plane, perpendicular to the optical axis using a beam spread function (BSF) [8]. Based on the BSF, without considering the pointing error due to slight jitter of the transceiver, the link disorder when the receiver deviates is modeled as

The geometry of the underwater optical transmission is shown in Fig. 7.11. The light source is located at a certain depth underwater, and this position is set as the origin of the coordinate system. The light energy received by the receiver at a given location underwater is essential for calculating the link budget. Brandon, therefore, modeled the irradiance distribution of the light source in a plane perpendicular to the optical axis. This irradiance distribution is also known as the beam spread function (BSF). Assuming that the light attenuation properties (spectral scattering, absorption, attenuation coefficient) are depth-dependent only, the irradiance distribution at a given wavelength is expressed as

Fig. 7.11 Problem geometry
for determining the BSF

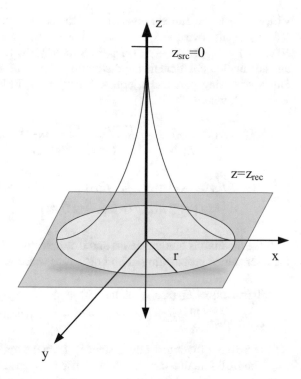

$$\text{BSF}\,(r, z_{\text{rec}}) = \frac{1}{2\pi} \int_0^\infty E_0\,(v, z_{\text{rec}})\,S\,(v, z_{\text{rec}})\,J_0(vr)v\,dv, \tag{7.21}$$

where (r, z_{rec}) denotes parameters in the spatial frequency domain and (v, z_{rec}) denotes parameters in the spatial coordinate system. The Hanel transform converts the parameters from the spatial frequency domain to the spatial coordinate system. $J_0(vr)$ denotes a Bessel function of order 0. $E_0\,(v, z_{\text{rec}})$ denotes the initial irradiance distribution of the beam in the spatial frequency domain. The irradiance of a laser light source follows a Gaussian distribution. Therefore,

$$E_0\,(v, z_{\text{rec}}) = P_0 \exp\left[-V_0\,(z_{\text{rec}})\,v^2/2\right], \tag{7.22}$$

where P_0 is the initial power of the light source. $V_0\,(z_{\text{rec}})$ is the variance of the Gaussian light source in free space.

$S\,(v, z_{\text{rec}})$ is the dielectric optical transfer function (OTF). OTF is the Fourier transform of the impulse response of the UWOC channel. OTF describes the spatial characteristics of the optical signal as it propagates through the water medium.

$$S\,(v, z_{\text{rec}}) = \exp\left\{-\int_{z_{\text{src}}}^{z_{\text{rec}}} [c(z) - b(z)p\,(v\,(z_{\text{rec}} - z))]dz\right\}, \tag{7.23}$$

where $p(v)$ is the Hankel transform of the scattering phase function. Decompose $S(v, z_{rec})$ into two parts containing the scattering and attenuation coefficients. $S(v, z_{rec}) = S_{NS}(z_{rec}) + S_S(v, z_{rec})$. Similarly, the terms containing the scattering and attenuation coefficients are distinguished and the BSF function is decomposed into a scattering part and a non-scattering part. BSF $(r, z_{rec}) = E_{NS}(r, z_{rec}) + E_S(r, z_{rec})$, where

$$
\begin{aligned}
E_{NS}(r, z_{rec}) &= \tfrac{1}{2\pi} \int_0^\infty E_0(v, z_{rec}) S_{NS}(v, z_{rec}) J_0(vr) \\
&= E_0(r, z_{rec}) \exp(-\int_{z_{src}}^{z_{rec}} c(z)dz),
\end{aligned}
\tag{7.24}
$$

$$
\begin{aligned}
E_S(r, z_{rec}) &= \tfrac{1}{2\pi} \int_0^\infty E_0(v, z_{rec}) \exp(-\int_{z_{src}}^{z_{rec}} c(z)) \\
&\times \left\{ \exp\left[\int_{z_{src}}^{z_{rec}} b(z) p\left(v(z_{rec} - z) \right) dz \right] - 1 \right\} J_0(vr) v\, dv.
\end{aligned}
\tag{7.25}
$$

If the medium is homogeneous and isotropic, then $\int_{z_{src}}^{z_{rec}} c(z)dz = cL$, the BSF of the laser source propagating in homogeneous water is

$$
\begin{aligned}
\text{BSF}(r, z_{rec}) &= E_0(r, z_{rec}) \exp(-cL) + \tfrac{1}{2\pi} \int_0^\infty E_0(v, z_{rec}) \exp(-cL) \\
&\times \left\{ \exp\left[\int_{z_{src}}^{z_{rec}} b(z) p\left(v(z_{rec} - z) \right) dz \right] - 1 \right\} J_0(vr) v\, dv.
\end{aligned}
\tag{7.26}
$$

The BSF is a function of the distance r from the optical axis and the link distance z_{rec}. The BSF function can therefore be used to measure the received optical energy in the presence of an offset of the receiver when the receiving plane is perpendicular to the optical axis. Note that there are some limitations of the BSF model. First, the BSF model requires that the scattering of photons in the water medium is dominated by forward scattering, and the receiving plane is perpendicular to the optical axis. Besides, Eq. (7.26) not consider the limitations of FOV on photon capture (i.e., the receiver field of view is set to 180°), receiver lens area, receiver pointing, etc.

7.5   Modeling of UWOC Link Turbulence

Absorption, scattering, and underwater optical turbulence (UOT) can have a serious negative impact on the reliability of UWOC systems. Among them absorption and scattering effects have been introduced in Sect. 2.1 of this chapter. By definition, UOT is an event that causes a rapid change in the refractive index of the water medium. Specifically, these events include temperature fluctuations, pressure changes, salinity changes, and bubbles, etc. UOT can distort the intensity and phase of the optical signal and degrade the performance of the UWOC system. In order to design a robust and reliable UWOC system, it is important to study the distribution pattern of the UOT-induced optical signal fluctuations [9].

Turbulence-induced changes in light intensity are similar to light scintillation. The light intensity perturbation can be quantitatively described by the scintillation index:

$$\sigma_I^2(r, d_0, \lambda) = \frac{\mathrm{E}\left[I^2(r, d_0, \lambda)\right] - \mathrm{E}^2\left[I(r, d_0, \lambda)\right]}{\mathrm{E}^2\left[I(r, d_0, \lambda)\right]}. \tag{7.27}$$

In the above equation $I(r, d_0, \lambda)$ denotes the instantaneous intensity at position (r, d_0). d_0 denotes the distance of the link. $\mathrm{E}[\cdot]$ denotes the expectation.

7.5.1 Atmospheric Turbulence Model

Considering that the physical properties of light propagation in the atmosphere as well as in water have some similarity. In the early studies of UOT, researchers applied atmospheric turbulence models to directly describe UOT. These atmospheric turbulence models include log-normal, K distribution, and Gamma-Gamma distribution.

(1) Lognormal distribution.

The most commonly used model to describe the fluctuations in irradiance caused by turbulence is the lognormal distribution model. The lognormal distribution is shown below.

$$f_{\tilde{h}}(\tilde{h}) = \frac{1}{2\tilde{h}\sqrt{2\pi\sigma_X^2}} \exp\left(-\frac{\left(\ln(\tilde{h}) - 2\mu_X\right)^2}{8\sigma_X^2}\right). \tag{7.28}$$

In the above equation, $\tilde{h}$ is the channel fading coefficient, which describes the variation in received light intensity; μ_X and σ_x^2 are the mean and variance of the log-normal distribution, respectively. To accurately describe the fading effect of turbulence, the effect of turbulence is separated from the scattering and absorption effect of the medium itself to ensure that the total energy of the optical signal remains unchanged. Set $\mathrm{E}[\tilde{h}] = 1$ so that $\mu_X = -\sigma_x^2$. The variance of log-normal fading at this point is related to the scintillation index, $\sigma_X^2 = 0.25 \ln\left(1 + \sigma_I^2\right)$. The log-normal fading model is a better fit for the weak atmospheric turbulent fading of $\sigma_I^2 \le 1$.

(2) K distribution.

In case of strong gas turbulence in $\sigma_I^2 \ge 1$, researchers often use the K distribution to describe the decay caused by turbulence.

$$f_{\tilde{h}}(\tilde{h}) = \frac{2\alpha}{\Gamma(\alpha)}(\alpha\tilde{h})^{(\alpha-1)/2}K_{\alpha-1}(2\sqrt{\alpha\tilde{h}}), \tag{7.29}$$

where $\Gamma(\cdot)$ is the Gamma function in the above equation. At $E[\tilde{h}] = 1$, α is a positive parameter related to the flicker index, $\alpha = 2/\left(\sigma_I^2 - 1\right)$ [exponential normal distribution model21]. $K_{\alpha-1}(\cdot)$ is a second class modified Bessel function of order $\alpha - 1$.

(3) Gamma-Gamma distribution.

The Gamma model can be used as an alternative to the log-normal decay model. Based on the Gamma model, the Gamma-Gamma model is proposed. The model describes the irradiance variation caused by turbulence through the product of two independent random variables, each of which obeys the Gamma distribution. The model can be used to describe atmospheric fluctuations with high scintillation index as well as low scintillation index.

$$f_{\tilde{h}}(\tilde{h}) = \frac{2(\alpha\beta)^{(\alpha+\beta)/2}}{\Gamma(\alpha)\Gamma(\beta)} (\tilde{h})^{\frac{(\alpha+\beta)}{2}-1} K_{\alpha-\beta}\left(2\sqrt{\alpha\beta\tilde{h}}\right), \qquad (7.30)$$

where α and β are parameters related to atmospheric conditions. At an expectation of 1, the scintillation index is $\sigma_I^2 = 1/\alpha + 1/\beta + 1/\alpha\beta$.

Atmospheric turbulence fading models such as lognormal, K distribution, and Gamma-Gamma distributions advanced early UOT research. It was shown that most atmospheric turbulence fading models could also be used to describe underwater optical channels. However, the statistical characteristics of the refractive index variation in water differ from those in the atmosphere. This makes it difficult for the log-normal distribution to describe the irradiance fluctuations caused by UOT. Therefore, the model of light energy decay caused by UOT remains to be studied.

7.5.2   Underwater Turbulence Model

In the presence of bubbles in the medium, typical single distributions such as the lognormal or Gamma distributions do not fit the measured data well, and a hybrid statistical model is needed to characterize the fading statistics due to UOT when the scintillation index takes different values.

(1) Mixture Exponential-Logarithmic distribution.

Mixture Exponential-Logarithmic distribution has been proposed to describe the statistical distribution of light intensity fluctuations in salty water containing air bubbles for the study.

$$f_{\tilde{h}}(\tilde{h}) = \frac{k}{\Gamma} \exp(-\tilde{h}/\Gamma) + \frac{(1-k)\exp\left(-(\ln(\tilde{h}) - \mu)^2/2\sigma^2\right)}{\tilde{h}\sqrt{2\pi\sigma^2}}, \qquad (7.31)$$

where k determines the weighting of the exponential distribution to the log-normal distribution, Γ is mean of exponential distribution, and μ and σ^2 are the

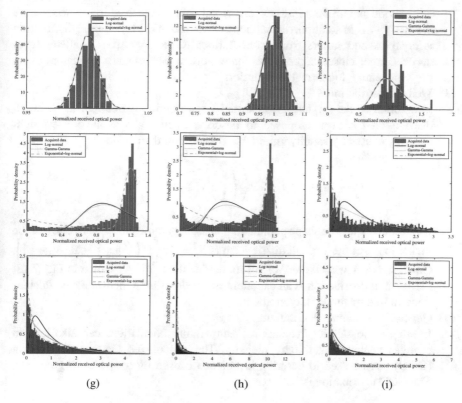

Fig. 7.12 Fitting effects of Lognormal, K distribution, Gamma-Gamma, and mixture Exponential-Logarithmic distribution on turbulent fading of different bubbles and salinities. (**a**) $\sigma_I^2 = 7.175 \times 10^{-5}$, (**b**) $\sigma_I^2 = 0.0011$, (**c**) $\sigma_I^2 = 0.0496$, (**d**) $\sigma_I^2 = 0.1015$, (**e**) $\sigma_I^2 = 0.2453$, (**f**) $\sigma_I^2 = 0.7885$, (**g**) $\sigma_I^2 = 1.0652$, (**h**) $\sigma_I^2 = 3.5695$, (**i**) $\sigma_I^2 = 2.1776$

mean and variance of the log-normal distribution. When $E[\tilde{h}] = k\Gamma + (1 - k)\exp(\mu + \sigma^2/2) = 1$, the scintillation exponent is equivalent to the second order moment of the decay distribution minus one, i.e., $\sigma_I^2 = 2k\Gamma^2 + (1 - k)\exp(2\mu + 2\sigma^2) - 1$. To verify the fitting effect of atmospheric decay model on UOT, Jamali conducted experiments on UOT with different bubble densities and salinities. The experimental results are shown in Fig. 7.12.

The experimental results show that the Lognormal distribution achieves an excellent fit to the fading data when the scintillation index σ_I^2 is less than 0.1. The Gamma-Gamma and K distributions describe the UOT at $\sigma_I^2 \geq 1$. None of these three distributions describes the received irradiance of the UWOC at time $0.1 \leq \sigma_I^2 \leq 1$. A combined exponential and lognormal distributions model describes the received irradiance PDF at $0.1 \leq \sigma_I^2 \leq 1$ perfectly.

However, the mixture exponential-lognormal distribution model does not consider temperature or salinity gradients in the watercourse, and the models do not provide an excellent fit to experimental data on received irradiance at time $0.1 \leq \sigma_I^2 \leq 1$. In addition, the mathematical form of the log-normal distribution is complex and suffers from computational difficulties. It is challenging to derive further closed expressions for important performance indicators using the exponential, lognormal distribution.

(2) Weibull distribution.

In addition to the mixture Exponential-Logarithmic distribution, the Weibull distribution has been proposed to describe fluctuations in underwater light intensity caused by salinity gradients. The Weibull distribution is

$$f_{\tilde{h}}(\tilde{h}) = \frac{k}{\lambda}\left(\frac{\tilde{h}}{\lambda}\right)^{k-1} e^{-\left(\frac{\tilde{h}}{\lambda}\right)^{k}}, \tag{7.32}$$

where k and λ are the shape parameter and scale parameter, respectively, and the flicker index of the Weibull distribution is $\sigma_I^2 = \lambda^2\left[\Gamma\left(1+\frac{2}{\lambda}\right) - \Gamma\left(1+\frac{1}{\lambda}\right)\right]$. Qubei shows how well the Weibull model fits the UOT. As shown in Fig. 7.13, the Weibull distribution is an excellent fit for describing the irradiance fluctuations caused by the salinity gradient.

(3) Generalized Gamma distribution.

In addition to salinity fluctuations, temperature fluctuations can also lead to changes in underwater light intensity. The generalized Gamma distribution (GGD) can be used to describe turbulence caused by temperature gradients. The GGD distribution is

$$f_{\tilde{h}}(\tilde{h}) = \frac{c\tilde{h}^{ac-1}}{b^{ac}}\frac{\exp\left(-\left(\frac{\tilde{h}}{b}\right)^{c}\right)}{\Gamma}, \tag{7.33}$$

where a and c are the shape parameters and b is the scale parameter. In turbulence with a generalized Gamma distribution, the scintillation index is $\sigma_I^2 = \frac{\Gamma(a)\Gamma\left(a+\frac{2}{c}\right)}{\Gamma\left(a+\frac{1}{c}\right)} - 1$.

As shown in Fig. 7.14, the experiments show that the GGD distribution achieves the best fit to the measured UOT data for all weak temperature gradients compared to the Weibull and Gamma distributions.

(4) Mixture Exponential-Gamma distribution.

In order to solve the problem of difficult calculation of Lognormal distribution, the researchers proposed a mixture Exponential-Gamma (EG) distribution to characterize the fluctuation of optical signal irradiance in the underwater channel due to bubbles and salinity. The EG model is

$$f_{\tilde{h}}(\tilde{h}) = \frac{w}{\lambda}\exp\left(-\frac{\tilde{h}}{\lambda}\right) + (1-w)I^{\alpha-1}\frac{\exp\left(-\frac{\tilde{h}}{\beta}\right)}{\beta^{\alpha}\Gamma(\alpha)}. \tag{7.34}$$

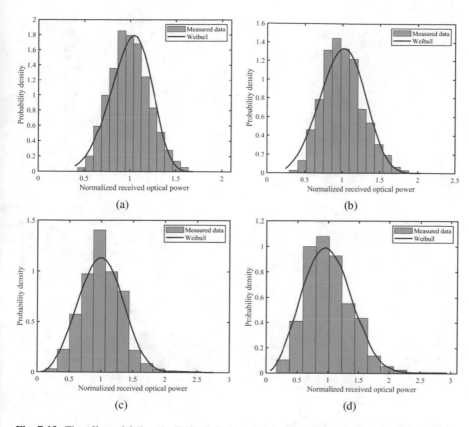

Fig. 7.13 The effect of fitting the Weibull distribution to the turbulent fading caused by salinity changes. (a) $\sigma_I^2 = 0.0487$, (b) $\sigma_I^2 = 0.0829$, (c) $\sigma_I^2 = 0.1134$, (d) $\sigma_I^2 = 0.1484$

In the Eq. (7.34), w determines the weighting of the exponential and Gamma distributions. λ is the parameter of the exponential distribution, and α and β represent the shape parameters and scale parameter of the Gamma distribution, respectively. The scintillation index is $\sigma_I^2 = 2w\lambda^2 + (1 - w)\alpha\beta^2(1 + \alpha) - 1$. As shown in Fig. 7.15, the EG model provides a perfect fit for laboratory measurements at different salinities and bubble densities. The model has a simple mathematical form, which facilitates the derivation of the equations.

(5) Mixture Exponential-Generalized Gamma.

To comprehensively describe the statistical model of the decay of UOT caused by changes in salinity, temperature, and bubble density, the mixture Exponential-Generalized Gamma (EGG) distribution is proposed. The EGG distribution is

$$f_{\tilde{h}}(\tilde{h}) = \omega\frac{1}{\lambda}\exp\left(-\frac{\tilde{h}}{\lambda}\right) + c\frac{\tilde{h}^{ac-1}\exp\left(-\left(\frac{\tilde{h}}{b}\right)^c\right)}{b^{ac}}\frac{}{\Gamma(a)}, \qquad (7.35)$$

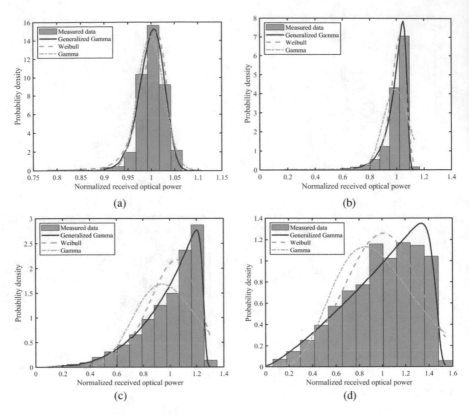

Fig. 7.14 Fitting effect of Generalized Gamma distribution, Weibull distribution, and Gamma distribution on turbulent decay caused by temperature gradient (TG). (**a**) TG=0.05°C · cm^{-1}, (**b**) TG=0.10°C · cm^{-1}, (**c**) TG=0.15°C · cm^{-1}, (**d**) TG=0.20°C · cm^{-1}

where w determines the weight of the exponential and generalized Gamma distributions in Eq. (7.35). a and c are the shape parameters and b is the scale parameter. The scintillation index is $\sigma_I^2 = \dfrac{2\omega\lambda^2 + (1-\omega)b^2\frac{\Gamma\left(a+\frac{2}{c}\right)}{\Gamma(a)}}{\left[\omega\lambda + (1-\omega)\frac{b\Gamma\left(a+\frac{1}{c}\right)}{\Gamma(a)}\right]^2} - 1$. Figures 7.16 and 7.17 show the effectiveness of the EGG model for fitting turbulent fading with different temperature gradients, bubble concentrations, and salinity, in fresh water and salty water, respectively. The results show that the EGG model is an excellent fit for turbulence induced by temperature, bubbles, and salinity and can be applied to turbulent flows of different intensities.

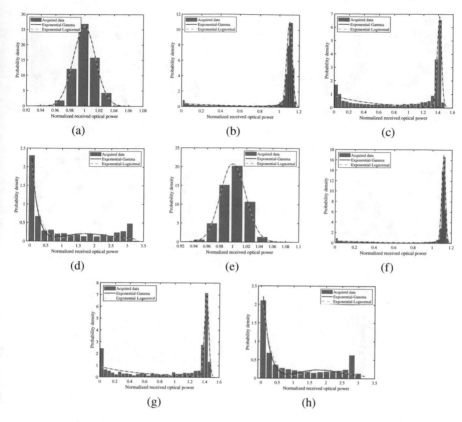

Fig. 7.15 Fitting effect of EG distribution and mixture Exponential-Lognormal distribution on turbulent decay caused by salinity and bubbles. (**a**) salt water, bubbles level (BL) = 0 L/min, (**b**) salt water, BL = 2.4 L/min, (**c**) salt water, BL = 7.1 L/min, (**d**) salt water, BL = 16.5 L/min, (**e**) fresh water, BL = 0 L/min, (**f**) fresh water, BL = 2.4 L/min, (**g**) fresh water, BL = 7.1 L/min, (**h**) fresh water, BL = 16.5 L/min

7.6   Noise in the UWOC Channel

Background noise is a crucial factor in the performance of UWOC systems. Noise is related to the geographical location of the UWOC systems. This section describes the modeling of noise, including background noise, dark current noise, thermal noise, and scattered particle noise. The noise variance is also introduced. The variance of the noise can be converted into a noise power spectrum and used to evaluate the UWOC system performance [10].

(1) Background noise.

In the oceanic transmissive zone (tens of meters deep), solar radiation is the main source of background noise. In the ocean depths, the main source of background noise shifts to blackbody radiation with bioluminescence. The

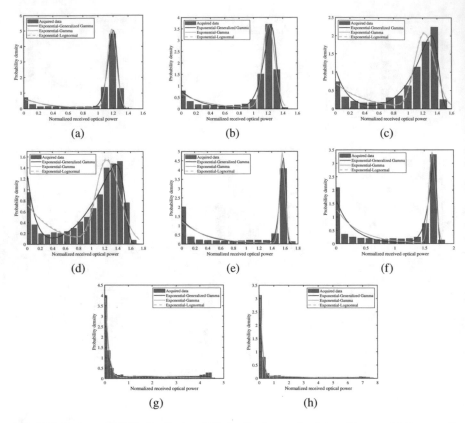

Fig. 7.16 Fitting of EGG distribution, EG distribution, and mixture Exponential-Lognormal distribution to turbulent decay caused by bubbles and temperature gradients in fresh water. (**a**) BL $=$ 2.4 L/min, TG $=$ 0.05°C.cm^{-1}, (**b**) BL $=$ 2.4 L/min, TG $=$ 0.10°C.cm^{-1}, (**c**) BL $=$ 2.4 L/min, TG $=$ 0.15°C.cm^{-1}, (**d**) BL $=$ 2.4 L/min, TG $=$ 0.20°C.cm^{-1}, (**e**) BL $=$ 4.7 L/min, TG $=$ 0.05°C.cm^{-1}, (**f**) BL $=$ 4.7 L/min, TG $=$ 0.10°C.cm^{-1}, (**g**) BL $=$ 16.5 L/min, TG $=$ 0.22°C.cm^{-1}, (**h**) BL $=$ 23.6 L/min, TG $=$ 0.22°C.cm^{-1}

wavelengths of light emitted by deep-sea organisms are concentrated in the blue-green region. The variance of the background noise is

$$\sigma_{BG}^2 = 2q\mathfrak{R}P_{BG}B, \tag{7.36}$$

where $\mathfrak{R}$ is the responsivity, q is the electron charge for $q = 1.6 \times 10^{-19}$ coulomb, and B is the electronic bandwidth.

The total background noise is equal to the set of solar radiation noise and blackbody radiation noise

$$P_{BG} = P_{BG\_\text{sol}} + P_{BG\_\text{blackbody}}. \tag{7.37}$$

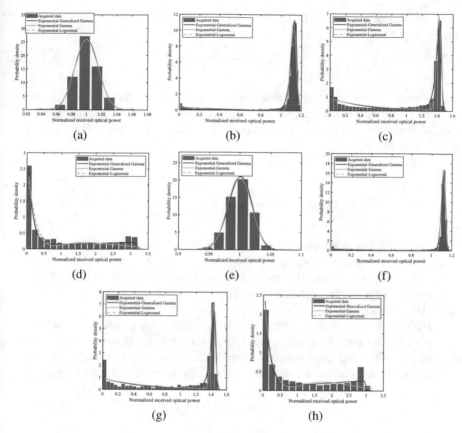

Fig. 7.17 Fitting of exponential-generalized Gamma mixing distribution, exponential-Gamma distribution, and exponential normal distribution to turbulent decay caused by bubbles in salt water. (**a**) BL = 2.4 L/min, TG = 0.05°C.cm^{-1}, (**b**) BL = 2.4 L/min, TG = 0.10°C.cm^{-1}, (**c**) BL = 2.4 L/min, TG = 0.15°C.cm^{-1}, (**d**) BL = 2.4 L/min, TG = 0.20°C.cm^{-1}, (**e**) BL = 4.7 L/min, TG = 0.05°C.cm^{-1}, (**f**) BL = 4.7 L/min, TG = 0.10°C.cm^{-1}, (**g**) BL = 16.5 L/min, TG = 0.22°C.cm^{-1}, (**h**) BL = 23.6 L/min, TG = 0.22°C.cm^{-1}

In the above equation, P_{BG_sol} represents the power of the solar background noise and P_{BG_sol} is

$$P_{BG-solar} = A_R(\pi FOV)^2 \Delta\lambda T_F L_{sol}, \tag{7.38}$$

where A_R is the received area, $\Delta\lambda$ is the filter bandwidth, and T_F is the filter transmittance.

Solar radiation $L_{sol}(W/m^2)$ is

$$L_{sol} = \frac{ERL_f e^{-Kd}}{\pi},$$ (7.39)

where E is downwelling irradiance (W/m^2), R is the underwater reflectance of downwelling irradiance, L_f is the factor describing the directional dependence of underwater radiance, K is the diffuse attenuation coefficient, d is the underwater depth.

Blackbody radiation noise is

$$P_{BG\_blackbody} = \frac{2hc^2 \Gamma A_R (\pi FOV)^2 \Delta\lambda T_A T_F}{\lambda^5 \left[e^{(hc/\lambda kT)} - 1 \right]},$$ (7.40)

where c is the speed of light in water, h is Planck's constant, $T_A = \exp(-\tau_0)$ is the transmittance in water, k is Boltzmann's constant, and Γ is the radiation absorption coefficient. The use of a narrowband spectral filter in front of the detection system helps to reduce background noise.

(2) Dark current noise.

Dark current noise is the small current generated on a component when no photons are passing through the optical sensor (e.g., photomultiplier tubes, photodiodes, and photocoupled elements). In non-optical components, it is called leakage current during reverse bias. The variance of the dark current is

$$\sigma_{DC}^2 = 2q I_{DC} B,$$ (7.41)

where I_{DC} is the dark current in amperes.

(3) Thermal noise.

Thermal noise is also known as Johnson noise and resistance noise. Thermal noise refers to the noise caused by passive devices such as resistors and feeders in communication equipment due to electron Brownian motion. The variance of the thermal noise current is

$$\sigma_{TH}^2 = \frac{4k T_e F B}{R_L},$$ (7.42)

where T_e is the equivalent temperature, F is the noise factor of the system, and R_L is the load resistance.

(4) Current shot noise.

Scattered particle noise is caused by the unevenness of electron emission inactive devices (such as electric vacuum tubes) in communication equipment. It is also known as scattering noise. The variance of the scattering noise current is

$$\sigma_{ss}^2 = 2q\,\Re P_s B, \tag{7.43}$$

where P_s is the power of the signal.

The total noise is the combination of all the noise in the communication system. Thus, in the absence of any optical signals. The variance of the total noise of the UWOC system σ_n^2 is

$$\sigma_n^2 = \sigma_{TH}^2 + \sigma_{DC}^2 + \sigma_{BG}^2 + \sigma_{ss}^2. \tag{7.44}$$

? Questions

1. How to model the beam attenuation characteristics of non-uniform media?
2. Light has wave-particle duality, so what are the ways to describe light attenuation in terms of light fields and particles? What are the advantages and disadvantages of each of them?
3. Why does Bill Lambert's law produce large errors when measuring the attenuation of light energy in media with high attenuation coefficients?
4. What assumptions need to be satisfied to simulate UWOC channels by the MC simulation method? Do these assumptions negatively affect the accuracy of MC simulation results?
5. The Log-Normal, K, Gamma-Gamma, Exponential-Log, Weibull, Generalized Gamma, mixed Exponential-Gamma, and mixed Exponential-Generalized Gamma distributions are suitable for describing which type of underwater turbulence?
6. Suppose $g = 0.924$, please calculate the scattering phase function by HG function and plot the curve of scattering with phase described by HG function using MATLAB and other software. And try to analyze the drawbacks of the HG function.
7. In the LOS UWOC system shown in following figure, the optical signal power reaching the receiver is often modeled as the product of the transmitter power, telescope gain, and path loss, i.e., $P_{Rx} = P_{Tx}\exp\left[-c(\lambda)\frac{d}{\cos(\theta)}\right]\frac{A_{Rec}\cos(\theta)}{2\pi d^2[1-\cos(\theta_0)]}$. where P_{Tx} is the average optical power of the transmitter. R_T and R_R are the sensitivity of the transmitter and receiver, respectively. d is the vertical distance between the transmitting and receiving planes. θ is the angle (misalignment angle) between the transmitting and receiving trajectories and the vertical line of the receiving plane. θ_0 is the half angle of the beam diffusion angle. A_{Rec} is the area of the receiver lens.

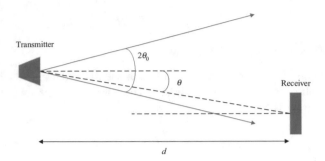

Assume that the laser has an average transmit power of 40 mW. The laser diffusion angle is 1 mrad. The transmitter and receiver sensitivity are both 1. The receiver aperture is 4 inches. The link distance is 10 m. There is a misalignment angle of 1 mrad between the transmitter and receiver. The link distance is 10 m. The transmitter-receiver has a misalignment angle of 1 mrad. The UWOC system is arranged in a turbid ocean with $c = 2.2$. Without considering environmental disturbances such as turbulence and noise, try to calculate the optical power that can be received by the UWOC system.

8. Suppose the UWOC system is subjected to turbulence that obeys a log-normal distribution and the received signal is appended with additive Gaussian white noise with mean 0 and variance σ^2. If the channel impulse response of the UWOC system is known and the signal is modulated by OOK, try to derive the BER expression of the UWOC system under the influence of inter-code crosstalk.

References

1. Z. Zeng, S. Fu, H. Zhang, et al., A survey of underwater optical wireless communications. IEEE Commun. Surv. Tutor. **19**(1), 204–238 (2016)
2. J.R.V. Zaneveld, *Light and Water: Radiative Transfer in Natural Waters* (Academic Press, San Diego, 1995)
3. R.W. Spinrad, K.L. Carder, M.J. Perry, *Ocean Optics* (Oxford University Press, 1994)
4. C. Mobley, E. Boss, C. Roesler, *Ocean Optics Web Book* (2010)
5. C. Gabriel, M.A. Khalighi, S. Bourennane, et al., Channel modeling for underwater optical communication, in *2011 IEEE GLOBECOM Workshops (GC Wkshps)* (IEEE, 2011), pp. 833–837
6. S. Arnon, J. Barry, G. Karagiannidis, R. Schober, M. Uysal, *Advanced Optical Wireless Communication Systems* (Cambridge University Press, 2012)
7. S. Tang, Y. Dong, X. Zhang, Impulse response modeling for underwater wireless optical communication links. IEEE Trans. Commun. **62**(1), 226–234 (2013)
8. S. Arnon, Underwater optical wireless communication network. Optical Engineering **49**(1), 015001 (2010)
9. E. Zedini, H.M. Oubei, A. Kammoun, et al., Unified statistical channel model for turbulence-induced fading in underwater wireless optical communication systems. IEEE Trans. Commun. **67**(4), 2893–2907 (2019)
10. H. Kaushal, G. Kaddoum, Underwater optical wireless communication. IEEE Access **4**, 1518–1547 (2016)

Chapter 8
Modulation Schemes

8.1 Introduction

Modulation is the use of a signal (modulating signal) to control another signal as a carrier signal, so that a parameter of the carrier signal changes with the modulating signal. Generally speaking, modulation techniques are divided into active modulation and passive modulation, active modulation is modulation at the optical transmitter, passive modulation is modulation at the other side, the modulated light source and the modulation process are separated. Active modulation has internal and external modulation: internal modulation is modulation of the light source parameters, and external modulation is modulation of a parameter of the light wave. UWOC channel modulation technology has a greater impact on system performance and has attracted widespread attention in recent years. Since UWOC can be regarded as FSO in an underwater environment, the traditional modulation technology used in the FSO communication system can also be applied to the UWOC system [1].

In UWOC, information (including language, text, images, symbols, etc.) carried by a laser beam is transmitted through an underwater optical channel to an optical receiver, which then identifies and recovers the information to its original form. This process of loading information into the light wave is called modulation, and the device that completes this process is called a modulator. This process of loading information into laser light waves is called laser modulation drive technology, where the laser is called the carrier, and the low-frequency information that plays a controlling role is called the modulating signal. Light waves are used as carriers for transmitting information in optical communications.

Modulation in UWOC can be classified according to a variety of criteria. According to the nature of modulation, underwater wireless optical modulation can be classified into amplitude modulation, intensity modulation, frequency modulation, phase modulation, and pulse modulation. According to the working principle of the modulator, it can be divided into electro-optical modulation, acoustic-optical

© The Author(s), under exclusive license to Springer Nature Switzerland AG 2022
Y. Lou, N. Ahmed, *Underwater Communications and Networks*, Textbooks in Telecommunication Engineering, https://doi.org/10.1007/978-3-030-86649-5_8

modulation, magneto-optical modulation, etc., and it can also be divided into analog modulation and digital modulation.

8.1.1   Analog and Digital Modulation

The original electrical signal from the source without modulation (spectral shift or transformation) is the baseband signal. It is featured by a very low frequency, a signal spectrum starting around zero frequency, and a low-pass shape. Based on the raw electrical signal characteristics, baseband signals can be categorized into digital baseband signals and analog baseband signals.

Analog modulation generally uses a sinusoidal signal as the carrier, and the spectrum of the baseband signal is transformed by compounding the carrier with the baseband signal for its transmission in the channel. The signal spectrum after linear modulation is a translation or linear transformation of the baseband signal spectrum (e.g., amplitude modulation). On the other hand, in a nonlinear modulation, there is no linear relationship between the modulated signal and the baseband signal (e.g., frequency modulation, phase modulation).

Digital modulation shifts the baseband signal spectrum to a higher frequency band more suitable for baseband signal transmission, while loading the baseband signal to one of the parameters of the high-frequency carrier. The noise introduced by analog modulation during transmission cannot be completely eliminated at the receiving end, but digital modulation has the potential to do so.

According to the number of levels of the baseband signal, digital modulation is divided into binary digital modulation and multi-decimal digital modulation; according to the different parameters of the carrier carrying information, digital modulation can be divided into ASK, FSK, PSK. Digital modulation can also combine several parameters mentioned above, such as QAM which is essentially a combination of ASK and PSK modulation.

8.1.2   Direct and Indirect Modulation

According to the relationship between modulating signal and light source, light modulation can be divided into direct modulation and indirect modulation.

(1) Direct modulation. Direct modulation controls the light source parameters (such as light intensity, etc.) directly by the modulating signal and obtains the light signal that changes with the modulating signal, respectively. The advantage of direct modulation is that the circuit is simple and easy to implement, but the transmission rate is low.

(2) Indirect modulation. Indirect modulation uses an external modulator to modulate a parameter of the optical carrier. The modulation object is the light wave
from the light source, and the parameters of the light source itself do not change.

8.1.3 Internal and External Modulation

Depending on the relative relationship between the modulator and the laser, internal
and external modulation can be classified. Internal modulation means that the
modulation signal is loaded during the laser oscillation process, and the modulation
signal is used to change the oscillation parameters of the laser, thus changing
the laser output characteristics to achieve modulation. The modulation signal
corresponds to the source in communication.

External modulation is the placement of a modulator on the optical path outside
the laser after laser generation, and the modulation signal is used to change the
physical characteristics of the modulator. When the laser beam passes through the
modulator, certain parameters of the light wave are modulated. External modulation
is usually achieved with electro-optic, acoustic-optic, or magneto-optic crystals.
External modulation is valued for its higher modulation rate (about an order of
magnitude) and much wider modulation bandwidth than internal modulation [2].

8.2 Selection Principles of Modulation Scheme

The choice of a more suitable modulation scheme depends on certain principles of
system requirements. For underwater wireless optical communication systems, the
two main principles are

- Power efficiency
- Bandwidth efficiency

In order to study the performance of any modulation form, the most important parameters are its bandwidth and the minimum power requirement for the
receiver detector to accurately detect the signal in the presence of noise. Since
the maintenance of equipment in UWOC is extremely difficult and the average
transmit optical power determines the visual safety and the power consumption at
the transmitter, it is very important to calculate the power requirements for each
modulation scheme. On the other hand, the bandwidth requirement is also important
because it determines whether the highest data rate can be achieved for a link with
a particular modulation form.

Power-efficient modulation schemes are relatively simple to implement because
in the case of line-of-sight UWOC, channel path loss at low data rates can
be effectively compensated. For short pulses, this modulation scheme meets the
requirement of low average power, but this modulation scheme requires a wider

bandwidth. On the other hand, for scattered UWOC scenarios with high data rates, bandwidth-efficient schemes (e.g., On-off keying (OOK), etc.) are more effective. The multipath distribution caused by the reflected light from floating objects and bubbles in the water makes the diffuse link have bandwidth-limited channel characteristics. Therefore, power efficiency schemes may not meet the bandwidth efficiency, for example, in these schemes PPM does not provide good performance. On the other hand, in bandwidth-efficient schemes, the power level limitation is significantly reduced at the edge of the link, thus limiting the usable range of the scheme. At this point, we need to choose a power-efficient modulation scheme, just like PPM and its deformation in bandwidth sacrifice makes the link performance improved at the same signal-to-noise ratio. Therefore, the choice of a modulation scheme must be a compromise between power and bandwidth.

8.3   Pulse Modulation

Pulse-like position modulation is a collective term for pulse position modulation methods. The main modulation methods are OOK, PPM, Digital Pulse Interval Modulation (DPIM), Differential Amplitude Pulse Interval Modulation (DAPIM), Pulse Amplitude and Position Modulation (PAPM), Dual-Header PIM (DHPIM), etc. [3–6]. All these modulation methods use the time interval between the light pulse and the reference point as the carrier of information transmission, and the receiver of the information uses the position of the light pulse in a certain period of time to determine the transmitted information, which is collectively called pulse-like position modulation. The modulation methods suitable for intensity modulation/direct detection of atmospheric wireless optical communication systems are mainly OOK, PPM, and multiplexed sub-carrier modulation. The average transmit power of OOK modulation is higher, while PPM reduces the average transmit power but increases the bandwidth requirement. The following is a systematic analysis of the symbol structure, bandwidth requirement, average transmit power, false time slot rate, and channel capacity of these modulation methods.

8.3.1   Binary On-Off Keying Modulation

OOK modulation is the most commonly used intensity modulation scheme in FSO systems. The modulation scheme can also be implemented in the UWOC system. OOK modulation is a binary switch keying modulation scheme. During the transmission of information using the OOK modulation method, the light pulse occupying part or the entire bit duration represents a single data bit 1, and when there is no light pulse, it represents a single data bit 0. Because the OOK modulation method is relatively simple and convenient to implement, people usually use this modulation method in UWOC. It modulates the transmitted binary information bit by bit. When the transmitted binary digital information bit is 1, the laser is continuously turned on and emits light pulses. When the transmitted binary digital information is 0, the laser is turned off. That is, binary data information is transmitted by controlling the presence or absence of light pulses.

In the OOK modulation scheme, there are two pulse formats: return-zero (RZ) and non-return-zero (NRZ) formats. In the RZ format, a pulse that only occupies a portion of the bit duration is defined as presenting 1; however, in the NRZ scheme, the pulse occupies the entire bit duration. RZ OOK has higher energy efficiency than NRZ OOK, but at the price of using more channel bandwidth. Due to significant absorption and scattering effects in the underwater environment for optical transmission, the transmitting OOK signal is subject to various channel attenuation and turbulence-induced fading effects. To mitigate these effects and achieve optimal OOK signal detection, dynamic thresholding (DT) techniques are widely used in receivers for UWOC systems using OOK. Several commonly used channel estimation techniques in free-space optical communication systems, such as the pilot symbol method, symbol-to-symbol maximum likelihood method, and ML sequence method, are also commonly used in UWOC systems. The two main shortcomings of UWOC system with OOK are low power efficiency and low bandwidth utilization. However, due to its low complexity, OOK-based UWOC systems are still the most popular ones until now.

8.3.2 Pulse Position Modulation

Although simple modulation schemes such as OOK are often used in UWOC systems, OOK technology has problems such as low power efficiency and low bandwidth utilization. Pulse position modulation (PPM) is another modulation technique commonly used in UWOC systems. Compared with OOK modulation technology, PPM modulation has higher energy efficiency and does not require dynamic thresholds, but at the expense of lower bandwidth utilization and more complex transceivers. The main disadvantage of PPM modulation is that the timing synchronization requirements are too high. Any timing jitter or asynchrony will reduce the system's bit error rate performance severely. In recent years, some researchers have studied the PPM technology's performance on the UWOC channel model. The study found that the PPM method's error rate is almost equal to the OOK modulation, but it has higher energy efficiency and spectrum utilization.

PPM technology is a kind of orthogonal modulation technology. Each transmitted M bit will be modulated into a pulse over 2^M time slots. The pulse position represents the transmitted information. In the PPM scheme, each time interval is T_s, 2^M time slots constitute a PPM frame. At the transmitting end, the signal will be emitted by light pulses to form a specific time slot. At the receiving end, the photodiode detects the light pulse, then judges its time slot and recovers the signal.

The important advantage of PPM over OOK is that it has more average energy efficiency. However, this comes at the cost of lower bandwidth efficiency, and the bandwidth required by LPPM technology increases as L increases. The LPPM modulation method modulates the binary information in units of groups, and the number of bits contained in each group (i.e., the number of binary digits) is the same, which is represented by N. Each group of binary numbers is mapped to 2^N time slots, and only one of these time slots has light pulses, and the position of this pulse time slot corresponds to the decimal number value corresponding to the group of binary numbers. If an N-bit binary number is written as $K = (k_1, k_2, \ldots, k_N)$, and the position of the time slot with light pulses is marked as s, then their corresponding mapping relationship can be expressed as

$$s = k_1 + 2k_2 + \cdots + 2^{n-1}k_N \in \{0, 1, \ldots, N - 1\}. \tag{8.1}$$

For the LPPM modulation method, the bandwidth utilization rate increases, but symbol synchronization and time slot synchronization are required during demodulation. A larger L result in a higher Peak-to-Average Power Ratio (PAPR). The higher switching speed of the electronic circuit makes it more difficult to synchronize the receiver.

PPM technology has the advantages of low transmit power and good anti-noise performance and is an attractive choice for wireless optical communication. However, the limitation of low bandwidth utilization makes PPM modulation reduce the efficiency of mass data transmission. More complex PPM, such as 8-PPM or 16-PPM, can achieve higher bandwidth utilization. In addition, there are some improved PPM schemes. These improved PPM can maintain power efficiency and anti-noise performance similar to traditional PPM but improve the system's bandwidth utilization on the original basis.

Differential pulse position modulation (DPPM) is differential version of PPM modulation. As long as all 0 time slots after the 1 time of PPM modulation are deleted, the corresponding DPPM signal can be obtained. Compared with PPM, DPPM symbols do not have serious symbol synchronization requirements, and more importantly, they can provide higher power utilization and bandwidth utilization.

8.3.3   Pulse Width Modulation

Similar to PPM, pulse width modulation (PWM) also uses pulses' relative position to represent data symbols. In L-PWM, light pulses will only appear in the first consecutive L time slot to represent a symbol, where L is the decimal number of the sign bit. Since PWM extends the total pulse time during one symbol transmission, each pulse's peak transmission power is reduced. The PWM scheme improves the spectral efficiency and the ability to resist intersymbol interference. However, PWM requires higher average power, and the average power requirement increases as the number of time slots per symbol increases.

8.3.4 Digital Pulse Interval Modulation

DPIM has also been widely used in UWOC systems. The DPIM modulation method is an improvement of the LPPM modulation method. The number of time slots between two consecutive optical pulse signals is used to transmit information. The number of time slots corresponding to each group of binary information is called a symbol. The symbol length is the decimal number value corresponding to each group of binary information, so the number of time slots it includes varies. It is usually divided into two situations, protected and unprotected time slots. In this modulation, an On optical pulse time slot is sent, followed by some Off time slots. The number of Off time slots depends on the decimal value of the transmitted symbol. An additional guard time slot is usually added to avoid continuous transmission of On pulses. Increasing the guard time slot can effectively reduce intersymbol interference.

If the first binary symbol sent is 000, it represents a decimal 0. According to the above modulation principle, an On time slot is followed by 0+1 Off time slots, and 1 time slots are added as guard time slots. Similarly, for the second transmitted binary symbol 001 representing a decimal 1, there should be 1+1 Off time slots after the On time slot. Compared with PPM and PWM, which require time slot and symbol level synchronization, DPIM is an asynchronous modulation scheme with variable symbol length. In addition, as the symbol length changes, DPIM also has higher bandwidth efficiency than PPM and PWM.

This book only discusses the situation where there is one guard slot. When a guard time slot is used, the number of time slots with the symbol S_k (k is the decimal number to be represented) is $k + 2$ (that is, one guard empty time slot is added after the initial pulse representing the symbol, and k empty time slots are added. Indicates the symbol information to be sent). When demodulating, subtract 1 from the number of received empty slots to get the corresponding sent decimal information.

The DPIM modulation method can be identified from its symbol structure that only the number of empty slots between two optical pulses needs to be calculated during demodulation. Therefore, only time slot synchronization is required, and symbol synchronization (i.e., frame synchronization) is not required. The modulation structures of OOK, LPPM and DPIM are shown in Fig. 8.1.

PIM modulation and PPM modulation's difference is only that PPM modulation uses the position of the optical pulse to express information, while PIM modulation uses the interval between two optical pulses to express information. PIM is an asynchronous modulation scheme with variable symbol length. However, the most critical problem of PIM is that it easily leads to error propagation. If an Off slot is demodulated to On during demodulation, all subsequent symbols will also be wrong.

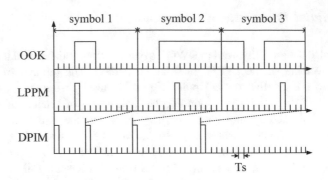

Fig. 8.1 OOK, LPPM, DPIM modulation structure diagram

8.3.5   *Performance Analysis of OOK, LPPM, DPIM Technology*

8.3.5.1   Transmit Power Performance Analysis

Taking into account the various limitations of underwater communication, it is required to increase the power utilization rate of UWOC as much as possible. Assuming that the three modulation methods' peak powers are the same (set to P), the average transmits power is analyzed.

For the OOK modulation method, when the binary numbers 0 and 1 appear with equal probability, the average transmit power can be expressed as

$$P_{OOK} = \frac{P}{2}. \tag{8.2}$$

For the LPPM modulation method, the number of time slots corresponding to each set of bits (i.e., the number of binary bits, let be N) is 2^N, and only one of these time slots has an optical pulse signal, then its average transmit power can be expressed as

$$P_{LPPM} = \frac{P}{2^N} = \frac{1}{2^{N-1}} P_{OOK}. \tag{8.3}$$

For the DPIM modulation method, the number of bits per group, the average number of time slots included is

$$\frac{1}{2^N} \sum_{k=1}^{2^N} (1 + k) = \frac{2^N + 3}{2}. \tag{8.4}$$

Moreover, each symbol of the DPIM modulation mode corresponds to an optical pulse signal, then the average transmit power is obtained from the above formula:

$$P_{DPIM} = \frac{P}{\frac{2^N+3}{2}} = \frac{4}{2^N+3} P_{OOK}. \tag{8.5}$$

The analysis shows that the average transmit power required by the LPPM modulation method is the smallest, the DPIM modulation method is slightly larger than that, and the average transmit power required by the OOK modulation method is the largest, and remains constant. As the value of N increases, the average transmits power required by LPPM and DPIM modulation methods gradually decreases and tends to zero. However, the value of N cannot be too large in practical applications, so N should be appropriately selected in practical applications to improve power efficiency.

8.3.5.2 Analysis of Bandwidth Requirements

In the UWOC systems, the receiver circuit's bandwidth is limited due to the capacitance in the receiver circuit. Therefore, the less bandwidth the system needs, the better. Assuming that the bit rates of the three modulation methods are same as R (bit/s), the number of bits (i.e., the number of binary digits) of each group of transmission information is the same, and all are N, and the bandwidth is represented by the reciprocal of the time slot width.

For the OOK modulation method, assuming that the time slot width is $\tau_0 = 1/R$, the required bandwidth is

$$B_{OOK} = \frac{1}{\tau_0} = R. \tag{8.6}$$

For the LPPM modulation method, the time slot width is $(N/2^N)\tau_0 = N/(2^N R)$, and the corresponding required bandwidth is

$$B_{LPPM} = \frac{2^N}{N} R = \frac{2^N}{N} B_{OOK}. \tag{8.7}$$

For the DPIM modulation method, the average time slot width is $2N\tau_0/(2^N + 3) = 2N/[(2^N + 3)R]$, and the corresponding bandwidth required is

$$B_{DPIM} = \frac{2^N + 3}{2N} R = \frac{2^N + 3}{2N} B_{OOK}. \tag{8.8}$$

According to the analysis, the OOK modulation method's bandwidth is low, and its value is constant. The bandwidth requirement of the LPPM modulation method is high, and as the value of N increases, the bandwidth required by the LPPM and DPIM modulation methods becomes larger and larger. Therefore, it is necessary to select N for LPPM and DPIM modulation methods appropriately.

8.3.5.3   Transmission Capacity Analysis

Transmission capacity refers to the maximum amount of signal that the system can transmit per unit time, and it reflects the strength of the system's information transmission capacity per unit time. Suppose that the number of bits (i.e., the number of binary digits) of each group of transmitted information is the same as N, and the slot width is the same as τ.

For the OOK modulation method, the transmission capacity can be expressed as

$$C_{OOK} = \frac{1}{\tau}. \tag{8.9}$$

For the LPPM modulation method, the corresponding length of each group of bits is $2^N \tau$, and the transmission capacity can be expressed as

$$C_{LPPM} = \frac{N}{2^N \tau} = \frac{N}{2^N} C_{OOK}. \tag{8.10}$$

For the DPIM modulation, the average length corresponding to the number of bits in each group is $[(2^N + 3)\tau]/2$, and its transmission capacity can be expressed as

$$C_{DPIM} = \frac{2N}{(2^N + 3)\tau} = \frac{2N}{2^N + 3} C_{OOK}. \tag{8.11}$$

8.3.5.4   Time Slot Error Rate Analysis

Time slot error rate is an essential indicator of the UWOC systems. For the convenience of discussion, we assume: (1) there is only additive white noise, the noise average is $n(t) = 0$, and the variance is $D = \sigma_n^2$; (2) the bandwidth of the transmitter and receiver of the system is broad.

The sampling arbiter obtains $\sqrt{S_t} + n(t)$, when the transmitted signal is 1 (with optical pulse), where S_t is the peak power when the sampling arbiter receives the signal; when the transmitted signal is 0 (without optical pulse), it obtains $n(t)$.

Suppose the decision threshold is b, then the probability of judging the signal 0 as 1 is

$$P_{01} = \frac{1}{2} \left\{ 1 + \text{erf} \left[\left(b - \sqrt{S_t} \right) / \sqrt{2\sigma_n^2} \right] \right\}. \tag{8.12}$$

The probability of judging the signal 1 as 0 is

$$P_{10} = \frac{1}{2} \left[1 - \text{erf} \left(b / \sqrt{2\sigma_n^2} \right) \right]. \tag{8.13}$$

Table 8.1 Comparison of modulation schemes

Modulation schemes	Modulation bandwidth (B)	The power required to achieve the same BER as OOK
OOK	R_b	P_{OOK}
L-PPM	$\frac{L}{\log_2 L} R_b$	$\sqrt{\frac{2}{L \log_2 L}} P_{OOK}$
L-PWM	$\frac{L}{\log_2 L} R_b$	$\frac{L+1}{\log_2 L} P_{OOK}$
L-DPIM	$\frac{L+3}{\log_2 L} R_b$	$\frac{\sqrt[4]{\frac{L+1}{\log_2 L}}}{(L+3)\sqrt{2}} P_{OOK}$

The total slot error rate of the arbiter can be written as $P_{se} = P_0 P_{10} + P_1 P_{01}$. Among them, P_0 and P_1 are the probability of sending signals 0 and 1, respectively, and $P_0 + P_1 = 1$.

Table 8.1 shows the modulation bandwidth B of OOK, L-PPM, L-PWM, and L-DPIM, and its transmission rate is R_b. Bandwidth efficiency can be obtained simply by R_b/B. The table above also shows the power required to obtain the same BER as the OOK modulation method. When $L = 2$, 2-PPM has the same power efficiency as OOK, but the required bandwidth is doubled. When L is large, L-PPM achieves the highest power efficiency, but the bandwidth efficiency is low. It can be seen that the power efficiency of L-DPIM is slightly lower than that of L-PPM, but the bandwidth efficiency of L-DPIM is much higher than that of L-PPM.

8.4 Sub-Carrier Intensity Modulation

Another modulation scheme implemented in UWOC is sub-carrier intensity modulation (SIM). In the SIM method, the optical sub-carrier signal and the information signal are pre-modulated. The optical sub-carrier can be modulated by any modulation method, such as 2PSK, QPSK, QAM, AM, FM, etc. [7]. The pre-modulation signal is used to modulate the intensity of optical carriers. Direct detection is used to recover the signal at the receiving end, just like in the IM/DD system. It does not need to adapt the threshold as the OOK modulation method, and it improves the bandwidth utilization rate than the PPM modulation method. Therefore, optical sub-carrier intensity modulation makes the system implementation easier.

SIM technology allows the optical link to transmit multiple information signals simultaneously. Sub-carrier multiplexing can be achieved using FDM to combine different modulated sub-carrier signals and then modulate the intensity of a continuous laser source as an optical carrier. Figure 8.2 shows the principle of the SIM optical system of the FSO link, which is also applicable to the UWOC systems. The disadvantage of this multiplexing scheme is tight synchronization and design complexity at the receiver. In short, the advantage of SIM technology is that it has high spectral efficiency. However, complex modulation/demodulation equipment is

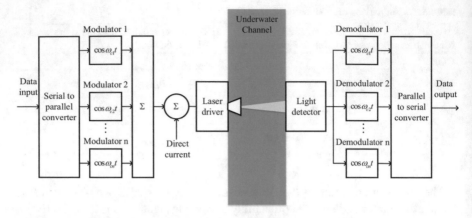

Fig. 8.2 SIM technology principle block diagram

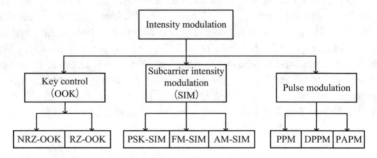

Fig. 8.3 Intensity modulation of UWOC systems

required, and the average power efficiency is poor. Using SIM, OFDM technology can also be implemented in the UWOC systems. For details, see the next section.

Both baseband and SIM signals can be demodulated using direct detection or incoherent detection technology. They have low cost and low complexity and are widely used in UWOC systems. Direct detection technology can also be used for analog modulation of optical carriers. However, it has not been widely used because it limits linearity to laser sources and modulation techniques, which is difficult to achieve. In short, in the UWOC systems, the most commonly used modulation scheme is intensity modulation, as shown in Fig. 8.3. Choosing an appropriate modulation scheme requires a trade-off between power efficiency, bandwidth requirements for a given data rate, and implementation complexity.

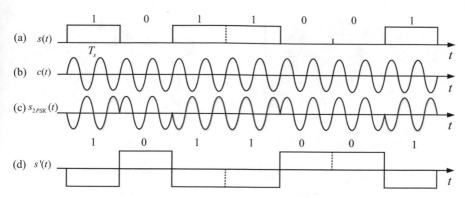

Fig. 8.4 2PSK signal waveform

8.4.1   PSK

Phase modulation uses binary digital information to control the phase of the sinusoidal carrier so that the phase of the sinusoidal carrier changes with the change of the binary digital information. Because of the different methods of controlling the carrier phase with binary digital information, binary digital phase modulation is divided into binary absolute phase modulation (2PSK) and binary differential phase modulation (2DPSK) [8].

8.4.1.1   2PSK

Binary absolute phase modulation is that digital information directly controls the phase of the carrier. For example, when the digital information is 1, the carrier is reversed (that is, 180° changes occur); when the digital information is 0, the phase of the carrier remains unchanged. Figure 8.4 shows the waveform of the 2PSK signal (for ease of drawing, draw two periods of the carrier within one symbol period).

In Fig. 8.4, (a) is digital information, (b) is carrier wave, (c) is 2PSK waveform, and (d) is the bipolar digital baseband signal. It can be seen from Fig. 8.4 that the 2PSK signal can be regarded as a bipolar baseband signal multiplied by the carrier, namely

$$s_{2PSK}(t) = s'(t) A \cos 2\pi f_c t. \tag{8.14}$$

It should be noted that the phase of the 2PSK waveform is relative to the phase of the carrier. Therefore, when drawing a 2PSK waveform, the carrier must be drawn first, and then the waveform of the 2PSK signal must be drawn based on the corresponding relationship between the digital information and the phase of the carrier. When the symbol width is not an integer multiple of the carrier period, draw the carrier first, and then draw the 2PSK waveform.

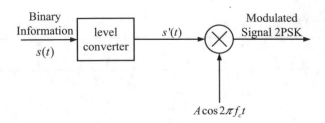

Fig. 8.5 2PSK modulator

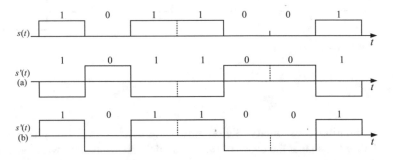

Fig. 8.6 Level converter input/output waveform

After understanding the waveform of the 2PSK signal, next we describe the components that generate the 2PSK signal, namely the 2PSK modulator. It can be seen from Eq. (8.14) that a multiplier can realize the 2PSK modulator. The block diagram is shown in Fig. 8.5.

Figure 8.5 shows that the level converter's function is to transform the digital input information into a bipolar full-duty digital baseband signal $s'(t)$. However, it should be noted that the same digital information can be transformed into two full-duty digital baseband signals with opposite polarities, as shown in Fig. 8.6. In a modulator, only one of these transformations can be used. As to which transformation is used, it is completely determined by the modulation rule. If the modulation rule that 1 becomes 0 is used, the level converter will change the digital information. 1 is transformed into a negative full-duty rectangular pulse, and digital information 0 is transformed into a positive full-duty rectangular pulse, as shown in Fig. 8.6a. The modulation rule corresponding to the waveform in Fig. 8.6b is that 0 changes to 1 unchanged.

It can be seen from Eq. (8.14) and Fig. 8.6. that a bipolar full-duty digital baseband signal of $s'(t)$ times $A \cos 2\pi f_c t$ produces a 2PSK signal. Therefore, according to the principle of spectrum transformation, the power spectrum of the 2PSK signal is

$$P_{2PSK}(f) = \frac{A^2}{4}[P_{s'}(f - f_c) + P_{s'}(f + f_c)]. \tag{8.15}$$

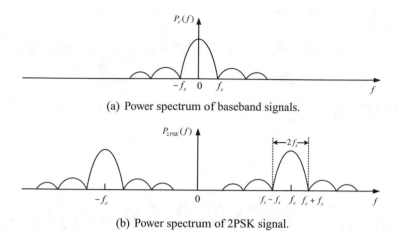

(a) Power spectrum of baseband signals.

(b) Power spectrum of 2PSK signal.

Fig. 8.7 The power spectrum of 2PSK signal. (**a**) Power spectrum of baseband signals. (**b**) Power spectrum of 2PSK signal

Among them, $P_{s'}(f)$ is the power spectrum of the bipolar full-duty rectangular pulse sequence $s'(t)$. The schematic diagram of power spectrum $P_{s'}(f)$ and $P_{2PSK}(f)$ is shown in Fig. 8.7. The power spectrum of the 2PSK signal has the same shape as the power spectrum of the 2ASK signal, except that a discrete carrier component is missing. This is because when the probability of 1 and 0 is equal to that of the bipolar digital baseband signal, the DC component is equal to zero. This is because when the probability of 1 and 0 is equal to that of the bipolar digital baseband signal, the DC component is equal to zero.

It can be seen from Fig. 8.7 that the bandwidth of the 2PSK signal is

$$B_{2PSK} = 2f_s. \qquad (8.16)$$

That is, the bandwidth of the 2PSK signal is twice the value of the digital information symbol rate.

8.4.1.2   2DPSK

Binary differential phase modulation uses binary digital information to control the phase difference between two adjacent symbols of the carrier so that the phase difference between two adjacent symbols of the carrier changes with the binary digital information. The phase difference between two adjacent symbols of the carrier is defined as

$$\Delta\varphi_n = \varphi_n - \varphi_{n-1}, \qquad (8.17)$$

"1" code $\longrightarrow$ $\Delta\varphi_n = 180°$ "1" code $\longrightarrow$ $\Delta\varphi_n = 0°$

"0" code $\longrightarrow$ $\Delta\varphi_n = 0°$ "0" code $\longrightarrow$ $\Delta\varphi_n = 180°$

(1)   (2)

Fig. 8.8 The relationship between binary digital information and phase difference between adjacent symbols of the carrier

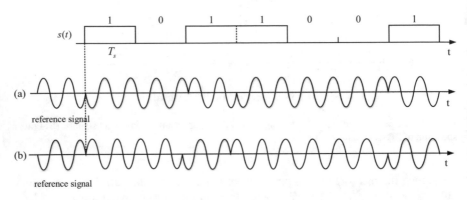

Fig. 8.9 2DPSK waveform

where φ_n and φ_{n-1}, respectively, representing the initial phases of the carrier of the nth and $n - 1$th symbols. Since the binary digital information has only two different symbols, 1 and 0, the carrier phase difference $\Delta\varphi_n$ controlled by the binary digital information also has only two different values. Two values of 0° and 180° are usually used. There are two one-to-one correspondences between the 1 code, 0 code, and 0° and 180°, as shown in Fig. 8.8.

The two corresponding relationships in Fig. 8.8 can be used for 2DPSK modulation. In the following discussion, the 2DPSK modulation adopts the corresponding relationship (1) in Fig. 8.8, that is, when the nth digital information is a 1 code, the control phase difference is $\Delta\varphi_n = 180°$, that is, the initial phase of the carrier of the nth symbol is relative to the $n - 1$th one. The carrier phase of the symbol changes by 180°; when the nth digital information is a 0 code, control $\Delta\varphi_n = 0°$, that is, the carrier phase of the nth symbol does not change relative to the carrier phase of the $n - 1$th symbol. Therefore, this correspondence is also called the 1 to 0 invariant rule. The 2DPSK waveform is shown in Fig. 8.9. Let $T_s = 2T_c$ be the carrier of two periods within one symbol width.

Since the variable and invariant in the 2DPSK modulation rules are relative to the initial carrier phase of the previous code, it is not necessary to draw the waveform of the modulated carrier wave when drawing the 2DPSK waveform, but the initial reference signal must be drawn, as shown in Fig. 8.9. The initial phase of the reference signal can be set arbitrarily. Figure 8.9a shows that the reference signal's initial phase is set to 0°, and the initial phase of the reference signal is set to 180° in

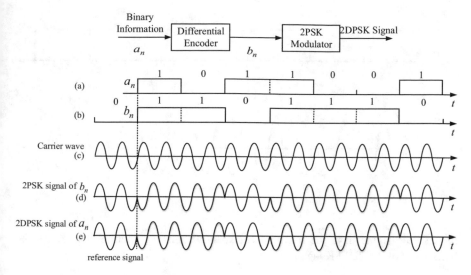

Fig. 8.10   2DPSK modulator and waveform

(b). The two waveforms look different on the surface, but the rules of changing and changing the carrier phases of the symbols before and after are exactly the same, so these two waveforms carry the same digital information.

The process of 2DPSK signal generation is first to perform differential encoding on the digital baseband signal, and then perform 2PSK modulation. Based on this formation process, binary relative phase modulation is also called binary differential phase modulation. The block diagram and waveform of the 2DPSK modulator are shown in Fig. 8.10.

The commonly used differential coding rule is

$$b_n = a_n \oplus b_{n-1}, \tag{8.18}$$

where $\oplus$ is modulo 2 plus; b_{n-1} is the previous symbol of b_n, and the initial b_{n-1} can be set arbitrarily. It can be seen from the modulated waveform that when using the formula (3.18) differential coding rule, the corresponding 2DPSK modulation rule is 1 changes and 0 unchanged.

For the convenience of comparison, the 2DPSK waveform obtained according to the definition is also shown in Fig. 8.10. As shown in Fig. 8.10e, this 2DPSK waveform is exactly the same as the waveform (d) obtained by 2PSK modulation by the differential code b_n. This shows that the modulator shown in Fig. 8.10 can be used to generate a 2DPSK signal. If researchers want to obtain a 2DPSK signal with an initial phase of 180°, just set the initial symbol to 1 during differential encoding.

It can be seen from Fig. 8.10 that the output signal of the 2DPSK modulator is a 2DPSK signal for the input information a_n, but a 2PSK signal for the differential code b_n, so for the same digital information sequence, the 2PSK signal, and

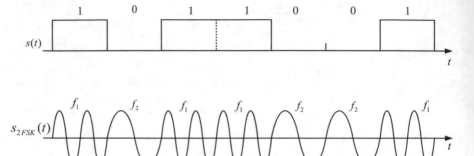

Fig. 8.11 2FSK signal

the 2DPSK signal have the same power spectral density function. Therefore, the bandwidth of the 2DPSK signal is

$$B_{2DPSK} = 2f_s, \tag{8.19}$$

where f_s is numerically equal to the symbol rate of digital information.

8.4.2 FSK

Binary FSK (2FSK) is to control the frequency of a sine wave with binary numbers so that the frequency of the sine wave changes with the change of the binary digital information. Since the binary digital information has only two different symbols, the modulated signal after modulation has two different frequencies f_1 and f_2. f_1 corresponds to the digital information 1, and f_2 corresponds to the digital information 0. The binary digital information and the modulated carrier are shown in Fig. 8.11 [9].

In a 2FSK signal, when the carrier frequency changes, the carrier phase is generally discontinuous. This signal is called a phase discontinuous 2FSK signal. 2FSK with discontinuous phase is usually generated by frequency selection method, as shown in Fig. 8.12. Two independent oscillators act as two frequency carrier generators, and the input binary signal controls them. The binary signal passes through two AND gate circuits, controlling one of the carrier waves to pass. The waveform at each point of the modulator is shown in Fig. 8.13.

The time-domain expression for a 2FSK signal is

$$f_{2FSK}(t) = b(t) \cos(\omega_1 t + \varphi_1) + \overline{b(t)} \cos(\omega_2 t + \varphi_2), \tag{8.20}$$

where $b(t)$ is the baseband signal and is expressed as

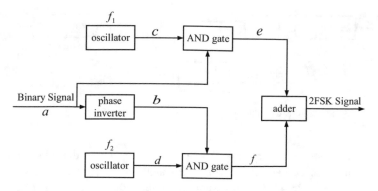

Fig. 8.12 2FSK signal modulator

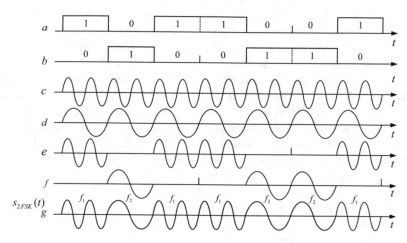

Fig. 8.13 2FSK modulator at each point waveform

$$b(t) = \sum_{n=-\infty}^{\infty} a_n g \left(t - n T_s \right). \tag{8.21}$$

As shown in Fig. 8.13, waveform g is the superposition of waveforms e and f. Therefore, the binary frequency modulation signal 2FSK can be regarded as the sum of two 2ASK signals with carrier frequencies of f_1 and f_2, respectively.

8.4.3 Polarization Shift Keying (PolSK) Modulation

The light wave is a transverse wave, and its vibration direction is perpendicular to the propagation direction. We often use the electric vector as the representative of the vibration vector in the light wave. The transverse wave property of light indicates

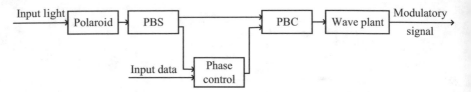

Fig. 8.14 Diagram of PolSK modulation

that the electric vector is perpendicular to light's propagation direction. In the two-dimensional space perpendicular to the propagation direction, the electric vector may have various vibration states, which is called the polarization of light. The polarization of light can be modulated to send data.

The traditional modulation methods are mostly intensity modulation and direct detection. The transmitting end directly modulates the intensity of the optical carrier to obtain a modulated signal. The receiving end uses a photodetector to convert the optical signal into an electrical signal. At this time, the noise of the channel, the noise of the background light and system itself have a significant impact on the performance of the communication systems. Therefore, the traditional modulation method has great limitations on the speed and communication distance of optical communication, while PolSK does not rely on information such as the intensity of the beam to demodulate. However, demodulate according to the light wave's polarization state, which well suppresses the channel's influence and provides a new way for high-speed and long-distance optical communication.

PolSK is a new digital optical modulation technology that has been widely discussed in recent years. Compared with traditional modulation methods, it has many advantages. The polarization state is the most stable characteristic during beam transmission. Polarization modulation can obtain stable output optical power, which is very important for peak power limiting systems.

Polarization modulation modulates the digital signal to the orthogonal polarization state of the optical carrier. Linearly polarized light with an azimuth angle of $-45°$ represents the digital signal 1, and linearly polarized light with an azimuth angle of $45°$ represents the digital signal 0.

Figure 8.14 is a schematic diagram of the PolSK modulation process, where PBS denotes polarization beam splitter, and PBC denotes polarization beam combiner [10]. The incident light is adjusted to linearly polarized light with an azimuth angle of $45°$ after passing through the polarizer. The linearly polarized light passes through the polarizing beam splitter and is divided into two linearly polarized lights with polarization states orthogonal to each other. The modulation information is used to modulate one of the light beams. That is, the phase of the light beam is controlled according to the modulation information, and then the polarization state of the light wave is modulated by the polarization beam combiner.

An ideal polarizer is shown in Fig. 8.15 can adjust the polarization state of incident light to linearly polarized light that is consistent with the direction of the polarizer's transmission axis.

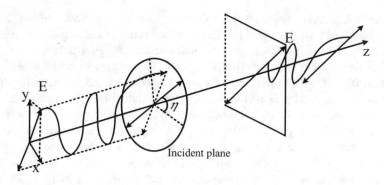

Fig. 8.15 Ideal polarizer

The polarization beam splitter's output is linearly polarized light whose polarization directions are along with the x-axis and y-axis directions. The phase of the y component is modulated according to the transmitted data stream. The output signal E_{out} of the phase modulator is given by

$$E_{out} = E_{in} \exp[j \Delta \varphi s_p(t)], \qquad (8.22)$$

where E_{in} is the incident light signal of the phase controller; $\Delta \varphi = 180°$ is the phase difference between mark and space; $s_p(t)$ is the digital signal 0 or 1. The polarization beam combiner combines the x component and the modulated y component into a beam of light. When the digital input signal is 0, no phase modulation is performed, and 45° linearly polarized light is output; when the digital input signal is 1, the y component is phase-shifted by 180°, and the output is −45° linearly polarized light.

By adjusting the phase controller, the modulation of any mutually orthogonal polarization states can be realized. For example, by using a phase controller, the phase difference between two orthogonal linearly polarized lights can be $\pi/2$ or $-\pi/2$, that is, the ellipticity is $\varepsilon = 45°$ and the azimuth angle is $\eta = 0°$. At this time, it is a special form of circularly polarized light—elliptically polarized light. The left-handed circular polarization state represents the 0 signal, and the right-handed circular polarization state represents the 1 signal.

Communication systems using PolSK modulation technology can be divided into single-signal polarization shift keying (SPolSK) systems and dual-signal polarization shift keying (DPolSK) systems. The SPolSK system refers to a system in which the receiving system only needs to receive one polarization state. For example, it can achieve demodulation as long as it receives the polarized light whose transmission signal is 1. The DPolSK system uses two photodiodes to detect the signals 0 and 1. The transmission system of DPolSK is the same as SPolSK. In the receiving system, there are two branches. The first branch uses a polarizer with azimuth to receive the transmitted digital signal 1; the second branch uses the azimuth. It is −45° polarizer to receive the transmitted digital signal 0. The polarized light emitted through the polarizer is detected by the photodiode through the trans-impedance amplifier. Then the two branches pass through a subtractor, which is used to eliminate the common-mode noise of 1 and 0. The amplitude of the received signal is twice that of the SPolSK system. Of course, the differential mode signal noise of the DPolSK system is also twice that of SPolSK. Since the two photodiodes and the trans-impedance amplifier are affected by dark current, shot noise, and thermal noise, random noise increases compared with single-branch demodulation. Although the two-way noise is different after passing through the subtractor, the two channels' noise is different. The noise ratio has a certain influence, thus affecting the bit error rate.

Due to the stability of the beam's polarization state during the transmission process, compared with other characteristics such as light intensity, it can better resist other channel interferences such as underwater turbulence and ambient light and has the better anti-interference ability to the phase noise of the laser. Therefore, the use of PolSK modulation technology is ideal for UWOC in a low signal-to-noise ratio environment. Although PolSK technology has strong anti-interference ability, it still has the problems of short transmission distance and low data rate. To overcome these problems, an improved polarization modulation technique is polarization pulse position modulation (P-PPM). This modulation scheme combines traditional PPM and PolSK, and transmits a series of PPM symbols in different polarization directions. P-PPM has the advantages of PPM and PolSK, and can increase the transmission bandwidth and distance of the UWOC systems based on anti-interference.

8.5   Multi-Carrier Modulation

Multipath induced fading is one of the most detrimental factors in UWOC and can significantly degrade UWOC systems performance. The amplitude and phase of the transmitted signal change after each absorption, scattering, and diffraction, so that different signals with different amplitudes and phases are received at the same time, resulting in Inter Symbol Interference (ISI). To mitigate ISI in single-carrier modulation schemes, the symbol duration of the transmitted signal should be less than or equal to the maximum delay spread of the UWOC channel. In a multi-carrier modulation scheme, the available signal bandwidth is divided into a number of narrow bandwidth slots (sub-carriers) and each sub-carrier is smaller than the coherent bandwidth. In this way, the symbol duration of each sub-carrier is larger than the delay spread of the UWOC channel, which can effectively reduce the ISI [11].

8.5.1   Basic Principles of OFDM

OFDM can be used to increase the communication rate of UWOC systems by allocating different amounts of data and power to each sub-carrier so that the system's communication performance is close to the channel capacity. Therefore, OFDM-based communication system is suitable for UWOC [12].

Optical OFDM belongs to the category of multi-carrier modulation, and data information is transmitted on multiple low-rate sub-carriers. The realization of OFDM with a wireless optical system provides a very economical and effective solution for improving its performance. OFDM allows the high data rate to be divided into multiple low data rates and transmitted in parallel. The main purpose of using this multi-carrier modulation scheme is to reduce the symbol rate and improve the UWOC system's performance.

Although OFDM technology has many advantages, optical OFDM is different from traditional radio frequency communication. Optical wireless communication usually directly uses the light intensity as the modulation object and uses the light intensity modulation technology to carry the data information that needs to be transmitted. Since the light intensity is all non-negative, the input light intensity signal must be a unipolar positive real number signal. There are currently two ways to ensure that the signal is a unipolar signal, one is through the use of DC bias, and the other is through the use of asymmetric limiting technology. However, adding a DC bias will greatly increase the average optical power of the transmitter and introduce limiting noise, which affects channel estimation accuracy. Because of the problem that the average optical power will limit the optical wireless communication system's transmission performance, the current ACO-OFDM modulation technology is widely used in the optical wireless communication system, and this technology can effectively improve the use efficiency of the optical power.

The specific implementation block diagram of ACO-OFDM modulation technology is shown in Fig. 8.16. In the optical wireless communication system, it is necessary to use the optical emphasis system technology to carry the optical signal's transmission data information. However, because the traditional baseband OFDM signal is a bipolar signal, it does not meet the optical modulator's data signal input requirements in the optical emphasis system. To solve this problem, in ACO-OFDM

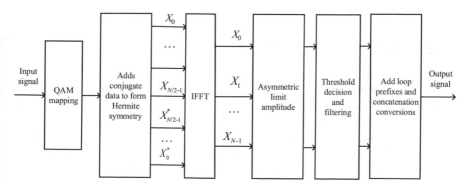

Fig. 8.16 ACO-OFDM modulation principle structure diagram

modulation, the optical signal needs to be processed symmetrically by Hermite before IFFT processing and then asymmetrically limited to ensure that the input signal is a real unipolar signal.

According to Fig. 8.16, a binary data symbol of length N is first converted into a bipolar complex vector through QAM constellation mapping. After serial-to-parallel conversion, the original single-channel signal of length N is converted into $N/4$ sub-carrier signals. They are transmitted separately. Then, the obtained complex vector undergoes Hermite symmetric transformation and is mapped into N sub-carrier transmission data vectors, and the center conjugate symmetry characteristic is satisfied. However, the signal will be transformed into a bipolar real signal after IFFT. In order to obtain a non-negative unipolar real signal, the obtained signal after IFFT also needs to adopt asymmetric amplitude limiting. The signal whose signal amplitude is greater than or equal to zero remains unchanged, while the signal whose signal amplitude is less than zero is set to zero directly. The transmission signal at this time satisfies: (1) when the number of sub-carriers in the ACO-OFDM system is even, there is $X(n) = 0$, that is, the system even carrier does not transmit data signals. Although the reliability of the input signal is guaranteed, it will reduce a certain spectrum utilization rate. (2) when the number of sub-carriers in the ACO-OFDM system is odd, there is $X(n) = X * (N - n)$, where $*$ denotes convolution, that is, only odd carriers with complex conjugate symmetry are used to transmit data, thus ensuring that the time-domain bipolar signal vectors obtained after IFFT are all real vector.

Compared with the DC bias OFDM modulation system, the advantage of ACO-OFDM lies in the application of asymmetric limiting technology and Hermitian symmetric processing in the modulation process. On the one hand, it can ensure that the optical modulator's data signal input is converted into a single polar signal. On the other hand, it can ensure that IFFT and output process the input signal as a real signal. The ACO-OFDM system can reduce the optical signal's sensitivity to the optical power. Because only the odd-numbered sub-carriers carry data information in the ACO-OFDM system and the even-numbered sub-carrier data information is all zero, without any loss of information data after asymmetric clipping.

8.5.2   Guard Interval and Cyclic Prefix

OFDM converts the incoming high-rate data streams into N parallel sub-channels, so that the data symbol period for each modulated sub-carrier is expanded by a factor of N of the original data symbol period, and the ratio of delay expansion to symbol period is reduced by a factor of N as well. To maximize the elimination of intersymbol interference, a Guard Interval (GI) can be inserted between two adjacent OFDM symbols, where the length of the GI should be larger than the maximum delay spread of the UWOC channel, so that the multipath component of one symbol does not cause interference to the following one symbol. During the GI cycle, no valid symbols can be transmitted. However, in this case, due to the

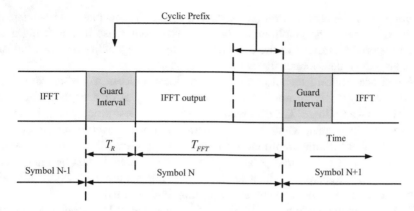

Fig. 8.17 Schematic diagram of cyclic prefix and guard interval

effect of multipath propagation, Inter Channel Interference (ICI) is generated, i.e., the orthogonality between sub-carriers is broken and interference occurs between different sub-carriers.

To eliminate ICI due to multipath, OFDM symbols need to fill in the protection interval with a cyclic prefix (CP), shown in Fig. 8.17. This ensures that the number of periods of the waveform contained in the delayed copy of the OFDM symbol is also integer within the FFT period. In this way, delayed signals with time delays smaller than the protection interval T_s do not generate ICI during the demodulation process.

The introduction of CP is one of the key technologies for OFDM systems. Under certain conditions, CP can completely eliminate the Inter Symbol Interference (ISI) caused by multipath propagation without destroying the sub-carrier orthogonality and suppressing the effect of inter-sub-channel interference. The length of the CP should be comparable to the length of the unit impulse response of the channel.

The CP acts as a protection interval, thus eliminating ISI, because its presence causes copies of the previous symbol multipath to fall within the CP of the latter symbol, thus eliminating interference between the first two symbols; due to the addition of the CP, a portion of each OFDM symbol becomes periodic, converting the linear convolution of the signal with the channel impulse response into a circular convolution, and it can be seen that the sub-carriers will remain orthogonal, thus preventing ICI.

8.5.3   Peak-to-Average Power Ratio

The output of a multi-carrier system is a superimposition of multiple sub-channel signals. If the phase of multiple signals is the same, the instantaneous power of the stacked signals will be significantly higher than the average power of the signals, resulting in a large peak-to-average power ratio (PAPR). This places a high demand on the linearity of the amplifier in the transmitter, otherwise it may bring about signal distortion and lead to the destruction of the orthogonality between the sub-channels, which deteriorates the system performance.

Amplitude limiting techniques use a nonlinear process to reduce the signal PAPR value by applying nonlinear operations directly at or near the peak of the OFDM signal amplitude. The disadvantage of the nonlinear process is that it causes distortion of the signal, which leads to degradation of the overall system error bit rate performance. Some amplitude limiting techniques are used to limit the signal after the IFT and before interpolation. The processed signal must be interpolated before the D/A transform, which leads to peak regeneration. To avoid this peak regeneration, the interpolated signal can be amplitude limited.

The set of signal code words available for transmission in code-like techniques is such that only those code words with amplitude peaks below a certain threshold are selected for transmission, thus avoiding signal peaks altogether. This type of technique is a linear process and does not suffer from the limiting noise found in amplitude limiting techniques. The starting point of these techniques is not to reduce the maximum value of the signal amplitude, but to reduce the probability of peaks.

8.6 Summary

OOK, PPM, DPIM, and PWM are the four major intensity modulation techniques implemented in the UWOC systems. OOK is the simplest to implement among the four modulation techniques, but the bandwidth efficiency is low. Integrating several auxiliary technologies such as DT technology and maximum likelihood method into the OOK scheme can improve its reliability. Compared with OOK, PPM has better power efficiency, but PPM requires tight timing synchronization and complex transceivers. Compared with PPM and PWM, DPIM has higher bandwidth efficiency, but DPIM has an error extension in the demodulation process.

Coherent modulation is superior to the IM/DD scheme, with higher receiver sensitivity, higher system spectrum efficiency, and better background noise suppression, but at the expense of higher implementation complexity and higher costs. In the actual implementation of UWOC, modulation technology's choice requires a comprehensive understanding of application scenarios. For discrete underwater sensor nodes, simple and low-cost modulation schemes such as OOK are preferred. However, for UWOC central nodes such as relay stations, multiple modulation technologies need to be integrated to improve the system's efficiency and reliability. Choosing the appropriate modulation technology must also cooperate with the appropriate encoding technology to improve the underwater optical application systems communication performance. Considering the severe optical attenuation effect in the underwater environment, the channel coding technology needs to maintain the system's performance. The simple block code has the advantages of low implementation cost, low complexity, and low energy consumption, but its error correction capability is also very limited. Therefore, they are suitable for compact underwater sensors that work in high signal-to-noise ratio underwater environments, such as clear deep-sea water. Complex channel coding technologies with high error correction capabilities, such as LDPC and Turbo coding, can be integrated into the underwater optical system's central data processing node to improve the robustness of the system.

Although most UWOC systems use OOK modulation technology because it is relatively simple to implement, OOK modulation has major disadvantages in terms of underwater optical power efficiency and bit error rate control capabilities. DPSK has good error control capabilities and high bandwidth, but it is more complicated to implement in embedded devices because of the use of optical interferometers in the receiving system. Since the transmitter in the FSK system is always ON, the power consumption is relatively high compared to other technologies, and it is not an ideal choice for submarine optical systems. Considering the bit error rate performance, required bandwidth and optical power in the underwater wireless optical communication system, and the complexity of implementation, PPM modulation technology is a good choice for system design.

In order to achieve a lower bit error rate, the value of L can be appropriately increased. For bandwidth efficiency, improved PPM modulation technology can improve its bandwidth performance.

? Questions

1. What is the main difference between using coherent demodulation and differential coherent demodulation for 2DPSK signals? What is the difference in BER performance?
2. What is the difference between intensity modulation and coherent modulation? What are the advantages and disadvantages of each of them?
3. What are the pulse modulation methods? What is PPM modulation? List the improved pulse modulation methods based on PPM?
4. In a UWOC system using the coherent modulation method, how does the receiver identify the phase of the optical signal?
5. What is the difference between the equipment of UWOC system with coherent modulation compared to UWOC system with non-coherent modulation?
6. The transmitted binary information sequence is 1011001, the code rate is 200 Baud, and the carrier signal is $\sin\left(8\pi \times 10^3 t\right)$, try to calculate:

   (1) How many carrier periods does each symbol contain?
   (2) Calculate the main lobe bandwidth of BPSK and BDPSK signals.

7. Knowing that the transmission rate of BPSK system is 2400 b/s, try to calculate:

   (1) The bandwidth efficiency of BPSK signal b/(s · Hz).

(2) If the baseband signal is preprocessed by $\alpha = 0.4$ cosine roll off filter and then modulated by BPSK, what is the channel bandwidth and bandwidth efficiency?

(3) If the transmission bandwidth remains unchanged and the transmission rate is increased to 7200 b/s, what changes should be made to the modulation system?

References

1. M.A. Khalighi, M. Uysal, Survey on free space optical communication: A communication theory perspective. IEEE Commun. Surv. Tutor. **16**(4), 2231–2258 (2014)
2. H. Kaushal, V.K. Jain, S. Kar, *Free Space Optical Communication* (Springer India, New Delhi, 2017)
3. X. He, J. Yan, Study on performance of M-ary PPM underwater optical communication systems using vector radiative transfer theory, in *ISAPE2012* (IEEE, 2012) pp. 566–570
4. M. Chen, S. Zhou, T. Li, The implementation of PPM in underwater laser communication system, in *2006 International Conference on Communications, Circuits and Systems*, vol. 3 (IEEE, 2006), pp. 1901–1903
5. D. Anguita, D. Brizzolara, G. Parodi, VHDL modeling of PHY and MAC Layer modules for underwater optical wireless communication, in *Proceedings of Papers 5th European Conference on Circuits and Systems for Communications (ECCSC'10)* (IEEE, 2010), pp. 185–188
6. C. Gabriel, M.A. Khalighi, S. Bourennane, et al., Investigation of suitable modulation techniques for underwater wireless optical communication, in *2012 International Workshop on Optical Wireless Communications (IWOW)* (IEEE, 2012), pp. 1–3
7. J.M. Kahn, J.R. Barry, Wireless infrared communications. Proc. IEEE **85**(2), 265–298 (1997)
8. W. Huang, J. Takayanagi, T. Sakanaka, et al., Atmospheric optical communication system using subcarrier PSK modulation. IEICE Trans. Commun. **76**(9), 1169–1177 (1993)
9. N. Chi, S. Yu, L. Xu, et al., Generation and transmission performance of 40 Gbit/s polarisation shift keying signal. Electronics Letters **41**(9), 547–549 (2005)
10. S. Benedetto, A. Djupsjobacka, B. Lagerstrom, et al., Multilevel polarization modulation using a specifically designed LiNbO/sub 3/device. IEEE Photon. Technol. Lett. **6**(8), 949–951 (1994)
11. F.M. Wu, C.T. Lin, C.C. Wei, et al., Performance comparison of OFDM signal and CAP signal over high capacity RGB-LED-based WDM visible light communication. IEEE Photon. J. **5**(4), 7901507–7901507 (2013)
12. H. Chun, S. Rajbhandari, G. Faulkner, et al., LED based wavelength division multiplexed 10 Gb/s visible light communications. J. Lightwave Technol. **34**(13), 3047–3052 (2016)

Chapter 9
Channel Coding

9.1 Introduction

Because the sea water has strong absorption and scattering effect on the optical signal, the transmitted optical signal suffers serious attenuation. This adverse effect increases the bit error rate of UWOC systems. In order to maintain low bit error rate in low SNR underwater environment, appropriate coding technology can be used to improve the efficiency and reliability of information transmission in UWOC systems.

The coding theory was first created in 1948 when Shannon, in his seminal paper "A Mathematical Theory of Communication," pointed out that any communication channel has a definite channel capacity C. If the required transmission rate of the communication system is $R < C$, there exists a coding method where the error probability of the message can be arbitrarily small when the code length n is sufficiently large and the maximum likelihood decoding (MLD) is applied. This is the famous noisy channel coding theorem, which lays the cornerstone of error correction coding theory [1].

Channel encoder is mainly used to deal with the transmission or storage of code words of the scrambled channel. The design and implementation are one of the main issues discussed in this chapter. The use of channel coding in UWOC systems can be represented in Fig. 9.1.

9.2 Channel Coding

9.2.1 Basic Idea of Channel Coding

The coding object of channel coding is the digital sequence m output from the source encoder, which is usually a sequence consisting of symbols 1 and 0, where symbols

© The Author(s), under exclusive license to Springer Nature Switzerland AG 2022
Y. Lou, N. Ahmed, *Underwater Communications and Networks*, Textbooks in Telecommunication Engineering, https://doi.org/10.1007/978-3-030-86649-5_9

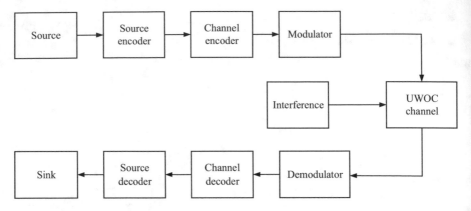

Fig. 9.1 Block diagram of UWOC communication system with coding

1 and 0 are independent with equal probability. The so-called channel coding is to add some extra code elements to the digital sequence m according to certain rules so that the information sequence m, which does not have regularity, is transformed into a digital sequence C with some regularity, which is also called a code sequence. In other words, the code sequence is correlated between each code element of the information sequence and the redundant code element. At the receiver side, the channel decoder uses this predicted coding rule to decode or check whether the received sequence of digits R conforms to the established rule, thereby finding whether there is an error in R or correcting the error in it. Detecting (finding) and correcting errors generated during transmission based on correlation are the basic idea of channel coding.

9.2.2 Classification of Coding

Coding rules can be classified into block codes and convolutional codes. If the rules of encoding are limited to this code group, i.e., the parity-check bits of this code group only depend on the information element of this code group, this type of code is called block code, and if the parity-check bits of this code group do not only depend on the information element of this code group but also depend on the information of the previous code group adjacent to this code group, this type of code is called convolutional code.

In block codes, each information code group to be encoded consists of k binary code element, and there are 2^k possible different information code groups. The channel encoder encodes each code group independently, and the value of 1 or 0 for each of the $n - k$ additional code elements is only relevant for the k code elements of the information code group. The encoder outputs a code group of length n, also known as a code word, and the number of code words is also 2^k. The set of 2^k code words is called a (n, k) block code. Since each code word of a block code depends only on the corresponding information code group, the encoder is memoryless. It can be implemented with combinational logic circuits. k/n is an important parameter, usually denoted by the symbol r, generally called the coding efficiency.

The encoder of a convolutional code also encodes an information code of k code element and outputs a code word of n code elements long. However, the parity-check bits that can be attached to each code not only depend on the information code set input to the encoder at this moment but also depend on the information code set input to the encoder in the first $N - 1 = m$ adjacent moments at this moment, so the encoder should contain $N - 1 = m$ levels of memory, and each level of memory stores k information elements input to the encoder in each of the first $N - 1 = m$ moments, so the convolutional code encoder has a memory, and it must use a sequential logic circuit to implement, $N - 1 = m$ is called the coded storage of the convolutional code, and N is called the constraint degree of the convolutional code. The convolutional code is often represented by (n, k, m).

If the coding rules can be represented by a linear system of equations, this can be called a linear code, otherwise called a nonlinear code. The structure of each code word after encoding can be divided into system codes and non-system codes. The first code element in each code word of the system code is consistent with the information code, while the code word of the non-system code does not have such structural characteristics.

According to the error type, error correcting codes can be divided into codes to correct random errors and codes to correct burst errors. Each code element's value in the code word can be divided into binary codes and multi-decimal codes. Since binary codes are most widely used, only binary codes are discussed here.

9.3 Linear Block Codes

In digital communication, in order to detect and correct errors occurred in message transmission at the receiving end, the transmitting end needs to encode the transmitted digital message sequence.

Assuming that the source output is a sequence of binary digits 0 and 1, in the block code, the binary message sequence is divided into groups of messages with fixed length, each group of messages consists of k message digit, denoted by U. There are 2^k different messages in total. The encoder transforms each message into a binary sequence of length n according to certain rules. If $n > k$, the sequence is called the code word (or code vector) of the message, there are 2^k, and the set of these 2^k code words is called the grouping code.

The main problem in channel coding research is to construct or discover good codes, and for large values of k and n, where there are more coding schemes, and coding devices can be complex, it is realistic to impose certain constraints on the coding schemes. Finding local optima in a subset requires focusing on achievable block codes, even if the construction of block codes is linear. A block code with this construction is called a linear block code [2].

9.3.1 Concepts Related to Linear Block Codes

9.3.1.1 Parity-Check Equations

A k-bit long information code group is transformed into an n-weight code word $(n > k)$ by a predefined linear operation. The set of 2^k code words compiled from 2^k information code groups is called linear block code, while the remaining $n - k$ code bits are parity-check bits, also called supervision bits.

An n-fold code word can be represented as a vector, i.e.,

$$(C_{n-1}, C_{n-2}, \ldots, C_1, C_0). \tag{9.1}$$

So a code word is also called a code vector. The linear code with information bits k and code length n is called (n, k) linear code.

For example, the message block length $k = 3$, followed by 4 parity-check bits in each message group, constitutes a $(7, 3)$ linear block code. Let the code word of the code be $(C_6 C_5 C_4 C_3 C_2 C_1 C_0)$, where $C_6 C_5 C_4$ is the information bits and $C_3 C_2 C_1 C_0$ is the parity-check bits, and each code element takes the value of 0 or 1. The parity-check element can be calculated by the following set of equations:

$$\left. \begin{array}{l} C_3 = C_6 + C_4 \\ C_2 = C_6 + C_5 + C_4 \\ C_1 = C_6 + C_5 \\ C_0 = C_5 + C_4 \end{array} \right\}. \tag{9.2}$$

Equation (9.2) is a linear system of equations, which determines the rules for obtaining parity-check bits from information bits, so it is called a parity-check equation or supervision equation. Since all code words are determined by the same rule, Eq. (9.2) is called a consistent parity-check equation or a consistent supervision equation, and the obtained parity-check bits are called consistent parity-check bits, and this coding method is called consistent supervisory coding or consistent parity-check coding. Since the consistent parity-check equation is linear, i.e., the relationship between the parity-check element and the information element is linear, the block code determined by the parity-check equation is a linear block code.

9.3.1.2 Parity-Check Matrix

For operational convenience, the system of parity-check Eq. (9.2) is written in matrix form to obtain

$$
\begin{bmatrix}
1 & 0 & 1 & 1 & 0 & 0 & 0 \\
1 & 1 & 1 & 0 & 1 & 0 & 0 \\
1 & 1 & 0 & 0 & 0 & 1 & 0 \\
0 & 1 & 1 & 0 & 0 & 0 & 1
\end{bmatrix}
\begin{bmatrix}
C_6 \\
C_5 \\
C_4 \\
C_3 \\
C_2 \\
C_1 \\
C_0
\end{bmatrix}
=
\begin{bmatrix}
0 \\
0 \\
0 \\
0
\end{bmatrix}.
\tag{9.3}
$$

Let $C = \begin{bmatrix} C_6 & C_5 & C_4 & C_3 & C_2 & C_1 & C_0 \end{bmatrix}, 0 = \begin{bmatrix} 0 & 0 & 0 & 0 \end{bmatrix}$, then

$$
\mathbf{H} =
\begin{bmatrix}
1 & 0 & 1 & 1 & 0 & 0 & 0 \\
1 & 1 & 1 & 0 & 1 & 0 & 0 \\
1 & 1 & 0 & 0 & 0 & 1 & 0 \\
0 & 1 & 1 & 0 & 0 & 0 & 1
\end{bmatrix}.
\tag{9.4}
$$

Then $\mathbf{H}$ is called the parity-check matrix.

9.3.1.3 The Generating Matrix

In digital circuits, a binary sequence of numbers is used to represent the information from the source, i.e., a 0–1 sequence. For this binary sequence, the symbols 0 and 1 define addition and multiplication as follows:

$$
\left.
\begin{array}{l}
0 \oplus 0 = 0, 0 \oplus 1 = 1, 1 \oplus 0 = 1, 1 \oplus 1 = 0 \\
0 \cdot 0 = 0, 0 \cdot 1 = 0, 1 \cdot 0 = 0, 1 \cdot 1 = 1
\end{array}
\right\}.
\tag{9.5}
$$

Such defined addition and multiplication are called modulo 2 addition and modulo 2 multiplication. Their corresponding inverse operations, subtraction, and division can be defined as follows:

$$
\begin{array}{l}
0 - 0 = 0, 0 - 1 = 1, 1 - 0 = 1, 1 - 1 = 0 \\
\frac{0}{1} = 0 \qquad \frac{1}{1} = 1
\end{array}
\tag{9.6}
$$

From Eqs. (9.5) and (9.6), it is clear that addition and subtraction are the same in modulo 2 arithmetic operations. And the laws of arithmetic, such as the law of union, the law of exchange, and the law of distribution, are still valid here. The set formed by two symbols 0 and 1 combined with the arithmetic operation it defines, Eqs. (9.5) and (9.6), is called the binary domain and is denoted as $GF(2)$.

Let the binary linear block code C_I (C_I denotes the set of code words) be defined by the parity-check matrix **H**. If U and V are any two of these code words, then $U + V$ must also be a code word in C_I. This shows that the sum of any two code words of a linear code is still a code word, and this property is called the closure of a linear code.

A binary sequence of length n can be considered as a point in an n-dimensional linear space on $GF(2)$, while the set of all 2^n vectors of length n constitutes an n-dimensional linear space V_n on $GF(2)$. The study of linear codes in linear spaces will simplify many problems and make them easier to solve. (n, k) linear code is a k-dimensional subspace V_k of an n-dimensional linear space V_n.

In the k-dimensional subspace of the linear space V_n consisting of (n, k) linear code, there must exist k linearly independent code words C that can all be expressed as these k code words: $g, g_1, g_2, \ldots, g_k$. Then, any other code word C in C_I can be expressed as a linear combination of these k code words. That is,

$$C = m_{k-1}g_1 + m_{k-2}g_2 +, \ldots, +m_0 g_k, \tag{9.7}$$

where $i = 0, 1, 2, \ldots, k - 1$.

Writing Eq. (9.7) in the form of a matrix yields

$$\mathbf{C} = \begin{bmatrix} m_{k-1} \; m_{k-2} \; \cdots \; m_0 \end{bmatrix} \begin{bmatrix} \mathbf{g}_1 \\ \mathbf{g}_2 \\ \vdots \\ \mathbf{g}_k \end{bmatrix} = m\mathbf{G}, \tag{9.8}$$

where $m = [\, m_{k-1} \; m_{k-2} \; \cdots \; m_0 \,]$ is the set of messages to be encoded and **G** is a matrix of order $k \times n$, i.e.,

$$\mathbf{G} = \begin{bmatrix} \mathbf{g}_1 \\ \mathbf{g}_2 \\ \vdots \\ \mathbf{g}_k \end{bmatrix} = \begin{bmatrix} g_{11} & g_{12} & \cdots & g_{1n} \\ g_{21} & g_{22} & \cdots & g_{2n} \\ \vdots & & \vdots & \\ g_{k1} & g_{k2} & \cdots & g_{kn} \end{bmatrix}. \tag{9.9}$$

Each row $g_i = g_{i1} \; g_{i2} \; \cdots \; g_{in}$ in **G** is a code word, and for each information group m, the code word corresponding to the (n, k) linear code can be obtained from matrix **G**, and the total number of code words is 2^k. Since matrix **G** generates (n, k) linear codes, matrix **G** is called the generation matrix of (n, k) linear codes.

9.3.2 Encoding of Linear Block Codes

(n, k) linear codes are coded by transforming a message group of length k into a code word of length n $(n > k)$ based on the parity-check matrix or the generation matrix of the linear code. The following is an example of a $(7, 3)$ system code to analyze how to construct the coding circuit of a $(7, 3)$ linear block code using the generation matrix and the parity-check matrix.

Let the information code group of $(7, 3)$ code be $m = (m_2 m_1 m_0)$, and the generation matrix is

$$\mathbf{G} = \begin{bmatrix} 1\,0\,0\,1\,1\,1\,0 \\ 0\,1\,0\,0\,1\,1\,1 \\ 0\,0\,1\,1\,1\,0\,1 \end{bmatrix}. \tag{9.10}$$

Substituting m and $\mathbf{G}$ according to $C = \begin{pmatrix} C_6 & C_5 & C_4 & C_3 & C_2 & C_1 & C_0 \end{pmatrix} = m\mathbf{G}$ gives

$$\begin{aligned} C_6 &= m_2 \\ C_5 &= m_1 \\ C_4 &= m_0 \\ C_3 &= m_2 + m_0 = C_6 + C_4 \\ C_2 &= m_2 + m_1 + m_0 = C_6 + C_5 + C_4 \\ C_1 &= m_2 + m_1 = C_6 + C_5 \\ C_0 &= m_1 + m_0 = C_5 + C_4. \end{aligned} \tag{9.11}$$

The parallel and serial encoding circuits of the $(7, 3)$-code can be drawn directly from the above set of equations, as shown in Fig. 9.2.

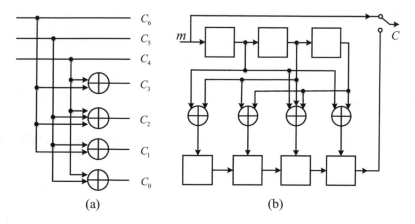

(a)   (b)

Fig. 9.2 The parallel and serial encoding circuits of the $(7, 3)$-code: (**a**) parallel coding circuits and (**b**) serial coding circuits

Similarly, the parity-check matrix can be used to encode, let the code word vector $C = (C_6\ C_5\ C_4\ C_3\ C_2\ C_1\ C_0)$, and the parity-check matrix of the code be

$$\mathbf{H} = \begin{bmatrix} 1 & 0 & 1 & 1 & 0 & 0 & 0 \\ 1 & 1 & 1 & 0 & 1 & 0 & 0 \\ 1 & 1 & 0 & 0 & 0 & 1 & 0 \\ 0 & 1 & 1 & 0 & 0 & 0 & 1 \end{bmatrix}. \tag{9.12}$$

From $HC^T = 0^T$, we have

$$\begin{aligned} C_3 &= C_6 + C_4 \\ C_2 &= C_6 + C_5 + C_4 \\ C_1 &= C_6 + C_5 \\ C_0 &= C_5 + C_4. \end{aligned} \tag{9.13}$$

It can be seen that the results obtained by calculating the parity-check bits by generating and parity-check matrices are the same, and the coding circuits are the same because generating and parity-check matrices describe the structure of the same code in different ways.

9.3.3   Decoding of Linear Block Codes

Since a linear block code can be encoded with a parity-check matrix, it can be decoded using a parity-check matrix. When word R is received, the parity-check matrix H can be used to check whether R satisfies the parity-check equation, i.e., whether $HR^T = 0^T$ holds. If the equivalence relation still holds, R is considered a code word; otherwise, it is judged that an error occurred in transmitting the code word. Therefore, whether the value of HR^T is 0 is the basis for checking whether the received code word is in error or not.

9.4   Cyclic Codes

Cyclic codes are an essential subset of linear block codes, which E. Prange first studied in 1957. Because of its cyclic property and excellent algebraic structure, it is possible to implement coding circuits and associated polynomial computation circuits with simple feedback shifters and use various simple and effective decoding methods. Since then, people's research on cyclic codes has made significant progress in theory and practice. Nowadays, cyclic codes have become the most intensively researched, theoretically mature, and widely used class of linear block codes.

For any code word of linear block code,

$$C = (C_{n-1}, C_{n-2}, \ldots, C_0). \tag{9.14}$$

The i-cycle shift of the resulting code word

$$C^{(i)} = (C_{n-1-i}, C_{n-2-i}, \ldots, C_0, C_{n-1}, \ldots, C_{n-i}) \tag{9.15}$$

is still a code word. The linear code is called an (n, k) cyclic code.

For the convenience of operation, each component of the code word can be used as the polynomial coefficient, and the code word is expressed as a polynomial, called code polynomial. Its general expression is

$$C(x) = C_{n-1}x^{n-1} + C_{n-2}x^{n-2} + \cdots + C_0. \tag{9.16}$$

The code polynomial of the code word $C^{(i)}$ obtained by cycling the code word C i times is

$$x = C_{n-1-i}x^{n-1} + C_{n-2-i}x^{n-2} + \cdots + C_0x^i + C_{n-1}x^{i-1} + \cdots + C_{n-i}. \tag{9.17}$$

Multiplying Eq. (9.16) by x and dividing by $x^n + 1$ give

$$\frac{xC(x)}{x^n+1} = C_{n-1} + \frac{C_{n-2}x^{n-1} + C_{n-3}x^{n-2} + \cdots + C_{n-1}}{x^n+1} = C_{n-1} + \frac{C^{(1)}(x)}{x^n+1}. \tag{9.18}$$

Equation (9.18) shows that the code polynomial $C^{(1)}(x)$ of the code word cycle once is the remainder of the original code polynomial $C(x)$ times x divided by $x^n + 1$, so write

$$C^{(1)}(x) \equiv xC(x) \,(\mathrm{mod}\, x^n + 1). \tag{9.19}$$

From this, it can be deduced that the i-cycle shift $C^{(i)}(x)$ of $C(x)$ is the remainder of $C(x)$ times x^i divided by $x^n + 1$, i.e.,

$$C^{(i)}(x) \equiv x^i C(x) \,(\mathrm{mod}\, x^n + 1). \tag{9.20}$$

9.4.1 Concepts Related to Cyclic Codes

9.4.1.1 Generating Polynomial and Generating Matrix

According to the cyclic property of cyclic codes, the other non-zero code words can be obtained by cyclic shifting of one code word. In the 2^k code words of (n, k) cyclic code, take the code polynomial $g(x)$ whose first $k - 1$ bits are 0 (its number is

$r = n-k$), and then by $k-1$ cyclic shifts, a total of k code polynomials are obtained: $g(x), xg(x), \ldots, x^{k-1}g(x)$. These k code polynomials are obviously independent of each other and can be used as k rows of the code generation matrix, so the (n, k) cyclic code generation matrix $\mathbf{G}(x)$ is obtained, i.e.,

$$\mathbf{G}(x) = \begin{bmatrix} x^{k-1}g(x) \\ x^{k-2}g(x) \\ \vdots \\ xg(x) \\ g(x) \end{bmatrix}. \tag{9.21}$$

Therefore, once the generation matrix of the code is determined, the code is determined. This shows that the (n, k) cyclic code can be determined by one of its $(n - k)$-code polynomials $g(x)$, so it can be considered that $g(x)$ generates an (n, k) cyclic code. Therefore, call $g(x)$ the generating polynomial of the code, i.e.,

$$g(x) = x^{n-k} + g_{n-k-1}x^{n-k-1} + \cdots + g_1 x + g_0, \tag{9.22}$$

and it can be seen that $g(x)$ is an $(n - k)$-first polynomial.

9.4.1.2   Parity-Check Polynomial and Parity-Check Matrix

Let $g(x)$ be the generating polynomial of the (n, k) cyclic code, then it must be a factorization of $x^n + 1$, i.e., there is

$$x^n + 1 = h(x) \cdot g(x), \tag{9.23}$$

where $h(x)$ is a k degree polynomial called the parity-check polynomial of the (n, k) cyclic code. Obviously, the (n, k) cyclic code can also be fully determined by its parity-check polynomial.

Now take the $(7, 3)$ code as an example.

$$x^7 + 1 = (x^3 + x + 1)(x^4 + x^2 + x + 1). \tag{9.24}$$

The 4th degree polynomial is the generating polynomial.

$$g(x) = x^4 + x^2 + x + 1 = g_4 x^4 + g_3 x^3 + g_2 x^2 + g_1 x + g_0. \tag{9.25}$$

The 3rd degree polynomial is then the parity-check polynomial

$$h(x) = x^3 + x + 1 = h_3 x^3 + h_2 x^2 + h_1 x + h_0. \tag{9.26}$$

From the equality of the coefficients of the same term at both ends of $x^7 + 1 = h(x) \cdot g(x)$, we get

$$
\begin{aligned}
g_3 h_0 + g_2 h_1 + g_1 h_2 + g_0 h_3 &= 0 \\
g_4 h_0 + g_3 h_1 + g_2 h_2 + g_1 h_3 &= 0 \\
g_4 h_1 + g_3 h_2 + g_2 h_3 &= 0 \\
g_4 h_2 + g_3 h_3 &= 0.
\end{aligned}
\tag{9.27}
$$

Write Eq. (9.27) in matrix form

$$
\begin{bmatrix}
0 & 0 & 0 & h_0 & h_1 & h_2 & h_3 \\
0 & 0 & h_0 & h_1 & h_2 & h_3 & 0 \\
0 & h_0 & h_1 & h_2 & h_3 & 0 & 0 \\
h_0 & h_1 & h_2 & h_3 & 0 & 0 & 0
\end{bmatrix}
\begin{bmatrix}
0 \\ 0 \\ g_4 \\ g_3 \\ g_2 \\ g_1 \\ g_0
\end{bmatrix}
=
\begin{bmatrix}
0 \\ 0 \\ 0 \\ 0
\end{bmatrix}
= \mathbf{0}^T,
\tag{9.28}
$$

where the elements of the column array are the coefficients of the generating polynomial $g(x)$, a code word, and the matrix equation $HC^T = \mathbf{0}^T$ of the linear block code shows that the first matrix is the parity-check matrix of the $(7, 3)$ cyclic code, i.e.,

$$
\mathbf{H}_{(7,3)} =
\begin{bmatrix}
0 & 0 & 0 & h_0 & h_1 & h_2 & h_3 \\
0 & 0 & h_0 & h_1 & h_2 & h_3 & 0 \\
0 & h_0 & h_1 & h_2 & h_3 & 0 & 0 \\
h_0 & h_1 & h_2 & h_3 & 0 & 0 & 0
\end{bmatrix}.
\tag{9.29}
$$

As seen in Eq. (9.29), the parity-check matrix's first row is an inverse order arrangement of the coefficients of the parity-check polynomial $h(x)$ of the code, while the second, third, and fourth rows are shifts of the first row. Therefore, the coefficients of the parity-check polynomials can be used to form the parity-check short array. Denoting the inverse polynomial of $h(x)$ by $h^*(x)$, we get

$$
\mathbf{H}_{(7,3)} =
\begin{bmatrix}
h^*(x) \\
x h^*(x) \\
x^2 h^*(x) \\
x^3 h^*(x)
\end{bmatrix}
=
\begin{bmatrix}
0 & 0 & 0 & 1 & 1 & 0 & 1 \\
0 & 0 & 1 & 1 & 0 & 1 & 0 \\
0 & 1 & 1 & 0 & 1 & 0 & 0 \\
1 & 1 & 0 & 1 & 0 & 0 & 0
\end{bmatrix}.
\tag{9.30}
$$

As $x^n + 1 = h(x) \cdot g(x)$, we have $h_0 = h_k = 1$. Generalizing Eqs. (9.29) and (9.30) to the general form, the parity-check matrix of the (n, k) cyclic code is obtained as

$$\mathbf{H} = \begin{bmatrix} h^*(x) \\ xh^*(x) \\ \vdots \\ x^{n-k-1}h^*(x) \end{bmatrix} = \begin{bmatrix} 0 & \cdots & 0 & 1 & h_1 \cdots h_{k-1} & 1 \\ 0 & \cdots & 1 & h_1 & h_{k-1} & 1 & 0 \\ \vdots & & & & & \\ 1 & h_1 \cdots h_{k-1} & 1 & 0 & & \cdots & 0 \end{bmatrix}. \tag{9.31}$$

We know $x^n + 1 = h(x) \cdot g(x)$, where $g(x)$ is an $n - k$ degree polynomial. Hence, with $g(x)$ as the generating polynomial, an (n, k) cyclic code is generated; with $h(x)$ as the generating polynomial, an $(n, n-k)$ cyclic code is generated. These two cyclic codes are dual to each other.

It should be noted here that the cyclic codes defined above are all non-systematic codes. The generation matrix G of the system code should satisfy

$$G = [I_k P], \tag{9.32}$$

where G is the $k \times k$ order unit square matrix. Since system codes are widely used in practical applications, only the encoding of system code forms is discussed below.

9.4.2   Encoding of Cyclic Codes

The generation matrix G of the system code is

$$G = [I_k P]. \tag{9.33}$$

The I_k on the left is the unit square, which corresponds to the information bits of the $n - 1$ to $n - k$ coefficients of the code word polynomial, while the rest are parity-check bits. Here let the information polynomial be

$$m(x) = m_{k-1}x^{k-1} + m_{k-2}x^{k-2} + \cdots + m_0. \tag{9.34}$$

Also let the parity-check numerical polynomial be

$$r(x) = r_{k-1}x^{k-1} + r_{k-2}x^{k-2} + \cdots + r_0, r = n - k. \tag{9.35}$$

The code polynomial of the (n, k) cyclic code is expressed as

$$C(x) = C_{k-1}x^{k-1} + C_{k-2}x^{k-2} + \cdots + C_{n-k}x^{n-k} + C_{n-k-1}x^{n-k-1} + \cdots + C_0. \tag{9.36}$$

The first k coefficients of $C(x)$ are information numbers and the last $r = n - k$ are parity-check bits, so there are

$$C_{n-1}x^{n-1} + C_{n-2}x^{n-1} + \cdots + C_{n-k}x^{n-k} = x^{n-k}(m_{k-1}x^{k-1} + \cdots + m_0) = x^{n-k}m(x), \tag{9.37}$$

and

$$C_{n-k-1}x^{n-k-1} + \cdots + C_0 = r_{r-1}x^{r-1} + \cdots + r_0 = r(x). \tag{9.38}$$

Substituting Eqs. (9.37) and (9.38) into Eq. (9.36) gives

$$C(x) = x^{n-k}m(x) + r(x). \tag{9.39}$$

Since $C(x)$ is a multiple of $g(x)$, such that $C(x) = q(x) \cdot g(x)$, then

$$q(x) \cdot g(x) = x^{n-k}m(x) + r(x). \tag{9.40}$$

Dividing both sides of the above equation by $g(x)$, we obtain

$$\frac{x^{n-k}m(x)}{g(x)} = q(x) + \frac{r(x)}{g(x)}. \tag{9.41}$$

Since $\partial^0 g(x) > \partial^0 r(x)$, Eq. (9.41) indicates that the parity-check digital polynomial $r(x)$ is the remainder of $x^{n-k}m(x)$ divided by $g(x)$. This means that the encoding of the cyclic code can be obtained by multiplying the information polynomial $m(x)$ by x^{n-k} and dividing it by the generating polynomial $g(x)$ to find the remainder, i.e., the parity-check digital polynomial of the code set $r(x)$. Thus, Eq. (9.41) can be rewritten as

$$r(x) \equiv x^{n-k}m(x) \,(\mathrm{mod}\ g(x)). \tag{9.42}$$

The coefficient of $r(x)$ is the parity-check bits. From Eq. (9.41), we get

$$g(x) \mid \left[x^{n-k}m(x) + r(x) \right]. \tag{9.43}$$

Therefore, the code word $C(x) = x^{n-k}m(x) + r(x)$ coded in this way is a multiple of $g(x)$, i.e., the set of code words is cyclic (Fig. 9.3).

From Eq. (9.38), we have $r = n - k = \partial^0 g(x)$, i.e., the number of parity-check bits is equal to the degree of $g(x)$. It can be seen that the encoding steps of (n, k) cyclic codes in the form of system codes are as follows: first, a suitable generating polynomial $g(x)$ is selected according to n and k, and then the encoding is performed according to the following steps, namely,

(1) $x^{n-k}m(x)$, increase the degree of $m(x)$ to $\leq n - 1$.
(2) Residual equation $r(x) \equiv x^{n-k}m(x)$, mod $g(x)$.
(3) $x^{n-k}m(x) + r(x) = C(x)$.

(1) All levels of the shift register are cleared, and the control gate is opened.

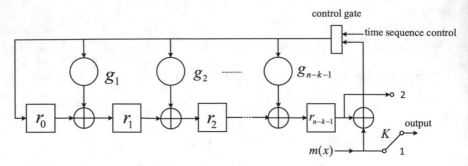

Fig. 9.3 Encoding circuit with $g(x)$ for $n - k$ level

(2) The k information bits $m_{k-1}, m_{k-2}, \ldots, m_0$ are sequentially inputted into the
coding circuit from the end and fed into the channel at the same time, and the
shifter is shifted once at each level when one information bit is added. When all
k information bits are inputted into the shift register, the $n - k$ bits in the shift
register are the parity-check bits.
(3) The control gate is closed, and the feedback is disconnected. Switch K is turned
from position 1 to position 2, and the stored numbers (parity-check bits) in the
register are sequentially shifted out and sent to the channel. k information bits
and r parity-check bits constitute a code word.

9.4.3   Decoding of Cyclic Codes

The decoding of linear codes is on the basis of the one-to-one correspondence
between the associated polynomial and the correctable error pattern, whereas the
error pattern is obtained from the associated polynomial. The decoding of cyclic
codes is a special subclass of linear codes, and the decoding steps of cyclic codes
are the same as those of linear codes, but its decoding is simpler and more accessible
due to the cyclic nature of cyclic codes.

Assuming that the received polynomial obtained at the input of the decoder after
the transmitted code polynomial $C(x) = C_{n-1}x^{n-1} + \cdots + C_1x + C_0$ enters the
channel be

$$R(x) = C(x) + E(x), \tag{9.44}$$

where $E(x) = e_{n-1}x^{n-1} + \cdots + e_1x + e_0$ is the error pattern generated by the
channel. The main task of the decoder is how to get the correct estimated germanium
error pattern $\hat{E}(x) = E(x)$ from $R(x)$, which in turn gives $C(x)$, and from that the
information code set $m(x)$.

Like the decoding process of linear grouping codes, the cyclic codes decoding
process consists of three steps.

(1) Calculating the received associated polynomial $R(x)$.
(2) Finding the estimate $\hat{E}(x)$ of the corresponding error pattern based on the associated polynomial.
(3) Calculate $R(x) - \hat{E}(x) = \hat{C}(x)$ to get the estimated code word $\hat{C}(x)$ output from the decoder, and send out the decoder to the user side, if $\hat{C}(x) = C$, then the decoding is correct, otherwise the decoding is wrong.

9.5 BCH Code

BCH codes can detect multiple random errors and correct the random errors generated by its unique generating matrix consisting of generating polynomials. So far, BCH codes have been the most thoroughly studied of all error correction codes, and a rigorous algebraic theory has been developed for BCH codes. Since there is a close algebraic relationship between the generating polynomial and the minimum code distance of BCH codes, researchers often construct BCH codes according to the actual required error correction capability of the code word. Since BCH codes are also a class of cyclic codes with their cyclic characteristics, BCH codes' coding circuit is relatively simple and easy to implement. Besides, BCH codes' decoder is also easy to implement, so it is one of the most widely used linear block codes [3].

9.5.1 Concepts Related to BCH Codes

9.5.1.1 Generating Polynomial and Generating Matrix

As one of the cyclic codes, BCH codes can be expressed by the roots of the generating polynomial $g(x)$: given any finite field $GF(q)$ and its expanding field $GF(q^m)$, where q is a prime number or a power of a prime number and m is a positive integer. A cyclic code generated by $g(x)$ is called a BCH code $R \supseteq \{a^{m_0}, a^{m_0+1}, \ldots, a^{m_0+\delta-2},\}$ in decimal if the code element is taken from a class of cyclic codes on $GF(q)$ whose set R of roots of the generating polynomial $g(x)$ contains the following $\delta - 1$ consecutive roots.

The generating polynomial of the BCH code is generated in the same way as generating the cyclic code. Let $m_i(x)$ and e_i be the smallest polynomial and order of a^{m_0+i} ($i = 0, 1, \ldots, \delta-2$) elements, respectively, then the generating polynomial and code length of the BCH code are

$$g(x) = \text{LCM} \left(m_0(x), m_1(x), \ldots, m_{\delta-2}(x) \right), \tag{9.45}$$

$$n = \text{LCM} \left(e_0, e_1, \ldots, e_{\delta-2}, \right). \tag{9.46}$$

If the root of the generating polynomial $g(x)$ has an element of the native domain in $GF(q^m)$, the length of the code word is $n = q^m - 1$.

For a BCH code of (n, k), let its generating polynomial be

$$g(x) = g_{n-k}x^{n-k} + g_{n-k-1}x^{n-k-1} + \cdots + g_1 x + g_0. \qquad (9.47)$$

Since $g(x), xg(x), \ldots, x^{k-1}g(x)$ a total of k code polynomials (they represent the $g(x)$ cyclic shifts 0, 1, $\ldots$, $k - 1$ times, respectively) must be linearly independent, they can be used to form a set of bases for the code, and the k linearly independent code vectors corresponding to these polynomials construct a generating matrix G of order $k \times n$, which is expressed in matrix form as follows:

$$G = \begin{bmatrix} g_{n-k} & g_{n-k-1} & \cdots & & g_1 & g_0 & 0 & \cdots & 0 \\ 0 & g_{n-k} & g_{n-k-1} & \cdots & & g_1 & g_0 & 0 & \cdots & 0 \\ \vdots & & & & & & & & \\ 0 & & \cdots & 0 & g_{n-k} & g_{n-k-1} & \cdots & g_1 & g_0 \end{bmatrix}. \qquad (9.48)$$

9.5.1.2   Parity-Check Matrix

$h(x)$ is usually used for the construction of generating matrices, and $h(x)$ is usually used for the construction of parity-check matrices. Assuming that $h(x)$ is also expressed in the form of $h(x)$, we have

$$h(x) = h_0 x^0 + h_1 x^1 + \cdots h_k. \qquad (9.49)$$

As in the construction of the generating matrix, the $h(x)$ polynomials are shifted $n-k-1$ times to have $x^{n-k-1}h(x), x^{n-k-2}h(x), \ldots, h(x)$. The parity-check matrix constructed at this point can be expressed as

$$H = \begin{bmatrix} h_0 & h_1 & \cdots & h_{k-1} & h_k & 0 & 0 & \cdots & & 0 \\ 0 & h_0 & h_1 & & h_{k-1} & h_k & 0 & 0 & \cdots & 0 \\ \vdots & & & & & & & & \\ 0 & \cdots & 0 & h_0 & h_1 & & h_{k-2} & h_{k-1} & h_k \end{bmatrix}. \qquad (9.50)$$

The role of the parity-check matrix of the BCH code is generally used for the decoding aspect by matrix multiplication with the received code word to obtain the associated formula, and the associated formula calculation obtains the error pattern. The obtained error pattern can be used to correct the code word error.

9.5.2 Encoding of BCH Codes

The BCH code encoding of (n, k) is closely related to its generating polynomial. As long as the target generating polynomial is found, a generating matrix can be generated based on the generating polynomial, and the (n, k) BCH code can be obtained by multiplying the sequence of message code words with the generating matrix. The basic steps are as follows:

1. In the first step, polynomial decomposition of equation $x^n - 1 = g(x)h(x)$ is performed according to polynomial $x^n - 1 = g(x)h(x)$.
2. In the second step, the $n-k$ polynomial of the decomposed polynomial is selected as the generating polynomial $g(x)$ of the (n, k) BCH code.
3. In the third step, the generated polynomial $g(x)$ is continuously shifted by k bits to obtain $g(x), xg(x), \ldots, x^{k-1}g(x)$ polynomials.
4. In the fourth step, the above k polynomials are used as the base to construct the corresponding generation matrix.

The result is obtained by multiplying the initial information sequence with the generation matrix in the fifth step.

9.5.3 Decoding of BCH Codes

The general decoding method of BCH codes is the same with that of linear block codes. First, the associated equation S is calculated from the information $R(x)$ received at the receiving end, then from the calculated associated equation S, the error pattern $\hat{E}(x)$ is found, and the received code word information $R(x)$ is subtracted from the error pattern to obtain the most probable code word $\hat{C}(x)$ to complete the decoding, i.e.,

$$\hat{C}(x) = R(x) - \hat{E}(x). \tag{9.51}$$

Let the length of the BCH code word be n, the length of the information bits be k, and the distance of the code word be d, on a finite field $GF(q)$, for any positive integer m_0, so that the (n, k, d) BCH code has $a^{m_0}, a^{m_0+1}, \ldots, a^{m_0+\delta-2}$ as its root, and then its generating polynomial can be expressed by Eq. (9.45) as

$$g(x) = (x - a^{m_0}) \ldots (x - a^{m_0+d-1}), \tag{9.52}$$

where $a \in GF(q^m)$, $d = 2t + 1$, t is the number of errors that can be corrected by the BCH code itself, and then its generating polynomial can be further expressed as

$$g(x) = (x - a^{m_0}) \ldots (x - a^{m_0+2t-1}). \tag{9.53}$$

Let the initial code word bit information be $q(x)$, then from the generating polynomial, $g(x)$ can be obtained from the transmitter after encoding the code word $C(x) = q(x)g(x)$, and the sent code word after modulation into the noise channel to get the code word can be expressed as $R(x) = C(x) + E(x)$, where $E(x)$ is the error pattern, expressed as

$$E(x) = Y_t x^{l_t} + Y_{t-1} x^{l_{t-1}} + \cdots + Y_1 x^{l_1} = \sum_{i=1}^{t} Y_i x^{l_i}, \qquad (9.54)$$

where Y_i refers to the error value, and x^{l_i} refers to the position in the code word where the error occurred in the code word. In the decoding process of BCH code, it is first calculated to obtain the associated equation S by received $R(x)$. When calculating the associated equation S, it is expressed by multiplying the parity-check matrix H of BCH code and the received matrix $R(x)$, i.e.,

$$S^T = H \cdot R^T = H \cdot (C + E)^T = H \cdot E^T. \qquad (9.55)$$

The associated equation S is expressed in matrix form as

$$S^T = \begin{bmatrix} S_{m_0} \\ S_{m_0+1} \\ \vdots \\ S_{m_0+2t+1} \end{bmatrix} = \begin{bmatrix} \sum_{k=1}^{t} Y_k x_k^{m_0} \\ \sum_{k=1}^{t} Y_k x_k^{m_0+1} \\ \vdots \\ \sum_{k=1}^{t} Y_k x_k^{m_0+2t+1} \end{bmatrix}. \qquad (9.56)$$

If the BCH code generates e ($e < t$) errors during transmission due to the channel, then for $1 \leq i \leq 2t$, there are

$$S(x) = e_{n-1} x^{n-1} + e_{n-2} x^{n-2} + \cdots + e_1 x + e_0. \qquad (9.57)$$

Here the error position polynomial $\sigma(x)$ is defined as

$$\sigma(x) = (1 - X_1 x)(1 - X_2 x) \cdots (1 - X_e x) = \sum_{i=0}^{e} \sigma_i x^i, \qquad (9.58)$$

where $X_1, X_2, \ldots, X_e$ is the location of the code word error. The simultaneous equation is used to solve the error pattern to determine the error in the code word, but the $2t$ equations above are nonlinear and therefore very difficult to solve.

In addition to solving the $2t$ nonlinear equations described above to obtain the polynomial roots and thus the error pattern, there are many other decoding methods for BCH codes, which are roughly divided into frequency-domain decoding methods and time-domain decoding methods. When BCH codes are decoded in the frequency domain, the BCH code word with noise is used as a digital signal, and the relevant information in the frequency domain is obtained through the discrete Fourier transform, and finally, the data is processed by using digital signal processing techniques, and the result is the decoding result; when BCH codes are decoded in the time domain, the special properties of BCH codes and their unique algebraic structure are usually used. The common time-domain decoding algorithms are the Peterson decoding algorithm and the BM iterative decoding algorithm.

9.6 RS Codes

RS code is a forward error correction channel code with excellent ability to correct random and burst errors and is very effective for data generated by over-sampling. When the receiver receives enough correct points, the original polynomial points can be restored even if many of the received polynomial points are distorted by noise interference.

RS code is a class of non-binary cyclic block codes proposed in 1960 [4]. Unlike Turbo codes and LDPC codes, RS codes do not require prior knowledge of channel characteristics for their implementation. Therefore, RS codes provide a good starting point for developing forwarding error correction codes due to their simple implementation and stable performance.

RS code is an algebraic geometry code with an excellent ability to correct random and burst errors. For RS codes with finite field $GF(2^m)$, the code element symbols are on this finite field, each code element corresponds to an m-bit multi-decimal number.

Due to various noise and interference factors, it is unavoidable that the transmitted information will generate errors during the transmission of information in practical applications. Generally speaking, the errors that occur can be divided into two types.

(1) The first type is random errors. That is, single-bit errors, each error bit is independent and uncorrelated, random errors are generally caused by thermal noise. Correcting such errors can be done with RS codes and convolutional codes. RS codes are linear block codes, and if there are fewer random errors generated by noise in a message block, the RS codes can correct all the random errors.

(2) The second type is burst errors. That is, the error bits appear continuously, but not for a long time. To correct such errors, RS codes are very effective. RS code is a maximum separation code, for a code length of n, message length of k RS code, its minimum distance is $d_{\min} = n - k + 1$, and the minimum distance of RS code is the largest of all linear packet codes, so it has the strongest error correction ability, up to $t = (n - k)/2$ errors can be corrected. One of the characteristics of RS codes is that an error of one bit within a symbol

is equivalent to an error of all bits. Hence, RS codes are more suitable for correcting burst errors.

9.6.1   Encoding of RS Codes

RS codes are a class of non-binary BCH codes in which the input signal is divided into groups of $k \cdot m$ bits, each group consisting of k symbols. Each symbol consists of m bits instead of one bit in a binary BCH code.

An RS code that corrects t symbol error has the following parameters:

- code length: $n = 2^m - 1$ symbols,
- information segment: k symbols,
- parity-check segment: $n - k = 2t$ symbols, and
- minimum code distance: $d = 2t + 1$ symbols.

The encoding of RS codes resembles that of BCH codes, and the information polynomial $m(x)$ to be encoded can be encoded using the generating polynomial $g(x)$; taking (7, 3) RS as an example, $m(x) = m_2 x^2 + m_1 x + m_0, m_i \in GF(2^3), i = 0, 1, 2, \ldots$. The encoding steps are

(1) $x^4 \cdot m(x)$.
(2) Use $g(x)$ to find the parity-check polynomial $r(x)$, $r(x) \equiv x \cdot m(x) \ldots \bmod g(x)$.
(3) Code word $c(x) = x^4 \cdot m(x) + r(x)$.

A division circuit of $g(x)$ can be used in finding the supervising polynomial $r(x)$; for (7, 3) RS codes, $n - k = 4$, the division circuit consists of a 4-stage shift, multiplier, and adder with feedback; unlike binary codes, these operations are operations on the $GF(2^3)$ domain, Still using (7, 3) RS codes as an example to illustrate the coding circuit of RS codes, as shown in Fig. 9.4.

In general, for (n, k) RS codes, the steps for system coding using $g(x)$ are as follows:

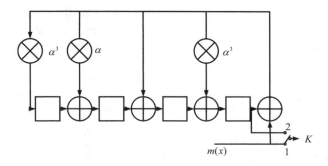

Fig. 9.4   (7,3) RS code encoder implemented with 4 stages

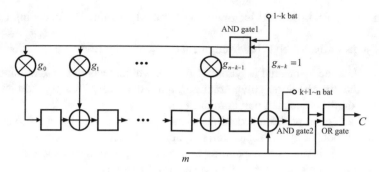

Fig. 9.5 General (n, k) RS code coding circuit

(1) $x^{n-k} \cdot m(x)$.
(2) Using $g(x)$ to find the parity-check polynomial $r(x)$, $r(x) \equiv x^{n-k} \cdot m(x) \ldots \mod g(x)$
(3) Code word $c(x) = x^{n-k} \cdot m(x) + r(x)$.

The coding circuit is shown in Fig. 9.5.

9.6.2 Decoding of RS Codes

The fundamental principles of RS decoding are to find the location of the error and the corresponding error value from the receiving polynomial $R(x)$, that is, to find the error polynomial $E(x)$, and then subtract the receiving polynomial $R(x)$ from the error polynomial $E(x)$ to find the code word polynomial $C(x)$, as follows:

$$C(x) = R(x) - E(x). \tag{9.59}$$

If $v \leq t$ errors are generated during transmission, then the error polynomial $E(x)$ can be expressed as

$$E(x) = Y_v x^{l_v} + Y_{v-1} x^{l_{v-1}} + \cdots + Y_1 x^{l_1} = \sum_{i=1}^{v} Y_i x^{l_i}, \tag{9.60}$$

where $Y_1, Y_2, \ldots, Y_v \in GF(2^m)$ is the error value and $X_i = \alpha^{l_i}$ is defined as the error location number, indicating that the error occurs at bit $n - l_i$ (the first coefficient of x^{n-1}) in $R(x)$. Therefore, the main work of RS decoding is to calculate the error position number X_i and the error value Y_i, totaling $2v$ unknown quantities.

The decoding algorithm of RS code can be divided into a hard-judgment decoding algorithm and a soft-judgment decoding algorithm. RS code soft-judgment decoding algorithm is more complicated to implement, so this chapter mainly introduces the RS code hard-judgment decoding algorithm. The decoding process is as follows.

The steps of the RS code hard-judgment decoding algorithm are as follows:

(1) Calculate the associated polynomial $\Lambda(x)$ from the receiving polynomial $S(x)$.
(2) Find the error position polynomial $\Lambda(x)$ and the error value polynomial $\Omega(x)$ based on the associated polynomial $S(x)$.
(3) Find each error position according to the error position polynomial $\Lambda(x)$, i.e., find the roots of the error position polynomial $\Lambda(x)$.
(4) Find the error value corresponding to each error position according to the error value polynomial $\Omega(x)$, i.e., find the error pattern $E(x)$.
(5) Subtract the error pattern from the received polynomial to obtain the decoding result and complete the decoding.

9.7   Convolutional Codes

The block codes discussed earlier are all grouped information code elements and then encoded independently, and there is no connection between the front and back code groups after encoding, and the decoding information is only obtained from this code group when decoding. Moreover, the shorter the information sequence block code word, the more information is lost. If the block code length increases to a maximum extent, the decoding code's complexity will increase exponentially. So researchers think that when the code length is certain, the relevant information of limited block is added to the code word, and the correlation of the code word is used to feed the information of the previous code back to the back for decoding reference. These ideas lead to the creation of convolutional codes.

Convolutional codes (also known as concatenated codes) were first proposed by Elias of MIT in 1955 [5]. Both convolutional codes and linear block codes are linear codes. The convolutional codes discussed in this section also compose k bits of information into n bits, but k and n are usually very small and particularly suitable for transmitting information in serial form with small delay. Unlike block codes, the n bits encoded in a convolutional code not only depend on the current k bits of information but also depend on the $(n-1)$ bits of information in the preceding segments, with Nn interrelated bits in the encoding process. The convolutional code's error correction capability increases as N increases, while the error rate decreases exponentially as N increases. The performance of convolutional codes is better than that of block codes for the same encoder complexity.

9.7.1   Encoding of Convolutional Codes

The general form of the convolutional code encoder is shown in Fig. 9.6, which consists of an input shift register consisting of N segments, each with k stages, for

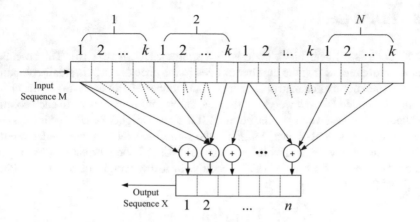

Fig. 9.6 General form of convolutional code

a total of Nk bits; a set of n modulo-2 adders; and an output shift register consisting of n stages. This corresponds to an input sequence of n bits per segment and an output of k bits. As shown in Fig. 9.6, the n output bits depend on the current k input bits and the previous $(N - 1)k$ input information bits. The encoding process can be seen as the convolution of the input information sequence with another sequence defined by the shift register and the modulo-2 sum connection type. The N is usually referred to as the constraint length. The convolutional code is often noted as (n, k, N), which has an encoding efficiency of (n, k, N).

9.7.2 Decoding of Convolutional Codes

The decoding methods of convolutional codes mainly include three kinds, which are threshold decoding, sequence decoding, and Viterbi decoding, proposed by Massey, Wozengraft, and Viterbi, respectively, thus promoting the development and application of the theory. Threshold decoding is an algebraic decoding method, and its main characteristics are simple algorithm, easy to implement, and the decoding operation time required for each information element is a constant, i.e., the decoding delay is fixed. Sequence decoding and Viterbi decoding are both probabilistic decoding. The delay time for sequential decoding is random and is related to the information interference situation. In contrast, the operation time of the maintenance ratio decoding code is fixed. The complexity of the decoding (either hardware implementation or software implementation) grows exponentially with m, but with the development of large-scale integrated circuit technology. These are no longer the key factors that prevent Viterbi decoding or sequence decoding from being widely used.

9.8 LDPC Codes

LDPC code is a linear block code with a sparse check matrix [6]. The so-called sparse nature means that the number of ones in the check matrix is small compared with the total number of matrix elements. The number of 1 in each row of the check matrix H is called the row weight of the row, and the number of ones in each column is called the column weight of the column. If the row weight of each row is the same and the column weight of each column is the same, the LDPC code is called regular LDPC code; otherwise, it is called a non-regular LDPC code. Equation (9.61) gives the check matrix of a regular LDPC code with a row weight of 4 and a column weight of 2.

$$\mathbf{H} = \begin{bmatrix} 1\,1\,1\,1\,0\,0\,0\,0\,0\,0 \\ 1\,0\,0\,0\,1\,1\,1\,0\,0\,0 \\ 0\,1\,0\,0\,1\,0\,0\,1\,1\,0 \\ 0\,0\,1\,0\,0\,1\,0\,1\,0\,1 \\ 0\,0\,0\,1\,0\,0\,1\,0\,1\,1 \end{bmatrix}. \tag{9.61}$$

In 1981, R.M. Tanner proposed that LDPC codes can be represented by bipartite graphs, one-to-one correspondence with the check matrix of LDPC codes. If a row is 1, there is an edge between the corresponding VN and CN, and if it is 0, it means that the corresponding VN and CN are not connected. In this book, CN_i is used to denote the i check node, and VN_j is used to denote the j variable node. Figure 9.7 represents the Tanner diagram corresponding to the equation's check matrix (4.61). The Tanner graph shown in Fig. 9.7 is regular, with the same number of connected edges for each VN and the same number of connected edges for each CN. The number of connected edges of a VN in the Tanner graph is called the degree of that VN, and accordingly, the number of connected edges of a CN is called the degree of that CN.

For non-regular LDPC codes, the column and row weights vary with the ranks, so the degree distribution polynomials are generally used to characterize the VN

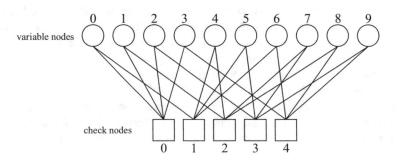

Fig. 9.7 Tanner diagram representation of the calibration matrix

and CN, denoted by $\lambda(X)$ and $\rho(X)$, respectively. In the VN degree distribution polynomial,

$$\lambda(X) = \sum_{d=1}^{d_v} \lambda_d X^{d-1},$$ (9.62)

where λ_d denotes the proportion of edges connected to the VN with degree d and d_v denotes the maximum VN degree value. Similarly, the CN degree distribution polynomial is

$$\rho(X) = \sum_{d=1}^{d_c} \rho_d X^{d-1},$$ (9.63)

where ρ_d denotes the proportion of edges connected to CN with degree d and d_c denotes the maximum CN degree value.

9.8.1 Encoding of LDPC Codes

LDPC code is a linear block code so that it can be systematically encoded like other linear block codes. Firstly, the checksum matrix H of $m \times n$ is converted into the following form by the Gaussian elimination:

$$H = \begin{bmatrix} P_{m \times k} & I_m \end{bmatrix},$$ (9.64)

where I_m is a unit matrix of order m, $k = n - m$, and then the system generation matrix can be expressed as

$$G = \begin{bmatrix} I_k P_{k \times m}^T \end{bmatrix}.$$ (9.65)

The calibration matrix H and the generation matrix G are mutually zero matrices, i.e., $G \cdot H^T = 0$. Let u be the sequence of information bits, then the encoded sequence is

$$c = u \cdot G = \begin{bmatrix} u\, u P_{k \times m}^T \end{bmatrix}.$$ (9.66)

9.8.2 Decoding of LDPC Codes

The decoding algorithm's choice directly determines whether the potential of the code itself can be exploited to the maximum extent. In particular, when decoding long codes, the complexity tends to increase exponentially with the code length, and a good decoding algorithm should improve the feasibility of the project based on guaranteed performance.

There are various decoding methods for LDPC codes, essentially iterative decoding algorithms for messages based on Tanner diagrams. According to the different forms of message transmission in the iterative process, the decoding methods of LDPC codes can be divided into hard-judgment decoding and soft-judgment decoding. If the message transmitted in the decoding process is a bit value, it is called hard-judgment decoding, and if the message transmitted in the decoding process is a message related to the posterior probability, it is called soft-judgment decoding, of which the confidence propagation iterative decoding algorithm is a representative, and in order to dry balance the performance and computational complexity, the two can be used in combination, called the hybrid decoding algorithm. Hard-judgment decoding is relatively simple to compute but has more unsatisfactory performance; soft-judgment decoding is relatively more complex to compute but has better performance, and the hybrid decoding algorithm is a compromise between the first two algorithms.

According to the number of bits used for information quantization in the iterative process, the decoding of LDPC codes can be divided into infinite bit quantization decoding and finite bit quantization decoding. Hard-judgment decoding is considered as 1-bit quantization decoding; soft-judgment decoding can be considered as infinite bit quantization decoding: hybrid decoding can be seen as variable bit quantization decoding.

9.9   Turbo Codes

Turbo code is a new coding method proposed by Berrou et al. [7]. The two-component code encoders are cascaded in parallel by an interleaver, and the decoder performs iterative decoding between the two-component decoders. The component decoders pass the external information of removing positive feedback between them, and the whole decoding process is similar to the Turbo work. The Turbo code's idea is to use shortcodes to construct long codes and pseudo-randomized the long codes interleaving to achieve a random code of maximum constraint length. Moreover, the long codes are transformed into shortcodes using iterative decoding, thus obtaining the performance close to the maximum likelihood decoding with less complexity.

Compared with the traditional coding methods, Turbo codes introduce interleavers with better performance, which are greatly developed in real engineering applications and profoundly impact the current stage of coding theory research. In the Turbo code coding process, a parallel cascade is used to form a long code with pseudo-random characteristics, and the internal structure of the cascade unit contains a pseudo-random interleaver. At the receiver side, the errors that occur centrally are also decentralized through corresponding interleaving and anti-interleaving, thus facilitating error correction and decoding. At the same time, multiple iterations between two soft input and soft output (SiSo) modules are used to strengthen the decoding information and gradually stabilize it, and finally, the pseudo-random decoding process is completed through specific judgments, which has better error correction performance. Turbo code's excellent characteristics make it a hot spot in the channel coding community and are applied in satellite communication, wireless communication, multimedia communication, mobile communication, and underwater communication. Shannon proposed and proved the famous noisy channel coding theorem in his "Mathematical Theory of Communication," and he cited three basic conditions in proving that the information rate reaches the information capacity to achieve error-free transmission.

(1) The use of a randomness compiled code.
(2) The coding length L tends to infinity, i.e., the packet's code group length is infinite.
(3) The decoding process uses the best maximum likelihood decoding scheme.

In the process of research and development of channel coding, basically, the last two conditions are the main direction. For condition (1), although a random selection of coding words can increase the probability of obtaining a good code, the maximum likelihood decoder's complexity increases with the number of words and when the coding length is large. The decoding is almost impossible to achieve. Therefore, for many years, random coding theory has been the main method for analyzing and proving coding theorems, but how to play a role in constructing codes has not attracted sufficient attention. In 1993, the discovery of Turbo code was regarded as a major breakthrough in the research of channel coding theory, which combines convolutional codes and random interleavers, while using soft output iterative decoding to approximate the maximum likelihood decoding code, and achieves extraordinary performance and surpasses the cutoff rate and directly approaches the Shannon limit. The Turbo code is a practical code that the channel coding community has been dreaming of, and its emergence marks a new stage in the research of channel coding theory.

Turbo codes have been greatly developed as far as they are concerned, and they are moving toward practical applications in all aspects. Meanwhile, the idea of iterative decoding has been widely used in coding, modulation, signal detection, and other fields.

9.9.1 Encoding of Turbo Codes

The structure of the coding side of the Turbo code is shown in Fig. 9.8. For the convenience of explanation and illustration, the modulation method of the transmitter side is 2PSK by default, i.e., the symbol bit is -1 when the information bit is 0, and the symbol bit is $+1$ when the information bit is 1, and the same demodulation method is used for the decoding link.

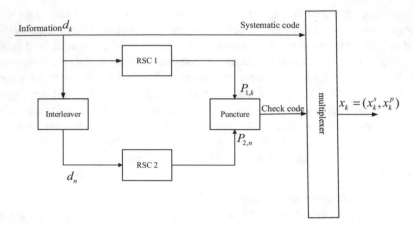

Fig. 9.8 Encoder structure of Turbo code

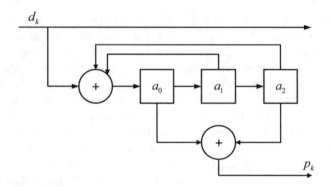

Fig. 9.9 (7, 5) RSC encoder

A parallel encoder consists of an interleaver, two-component encoders, a censoring module, and a multiplexer. The interleaver can disperse long burst errors to increase the reliability of decoding. The d_k represents the sequence of messages to be transmitted with length N, $P_{1,k}$ and $P_{2,n}$ represent the checksum codes generated by the two-component encoders, also known as component codes, respectively, x_k represents the symbols transmitted by the channel, where $x^s{}_k$ represents the system code, and $x^P{}_k$ represents the checksum code.

Since the structures of component encoder 1 and component encoder 2 are the same, they are introduced in the following parts. The component codes are generally chosen as Recursive Systematic Convolutional (RSC) codes, group codes, non-recursive convolutional codes, or non-convolutional codes, and only the more commonly used RSC codes are given here. Figure 9.9 shows the RSC encoder with the feedback generating polynomial of 111 (binary representation) and the forward generating polynomial of 101, i.e., the (7, 5) RSC code.

Fig. 9.10 Markov state transfer diagram for (7, 5) RSC encoder

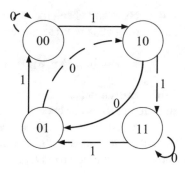

In Fig. 9.9, a_0 represents the input information at time k, a_1 and a_2 store the input information at times $k - 1$ and $k - 2$, and a_1 and a_2 together determine the current state of the RSC encoder. The RSC encoder has two registers, so the possible states of the encoder include four possibilities 00, 01, 10, and 11, and the check code p_k at the time the encoder state only determines k at time k and the input a_0 at time k, so the Markov process of transferring the encoder between the four states in Fig. 9.10 is the check code generation. Therefore, the Markov process of transferring the encoder between the four states in Fig. 9.10 is the check code's generation process. The state transfer diagram is shown in Fig. 9.10.

The solid line represents the case when the input information is 0, and the dotted line represents the case when the input information is 1. According to the transfer diagram, we can know the current encoder state at k. Suppose the encoder is in 01 state at k, if the information 0 is input to the encoder, the check code generated by the encoder will be 0, and the encoder will enter 10 state, and if the information input is 1, the encoder will generate check code 1 and enter 00 state; according to this way, the check code p_k can be generated.

Because the generated check digit p_k is redundant and not all code words need to be transmitted, the check digit p_k can be redundant to improve transmission efficiency encoding code rate.

According to the Turbo code encoder structure in Fig. 9.8, if the censors are not used, the length of the information code and the two check codes are both N. At this time, the useful information accounts for 1/3, and the code rate is 1/3, and the following transmit code word is obtained after multiplexing by the multiplexer:

$$x = \{d_0, P_{1,0}, P_{2,0}, d_1, P_{1,1}, P_{2,1}, d_2, P_{1,2}, P_{2,2}, \ldots, d_{N-1}, P_{1,N-1}, P_{2,N-1}\},$$
$$(9.67)$$

where d_k represents the system code, $P_{1,k}$ represents the check code output from component encoder 1, $P_{2,k}$ represents the check code output from component encoder 2, and there is $P_{2,k} = P_{2,n}, k = n$.

Suppose the check digit is deleted, for example. In that case, the check digit in the even position of the output $P_{1,k}$ of the component encoder 1 and the odd position of the output $P_{2,k}$ of the component encoder 2, the code rate increases to 1/2, and the system code, and check the multiplexer processes digit to obtain the corresponding transmitting code word sequence as follows:

$$x = \{d_0, P_{1,0}, d_1, P_{2,1}, d_2, P_{1,2}, \dots, d_{N-1}, P_{2,N-1}\}. \tag{9.68}$$

The transmitter code word can be processed on the transmitter side, and the encoding process has been completed. The above describes the Turbo code encoder structure and how the bits to be transmitted are processed to obtain the corresponding transmit data. The following section describes the Turbo code decoder corresponding to the receiver side.

9.9.2 Decoding of Turbo Codes

The block diagram of the Turbo decoder is shown in Fig. 9.11. The component decoder, interleaver, deinterleaver, and demultiplexer together constitute the Turbo code decoder, x_k represents the transmitted data, y_k represents the received data at the receiver, and w_k represents the additive Gaussian white noise with zero mean and variance of σ_w^2.

$$y_k = x_k + w_k. \tag{9.69}$$

In Fig. 9.11, $L_{D,1}^{pri}(x_k^s)$ and $L_{D,2}^{pri}(x_n^s)$ represent the a priori input information of SiSo decoder 1 and SiSo decoder 2, respectively; $L_{D,2}^{pos}(x_n^s)$ represents the a

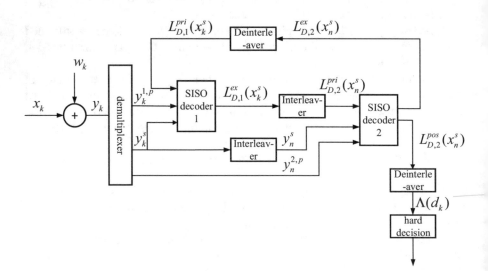

Fig. 9.11 Turbo code decoder structure

posteriori information output by SiSo decoder 2; and $L_{D,1}^{ex}(x_k^s)$ and $L_{D,2}^{ex}(x_n^s)$ represent the external information output by SiSo decoder 1 and SiSo decoder 2, respectively, where $L_{D,2}^{ex}(x_n^s)$ is obtained by subtracting the a priori information from the a posteriori information output by SiSo decoder 2, which is obtained as

$$L_{D,2}^{ex}(x_n^s) = L_{D,2}^{pos}(x_n^s) - L_{D,2}^{pri}(x_n^s). \tag{9.70}$$

It should be noted that the entire Turbo decoding process passes only the LLR values of the useful information, and the LLR values of the checksum information are not processed.

The workflow of Turbo code decoding is described in the following flowchart. The workflow diagram is shown in Fig. 9.12.

(1) The received data is demultiplexed. The valid data and the checksum data corresponding to component encoder 1 are extracted according to the corresponding multiplexing order at the transmitter, and the valid data is interleaved to obtain the checksum data corresponding to component encoder 2.
(2) Input the a priori information of the sum decoder 1 to the SiSo decoder 1, which should be set to 0 first if it is the first time.
(3) The external information output from SiSo decoder 1 is interleaved by the interleaver and input to SiSo decoder 2 together with the data for decoding.
(4) The end of the iteration is judged, on the one hand, whether the maximum number of iterations is reached, and on the other hand, if there is a stopping criterion, the corresponding judgment should be made. If the iteration is finished, the final result will be obtained by a likelihood ratio judgment, i.e., hard judgment. If the iteration is not finished, the a posteriori information is subtracted from the a priori information to obtain the external information, and then the a priori input information for the next iteration of SiSo decoder 1 is obtained by interleaving and then return to step 2 to continue the processing of the next iteration.

9.10 CRC Codes

In data communication and networks, the number of elements of information code is usually quite large, consisting of a frame of a thousand or even thousands of data bits, and then the generation polynomial of cyclic code is used to generate r-bit check bits. At this time, the code word composed of information code elements and check bits is not necessarily a strictly defined cyclic code, but a cyclic redundancy check code (CRC) widely used in data communication and networks, which is a shortened cyclic code derived from regular one and is a very important class of error detection codes. CRC is derived from cyclic codes and is a very important type of error checking code. The coded code word of the cyclic code can be generated

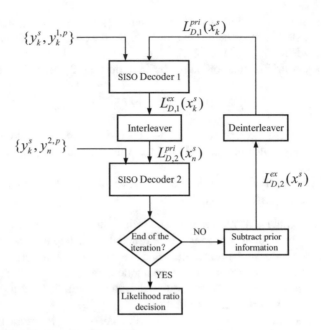

Fig. 9.12 Turbo code decoding flow chart

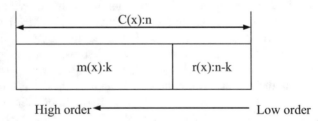

Fig. 9.13 Structure of CRC code

by polynomial $g(x)$ integer division. The receiver can use this feature for error detection. If the received code word cannot be divided by $g(x)$, then there is an error. The structure of CRC codes is shown in Fig. 9.13.

In Fig. 9.13, the k coefficients of $m(x)$ correspond to k information bits, and the $n - k$ coefficients of $r(x)$ correspond to $n - k$ parity bits. For the system loop code, on the transmitter side, $C(x) = x^{n-k}m(x) + r(x)$, where $r(x)$ is equal to the remainder of $x^{n-k}m(x)$ divided by $g(x)$.

If there is no error in the receive stream, the received code $R(x)$ should be equal to the transmit code $C(x)$, i.e., $C(x) = R(x) = x^{n-k}m(x) + r(x) = q(x)g(x)$, at which point the received code should be divisible by the generating polynomial $g(x)$. Conversely, if it is not divisible, there must be an error in the transmission.

The cycle of the cyclic redundancy check code shown in $g(x)$ is a cyclic code to generate polynomial, where redundancy is shown in the check digit $n - k$ length certain. Since it is a cyclic code, there should be $g(x)h(x) = x^n + 1$, that is, $n - k$ is a factor of n, once $n - k$ is fixed, then n is also fixed, or there are only a few values. But in the practical application of frame checksum, the frame length n is uncertain and can also change continuously. So the engineering application CRC code is not the original (n_0, k_0) cyclic code designed by $g(x)|(x^{n_0} + 1)$, but a shortened version of this code. That is, the cyclic code (n_0, k_0) is designed first, and then any i bits are shortened to get a CRC code of $(n_0 - i, k_0 - i)$. Due to the shortening, the CRC code loses the cyclic characteristics outside the cyclic code, but the intrinsic characteristics of the cyclic code still exist, and the cyclic code can analyze its error correction ability, and the codec circuit can also use the cyclic code to realize.

9.10.1 Encoding of CRC Codes

The CRC code applies a check digit for error checking, and at the transmitter, the check digit is calculated in the following form: the transmitter represents the original data (network data (ND), information bits) of length m as a binary polynomial $nd(x)$, and we have to select a degree r called the $g(x)$ of the generating polynomial. Polynomial $nd(x)$ is first multiplied by x^r and then divided by $g(x)$, and the resulting polynomial $fcs(x)$ sequence of length r is the check code FCS, which is appended to the ND after.

$$\left(nd(x) \cdot x^r\right) \bmod g(x) = fcs(x). \tag{9.71}$$

9.10.2 Decoding of CRC Codes

There are usually two requirements for decoding at the receiver side: error detection and error correction. The purpose of cyclic redundancy check is to detect errors, therefore, its decoding is very simple, i.e., judging whether the received codeword polynomial $T(x)$ can be divided by polynomial $g(x)$ or not.

When there is no error in transmission, i.e., the received $t'(x)$ is the same as the sent code group $t(x)$, i.e., $t'(x) = t(x)$, the received code word must be divisible by $g(x)$; if there is an error in transmission, $t'(x)$ is not divisible by $g(x)$, and therefore, it is possible to determine whether there is an error in the code word based on whether the remainder term is 0 or not. At the receiving end, the polynomial expression $t'(x)$ of the received message T' will be judged whether it can be divided by the generated polynomial, that is,

$$t'(x) \bmod g(x) = 0? \tag{9.72}$$

If Eq. (9.72) is not satisfied, then the received message is an error, and the error can be detected, and if Eq. (9.72) is satisfied, T is considered to be transmitted correctly. In fact, the ordinary CRC code itself is a very effective error detection method, but the CRC coding on top of the ordinary CRC code will further improve the error detection performance, and this has become a trend of research.

9.11   Luby Transform (LT) Codes

LT codes are the first practical digital fountain codes, and many subsequent fountain codes, such as Raptor codes, are proposed based on LT codes [8]. To further study network fountain code, it is necessary to analyze the compilation code principle of LT code. The degree distribution function plays an indispensable role in the process of compiling LT codes, so this section analyzes the degree distribution functions commonly used in LT codes and finally analyzes the performance of LT codes.

9.11.1   Encoding of LT Codes

LT codes are adaptive and can generate a theoretical source of coded groupings, and the coding process is as follows:

(1) Equating the original data into K input groupings, denoted as $\{S_1, S_2, \ldots, S_K\}$, each with the same number of bits.
(2) Randomly generating a degree value $d, d \leq K$ based on the degree distribution function Ω.
(3) Randomly and uniformly selecting d input groupings out of K input groupings, denoted as $\{S_1^{(i)}, S_2^{(i)}, \ldots, S_d^{(i)}\}$, and i denoting the i coded grouping.
(4) The selected d input groups are subjected to the iso-or operation to generate the coding groups $C_i, i = 1, 2, \ldots, i$ indicating the i coding group.

Repeat the above encoding steps to generate an infinite length of encoding groups. The specific encoding process of LT code can be represented by the Tanner diagram, as shown in Fig. 9.14. The nodes in the first row are variable nodes, which correspond to the input group. The nodes in the second row are called check nodes, which correspond to the encoded grouping. The variable node degree distribution determines the degree of the variable nodes, and the degree of the variable nodes indicates the number of coded groups with this input group involved in coding; the degree of the check nodes is selected from the check node degree distribution, and the check node degree indicates the number of input groups required to generate this coded group. Throughout this paper, if not specifically stated, the degree distribution specifically refers to the test node degree distribution.

In Fig. 9.14, the original data is equally divided into $K = 3$ input groupings, each containing 3 bits; based on the degree distribution function, a random degree

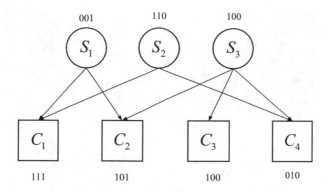

Fig. 9.14 Schematic diagram of LT code encoding

value d is generated, and d can be expressed as the number of edges connecting the coded groupings to the input groupings; then $d = 2$ groupings are randomly and uniformly selected among K input groupings, i.e., input groupings $S_1 = 001$ and $S_2 = 110$ are selected; finally, the coded grouping C_1 is generated by performing a dissimilarity operation between S_1 and S_2, i.e.,

$$C_1 = S_1 \oplus S_2 = 111. \tag{9.73}$$

9.11.2 Decoding of LT Codes

After the receiver receives the coded group, the BP decoding algorithm with low computational complexity is used to decode the coded group, and the specific decoding process is as follows:

(1) Find the coded grouping of $d = 1$, recover the input grouping connected to it directly by copying, and delete the connected edge between them.
(2) Generate a new coded group by performing a dissimilarity operation between the translated input group and its neighboring coded groups, replacing the original coded group with the generated new coded group, and deleting the connected edges of the input group all its connected coded groups.
(3) Repeat steps (1) and (2) until all input groups are successfully decoded.

The LT code decoding process is shown in Fig. 9.15. Firstly, the coded group of $d = 1$ is found, and the degree value of the third coded group $C_3 = 110$ is 1 as shown in Fig. 9.15a, so the input group S_3 connected with it can be recovered directly by copying, i.e., $S_3 = C_3 = 110$, and then the connecting edge between S_3 and C_3 is deleted, as shown in Fig. 9.15b. Then, S_3 and its connected code-groups $C_2 = 101$, $C_4 = 010$ are XORed to generate new code-groups $C_2 = S_3 \oplus C_2 = 001$

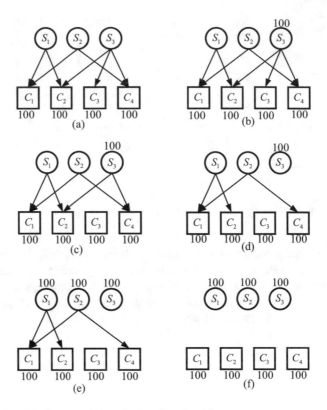

Fig. 9.15 Schematic diagram of LT code decoding algorithm

and $C_4 = S_3 \oplus C_4 = 110$, and replace the original C_2 and C_4 respectively. After that, the connect edges between S_3 and C_2, C_4 are deleted, as shown in Figs. 9.15c and 9.15d. The above decoding steps are repeated to find the coded group of $d = 1$, as shown in Fig. 9.15e, and the degree values of coded groups C_2 and C_4 are both shown in Fig. 9.15e and the degree values of encoding groups C_2 and C_4 are both 1, so the input groups S_1 and S_2 connected to them can be recovered directly by copying, and then the connected edges between them are deleted. At this point, the decoding of all the original data is completed.

As can be seen from Fig. 9.15, the decoding process of LT code starts from the set of coded groups with degree 1, which is called the output decodable set, and the set of input groups connected to it is called the decodable input set. If there is no coded group with degree 1, or the coded group with degree 1 disappears during the decoding process, the decoding will fail. The degree values of all coded groups are chosen randomly from the degree distribution function, so the suitable degree distribution function is selected to ensure the fountain code's decoding performance.

1. What is the significance of channel coding? What is the basis for the classification of channel coding methods?
2. What are the advantages of linear grouping codes?
3. What are the similarities and differences between RS codes and BCH codes?
4. How many decoding methods are there for convolutional codes? What are the advantages and disadvantages of each of these decoding methods?
5. Compare several commonly used coding methods for underwater optical communication and point out their advantages and disadvantages.
6. The generation matrix of a (7, 3) code is known to be

$$G = \begin{bmatrix} 1\,0\,0\,1\,1\,1\,0 \\ 0\,1\,0\,0\,1\,1\,1 \\ 0\,0\,1\,1\,1\,0\,1 \end{bmatrix}.$$

   Try to calculate its supervision matrix.
7. Knowing that

$$x^{15} + 1 = (x+1)\left(x^4 + x + 1\right)\left(x^4 + x^3 + 1\right)\left(x^4 + x^3 + x^2 + x + 1\right)\left(x^2 + x + 1\right),$$

   how many cyclic codes of length 15 can be made out of it? Please list their polynomials.

References

1. Z. Zeng, S. Fu, H. Zhang, et al., A survey of underwater optical wireless communications. IEEE Commun. Surv. Tutor. **19**(1), 204–238 (2016)
2. M.A. Khalighi, M. Uysal, Survey on free space optical communication: A communication theory perspective. IEEE Commun. Surv. Tutor. **16**(4), 2231–2258 (2014)
3. W.P. Wang, B. Zheng, The simulation design of LED-based close-range underwater optical communication system, in *2013 10th International Computer Conference on Wavelet Active Media Technology and Information Processing (ICCWAMTIP)* (IEEE, 2013), pp. 283–285
4. W.C. Cox, J.A. Simpson, C.P. Domizioli, et al., An underwater optical communication system implementing Reed-Solomon channel coding, in *OCEANS 2008* (IEEE, 2008), pp. 1–6
5. J.G. Proakis, M. Salehi, *Digital Communications* (McGraw-Hill, New York, 1995)
6. J.S. Everett, Forward-error correction coding for underwater free-space optical communication (2009)
7. S. Chen, J.L. Song, Z.M. Yuan, et al., Diver communication system based on underwater optical communication. *Applied Mechanics and Materials*, vol. 621 (Trans Tech Publications Ltd., 2014) pp. 259–263
8. M. Luby, LT codes, in *The 43rd Annual IEEE Symposium on Foundations of Computer Science. Proceedings* (IEEE Computer Society, 2002), pp. 271–271

Chapter 10
Link Performance Enhancement Techniques

10.1 Diversity Technology

10.1.1 Definition of Diversity

Diversity technology is widely used in UWOC systems, to reduce the effects of turbulence-induced fading. Diversity is achieved by receiving multiple copies of the transmitted signal at the receiving end, which carry the same information but are less correlated in terms of fading statistics. The basic idea is that the same information reaches the receiver through multiple statistically independent channels, and the signal carrying this information can have multiple copies that fade in independent or uncorrelated patterns. For example, some copies of the signal may be heavily faded, while others are minimally faded. The probability of multiple independent or incompletely correlated channels falling into deep fading at the same time is greatly reduced, resulting in improved performance. In general, the diversity order (diversity gain) determines the slope at which the error probability curve falls. The higher the diversity order, the faster the BER curve falls.

10.1.2 Classification of Diversity

Diversity techniques often require the transmitter to transmit multiple copies of the same signal, which and their copies are transmitted through mutually independent fading. Depending on the form of channel resources utilized, diversity techniques are divided into the following forms [1]:

(1) Time diversity: applicable to time-selective fading channels, time diversity can be obtained by coding and interleaving, after the information is coded, the coded symbols are dispersed at different coherent times so that different parts of the

code fade independently. A delay of a few milliseconds will be incurred to obtain the time diversity gain, which increases the communication delay and decreases the bit rate.

(2) Frequency diversity: it can be used when the bandwidth of the propagating signal is larger than the correlation bandwidth of the channel. The space–time coding can help exploiting the frequency diversity gain. However, to obtain the frequency diversity gain requires high-frequency broadband components that are not easily available in UWOC systems.

(3) Spatial diversity: it can be obtained by using multiple transmit or receive antennas placed far enough away from each other than the coherence distance.

(4) Cooperative diversity: it is also called relay-assisted diversity and uses relay nodes, thus obtaining a generalized distributed spatial diversity.

Spatial diversity is more commonly used in UWOC systems, while time, frequency, cooperative, or wavelength diversity is also feasible.

10.1.3 Spatial Diversity

Typically, spatial diversity is achieved at the transmitter/receiver end by an array of spatially separated apertures. These apertures are spatially separated by a certain distance to ensure that the received signals are uncorrelated or less correlated. In a spatial diversity set, the transmitted signals and copies arrive at the receiving end as redundant in the spatial domain. Spatial diversity does not require sacrificing the bandwidth utilization of the system.

Spatial diversity can be divided into receiving diversity and transmitting diversity according to whether the transmitter or receiver uses multiple apertures. Receiving diversity uses multiple apertures at the receiving side to receive the transmitted signal and its copies. A reasonable combination of signal copies can improve the total system receiving signal-to-noise ratio and reduce multipath fading. Transmitting diversity is the use of multiple apertures on the transmitting side, where the signal is processed by the transmitter and then transmitted through multiple antennas.

10.1.4 Cooperative Diversity

In relay-assisted UWOC systems, relay nodes can be distributed serially or in parallel. Increasing the number of serial relay nodes will increase the gain in path loss. Increasing the number of parallel relay nodes increases the diversity gain. Using the diffuse nature of light propagation in water, neighboring nodes (also called relays) are able to overhear the signals transmitted by the source node. The source and relay nodes act together in the transmission system to form a virtual transmitting array, although each node is equipped with a transmitting unit.

   Multi-hop transmission is a type of relay-assisted transmission that reduces absorption, scattering, and turbulence-induced fading effects and thus improves path loss gain if the relays are configured in a serial fashion, allowing relatively long distances to be split into multiple shorter distances. This helps to improve link coverage with limited transmission power.

   Using parallel relaying schemes, virtual transmit or receive arrays can be formed, and the diversity gain can then be obtained using conventional spatial diversity methods, which are effective against turbulence-induced fading.

10.1.5   Combining Schemes

In the previous section, we described that the transmitted signal diversity could be obtained by designing symbol redundancy for sending information in the time domain, spatial domain, frequency domain, etc. However, the performance of systems using diversity technology often depends on the way in which the receiver combines copies of the signals from multiple diversity branches, thus improving the overall received SNR. In the following parts, we discuss how the received diversity signals can be used to improve the system's BER performance. Currently, there are four main processing options for multiple diversity branched signals received by the receiver: selection combining, switching combining, maximum ratio combining, and equal gain combining.

(1) Selection combining.

   Selection combining scheme is simple, and a block diagram of it is shown in Fig. 10.1. Let the number of receiving apertures of the system be n_R. The system selects the diversity branch signal with the maximum instantaneous signal-to-noise ratio from the n_R received diversity branches as the judgment input signal. Therefore, the signal-to-noise ratio of the output signal is the best signal-to-noise ratio of the input signal. In practical systems, the sum of the measured signal and noise power is generally used instead of the signal-to-noise ratio, as it is difficult to measure the signal-to-noise ratio alone.

Fig. 10.1 Block diagram for
selection combining

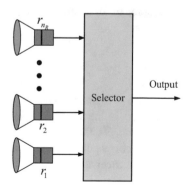

Fig. 10.2 Block diagram of
the switching combining

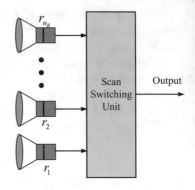

(2) Switching combining.

Replace the logical selector in Fig. 10.1 with a scan switching unit to obtain a switching and merging diversity system, as shown in Fig. 10.2.

For the systems shown in Figs. 10.1 and 10.2, the receiver scans all the diversity branches and compares them with the preset threshold and then selects the diversity branch signals whose SNR is above the preset threshold as the combined output signals. As soon as the signal-to-noise ratio of this signal drops below the set threshold, the receiver will start scanning again. The receiver will start scanning again until it finds a diversity bypass signal that meets the preset threshold, at which point the bypass switching is performed.

As can be seen from the switching combining scheme's operation, the system does not continuously select the instantaneous signal with the highest signal-to-noise ratio, so its performance is worse than that of the selective merging scheme. However, this scheme is much simpler as it does not require continuous monitoring of all diversity tributaries at the same time. For both the selection combining and switching combining schemes, the output signals of both are only one signal from all the diversity branches. So both schemes can be used for both coherent and incoherent modulation.

(3) Maximum ratio combining.

The schematic block diagram for the maximum ratio merge diversity is shown in Fig. 10.3. The combined output signal is a linear combination of weighted copies of all received signals:

$$r = \sum_{i=1}^{n_R} \alpha_i r_i, \tag{10.1}$$

where r_i is the received signal of the ith diversity branch (ith receive aperture) and α_i is the weighting factor of the corresponding receive aperture. In this scheme, each receiving aperture's weighting factor is chosen to be proportional to its corresponding signal voltage and noise power. Let A_i denote the amplitude of the received signal r_i and ϕ_i the phase of the received signal r_i. It is not

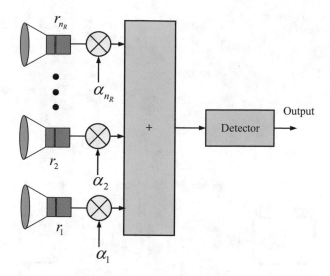

Fig. 10.3 Block diagram of the maximum ratio combining

necessary to assume that the average noise power of each aperture is the same so that the corresponding weighting factor is

$$\alpha_i = A_i e^{-j\phi_i}. \tag{10.2}$$

The maximum ratio combining scheme gives the maximum output signal-to-noise ratio, which is the optimal combining scheme, and Rappaport gives proof that the maximum output signal-to-noise ratio is equal to the sum of the instantaneous signal-to-noise ratios of the individual signals. In this scheme, the individual received signal must remain in phase and be weighted and summed by the same value as their respective amplitudes. Since this scheme requires knowledge of the turbulence-induced fading, amplitude of the channel, and the signal's phase information, it can only be used in case of coherent detection.

(4) Equal gain combining.

Equal gain combining is also known as phase equalization. Compared to maximum ratio combining, equal gain combining only corrects the channel's phase shift and does not process the amplitude. The principle block diagram of equal gain combining diversity is shown in Fig. 10.4.

For equal gain combining, the output signal-to-noise ratio of the merger is

$$r_{EGC} = \frac{\left(\sum_{i=1}^{n_R} \sqrt{r_i}\right)^2}{n_R}. \tag{10.3}$$

Fig. 10.4 Block diagram of
the equal gain combining

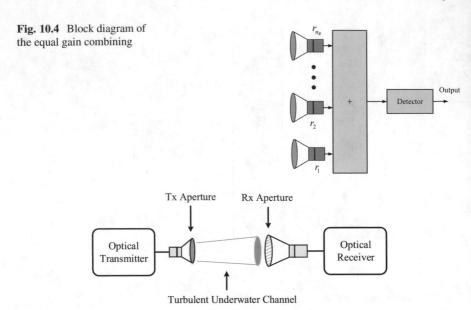

Fig. 10.5 Block diagram of aperture averaging of UWOC systems

Here it is assumed that the noise level is the same for all the diversity branches. This scheme is not the best way to combine in any sense and is only optimal to maximize the signal-to-noise ratio if the signal-to-noise ratios of all the branch signals are the same.

10.2 Aperture Averaging

The light intensity scintillation caused by turbulence in UWOC can cause the optical power received by the detector to fluctuate, where fluctuation variance is related to the size of the receiving aperture, the larger the receiving aperture, the smaller the optical power fluctuation variance, which is called the aperture averaging effect (Fig. 10.5).

The use of large aperture reception can effectively suppress the impact of communication optical flicker on the optical communication link, but large aperture reception can increase the weight of the receiver terminal, in applications that require miniaturization of the communication terminal, large aperture reception technology be somewhat limited [2]. When designing a system, there is often a trade-off between size, weight, and lightness of the communication terminal. In addition, the received optical power variance decreases nonlinearly with the increase of the received aperture size, so the optimal selection of the received aperture size becomes a key issue in the study of large aperture reception technology.

10.3 Acquisition, Pointing, and Tracking

In wireless communication, due to the high transmitting power of microwave aperture, the field effect of electromagnetic waves leads to a large diffusion angle of electromagnetic signals and therefore does not require very precise alignment of the antenna; however, in UWOC systems, especially when using laser light sources, because the light transmission has strong directivity, there will be the situation of misalignment. Therefore, acquisition, pointing, and tracking (APT) is a key technology to enhance the performance of UWOC systems [3].

(1) Acquisition. Judgment and identification of the target in the uncertain area, by scanning the uncertain area until the beacon light signal is received in the capture field of view, laying the foundation for the subsequent aiming and tracking.

(2) Pointing. (a) Make the view axis of the communication transmitting end and receiving aperture normal direction maintain very precise coaxiality during the communication process. (b) In long-range UWOC system, when the transmitting and receiving sides move tangentially at high relative speeds, over-sighting is required at the transmitting end.

(3) Tracking. Due to the relative motion between the transmitter and the receiver, the spot scintillation and wander caused by underwater turbulence, and the effect of platform vibration on the aiming accuracy, the aiming error is controlled within the allowable range by tracking when the aiming is completed.

The APT system works in two stages, as shown in Fig. 10.6, when the receiver has not yet captured the beacon light, the visual axis in the uncertain area for open-loop scanning, open-loop scanning of the uncertain area information imported by the guidance information, guidance information is generally each other's position coordinate information; when the main control computer receives the guidance information, calculate the relative position of the receiver and the transmitter both sides, and drive the tracking frame turntable pointing at the target. The speed measurement unit and the angle control unit are the feedback units of the system, which constitute a double closed-loop control system. The image processing unit, as the detection mechanism of the spot position, provides the spot position information for the system in real time. Due to the deviation of the guidance information and the error of the actuator, the turntable will have error after pointing and generally cannot aim at the target accurately.

After the turret pointing, due to the existence of errors, the target position as the center, the beam position may be distributed in the area called the uncertainty region. Since the angle of the uncertain area is larger than the divergence angle of the beacon beam, the beacon light should be scanned in the uncertain area until the beacon light spot appears in the receiver field of view to complete the capture of the beacon light; once the beacon light is captured successfully, the system will detect the deviation between the center axis of the receiving aperture field of view and the center axis of the beacon beam through the coarse tracking imaging unit and

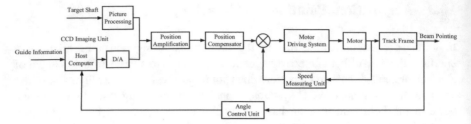

Fig. 10.6 Schematic diagram of APT structure of a typical UWOC system

continuously correct the angle of the receiver and transmitter by driving the motor to achieve precise aiming.

10.4 Multiplexing

Multiplexing means that multiple signals are transmitted independently using the same channel, where transmission of each signal is independent of each other and does not interfere with each other. In UWOC systems, multidimensional multiplexing is often used. The use of multidimensional multiplexing is an effective way to increase the capacity of the system, as it enables multiple signals to be transmitted in parallel on the one hand and overcomes modulation bandwidth limitations on the other. The primary methods of multiplexing are frequency division multiplexing (FDM), time division multiplexing (TDM), code division multiplexing (CDM), wavelength division multiplexing (WDM), sub-carrier multiplexing (SCM), and polarization division multiplexing (PDM) [4, 5]. This section describes these multiplexing techniques in principle.

10.4.1 Time Division Multiplexing

Dividing the channel into N time slots in time and transmitting N data in parallel are called time division multiplexing (TDM). A schematic of TDM with 4 time slots is given below, as shown in Fig. 10.7. The dynamic allocation of N time slots to multiple users is called time division multiple access (TDMA), as shown in Fig. 10.8.

After source coding, channel coding, interleaving, and other processing of multiplexed data in accordance with a certain timing relationship to modulate the carrier, TDM/TDMA can be achieved, and the principle of implementation is shown in Fig. 10.9.

TDM technology has two outstanding advantages: firstly, the convergence and splitting of multiple signals are digital circuits, simpler, and more reliable than the

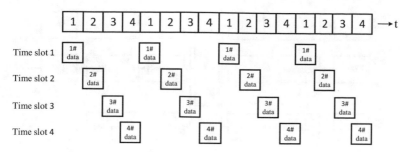

Fig. 10.7 Time division multiplexing diagram

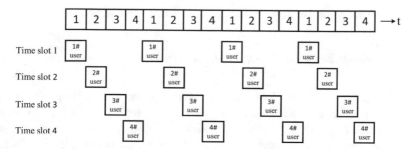

Fig. 10.8 Time division multiple access diagram

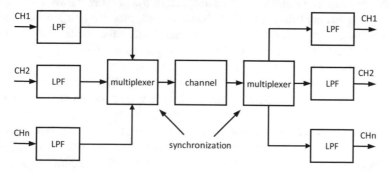

Fig. 10.9 Block diagram of time division multiplexing implementation

analog filter splitting of frequency division multiplexing. Secondly, the channel's nonlinearity can produce crosstalk distortion and high harmonics in the FDM system, causing inter-road crosstalk, and therefore, FDM requires high nonlinear distortion of the channel, while the nonlinear distortion requirement of the TDM system can be reduced.

However, TDM has high requirements for clock phase jitter in the channel and clock synchronization between the receiver and the transmitter. By synchronization, the receiver can correctly identify the individual serial numbers from the data stream. To do this, a frame synchronization signal must be added to each frame. This can be a specific code group or a pulse of a specific width.

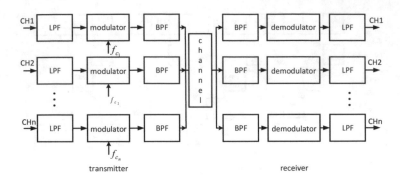

Fig. 10.10 Block diagram of frequency division multiplexing implementation

10.4.2 Frequency Division Multiplexing

Dividing the channel into N carriers by frequency and transmitting N channels of data in parallel are called frequency division multiplexing (FDM), as shown in Fig. 10.10. In FDM, the channel's available frequency bands are divided into several non-overlapping frequency bands, each of which is occupied by a signal and can therefore be demodulated and received separately by dividing them with an appropriate filter. The schematic block diagram of the FDM is shown in Fig. 10.10.

As the signals are often not strictly band-limited, they are first low-pass filtered and then linearly modulated at the transmitter. Before the modulated signals are summed and fed into the channel, they are also passed through a bandpass filter to avoid their spectra overlapping. The receiver first uses the bandpass filter to extract each signal separately, then demodulates, and low-pass filters the output.

The main problem in FDM is the mutual interference between the signals, which is called crosstalk. The main cause of crosstalk is the spreading of the modulated signal spectrum caused by system nonlinearity. The transmitting bandpass filter can partially eliminate the crosstalk caused by modulation nonlinearity, but the crosstalk caused by nonlinearity in the channel transmission cannot be eliminated. The requirements for system linearity in FDM are therefore very high. Rational selection of the carrier frequency and a certain protection interval between the tuned signal spectrum is also an effective measure to reduce crosstalk. N carriers' dynamic allocation to multiple users is called frequency division multiple access (FDMA). FDM/FDMA is achieved using modulation techniques to modulate multiple channels of data from multiple users onto multiple carriers. TDM is split in the time domain, but mixed in the frequency domain; similarly, FDM is split in the frequency domain, but mixed in the time domain. FDM is more difficult to integrate as it uses modems and filters.

10.4.3 Code Division Multiplexing

Dividing the channel into N code channels by code word and transmitting N data channels in parallel are called code division multiplexing (CDM). The dynamic allocation of N code channels to multiple users is code division multiple access (CDMA). The schematic diagram and the implementation of CDM/CDMA in UWOC systems are shown in Figs. 10.11 and 10.12.

Spread spectrum technology means spectrum expansion, i.e., expanding the spectral bandwidth of a signal. According to the Shannon formula,

$$C = B\log_2\left(1 + \frac{S}{N}\right). \tag{10.4}$$

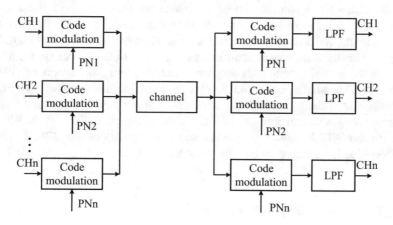

Fig. 10.11 Schematic diagram of the CDM principle

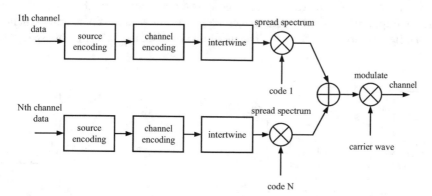

Fig. 10.12 Schematic diagram of CDM implementation

Increasing the bandwidth B reduces the requirement for SNR without changing the channel capacity. Spread spectrum allows normal communication at very low signal transmitting power, facilitating concealment, hence the original application of spread spectrum communication in military communications.

10.4.4 Wavelength Division Multiplexing

Wavelength division multiplexing (WDM) technology is a communication method in which multi-channel optical signals with different wavelengths are transmitted through underwater optical channel, optical carrier separation is carried out by different color filters at the receiving end, and signal processing is carried out by the optical receiver to recover the original information.

If the RGB-LED transmitter is used, i.e., the beams are combined and transmitted using RGB tricolor modulation, the three light signals of different wavelengths are loaded onto the RGB three color LEDs after an electrical amplifier and a DC bias, through which the three beams of different colors are coupled in space to form white light. On the receiving end, after focusing through a lens, the signals of different wavelengths are filtered out using filters corresponding to the three RGB colors, respectively, and finally a receiving circuit is used for signal acquisition and off-line processing; the principle of UWOC WDM is shown in Fig. 10.13. With the RGB tricolor WDM technology, the transmission capacity of the UWOC systems can theoretically be increased by a factor of three.

10.4.5 Sub-carrier Multiplexing

Sub-carrier multiplexing (SCM) technology is a multiplexing method which transmits multiple signals by the same wavelength transmitter in the underwater channel with different carrier modulation. The flexibility of SCM lies in its flexible spectrum

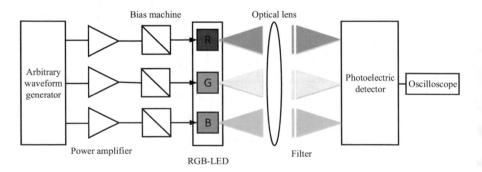

Fig. 10.13 Schematic diagram of the principle of WDM technology

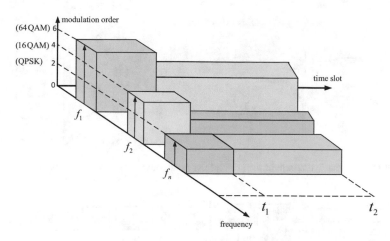

Fig. 10.14 Schematic diagram of the principle of sub-carrier multiplexing

allocation, as shown in Fig. 10.14. It can be seen, the modulation order, bandwidth and center frequency can be dynamically adjusted for different sub-carriers.

Specifically for UWOC systems, higher order symbols can be allocated in the low-frequency region to take advantage of multiplexing techniques to increase communication rates. Besides, the PAPR of OFDM systems usually increases with the number of sub-carriers. SCM can reduce the number of sub-carriers by flexibly deploying each sub-carrier modulation order, which ultimately reduces the PAPR and improves communication performance. According to the characteristics of SCM, it can be applied in multi-access or bi-directional transmission. In particular, for bi-directional transmission, the downlink transmission can be reserved for the low- and medium-frequency regions and 64QAM or 16QAM-OFDM modulation can be used to improve the downlink communication rate; in contrast, the uplink rate demand is relatively low and the high-frequency region can be used with OOK or QPSK modulation.

10.4.6   Polarization Division Multiplexing Technology

The light emitted from the LED is non-coherent. However, it can be used to obtain linearly polarized light using an external polarizer to achieve polarization division multiplexing (PDM). The transmitter end of the UWOC systems can be received using multiple starters with different polarization directions and a corresponding number of detectors. Under the same wavelength channel condition, two independent data are transmitted simultaneously through two mutually orthogonal polarization states of light to achieve doubling of the information transmission capacity without adding additional bandwidth resources and to improve spectrum

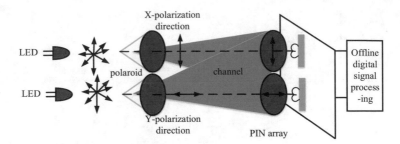

Fig. 10.15 Schematic diagram of the principle of polarization division multiplexing

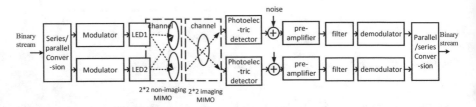

Fig. 10.16 Schematic diagram of the imaging MIMO and non-imaging MIMO principles

utilization. The schematic diagram of UWOC PDM technology is shown in Fig. 10.15.

For the digital signal processing on the receiving end, on the one hand, the sensitivity of the system can be reduced with the help of coherent detection technology to obtain information on the amplitude, frequency, phase, and polarization direction of the optical carrier and make full use of the high-order, multidimensional modulation format to increase the spectrum utilization of the system; on the other hand, the digital signal processing technology can be used to flexibly complete the digital polarization demultiplexing to obtain the ideal demultiplexing results.

10.4.7 MIMO Technology

UWOC MIMO technology uses multiple LEDs to transmit data and multiple receivers to receive data, which can significantly increase the communication capacity of systems. In general, according to the correspondence between the receiver and the transmitter, UWOC MIMO can be divided into two types: imaging MIMO and non-imaging MIMO, as shown in Fig. 10.16. This section focuses on the above two MIMO technologies and introduces MIMO technology principles and the corresponding UWOC MIMO implementation.

UWOC imaging MIMO technology means that the optical signals from multiple transmitters pass through an imaging prism and are received at the receiving end by multiple receivers corresponding to the transmitter's optical signals. Imaging MIMO requires that each receiver can only receive the optical signal from the corresponding transmitter. Therefore, there is no need for MIMO demultiplexing algorithms at the receiver to solve the optical signal crosstalk problem, which can significantly reduce the level of difficulty of realizing receiver integration.

Compared to imaging MIMO, non-imaging MIMO does not require precise alignment and is more practical for practical applications, requiring only a concentrator device to be placed in front of each receiver for complete construction. For non-imaging MIMO UWOC systems, each receiver receives signals from multiple transmitters, which can cause crosstalk between them, and how to recover the transmitted data stream from the crosstalk is a key priority for non-imaging MIMO technology. A post-equalization algorithm is used at the receiver to insert a training sequence into the header of each transmit data frame, and the training sequence is designed to be inserted into two adjacent time slots, thus ensuring that only one transmitter transmits the training sequence for each time slot.

Today, artificial intelligence (AI)-based deep neural network (DNN) and machine learning techniques have been frequently applied in classification, prediction, data mining, and pattern recognition. However, less research has been conducted on AI to solve spatial reuse in UWOC MIMO systems. With the continuous development of related technologies, the combination of AI and UWOC MIMO is expected to become a hot research field in the future. Compared to traditional MIMO demultiplexing algorithms and post-equalization algorithms, artificial intelligence techniques can provide more flexible and accurate estimation and compensation of UWOC channels and received symbols. Existing research shows that AI techniques can effectively improve the capacity and robustness of UWOC MIMO systems.

As deep neural networks, machine learning, and other theoretical methods continue to develop and deepen, they will inevitably have a profound impact the development of MIMO technology in the future, from traditional MIMO algorithms to artificial intelligence-based MIMO algorithms, will be the general trend.

10.5 Adaptive Technology

10.5.1 Adaptive Optics Technology

Adaptive optics (AO) is a technology that uses deformable mirrors to correct wavefront distortions caused by factors such as turbulence to improve the performance of optical systems. This technique reduces scintillation-induced signal attenuation by precorrecting the transmitting beam before it enters the channel [6]. The adaptive optics system has two important components: a deformable reflector for optical correction and a wavefront sensor that can measure turbulence hundreds of times per second. The reflector and sensor are connected by a digital signal processor.

The deformable mirror consists of a thin sheet of glass attached to the back of some devices. Currently, deformable mirrors are small piezoelectric actuators based

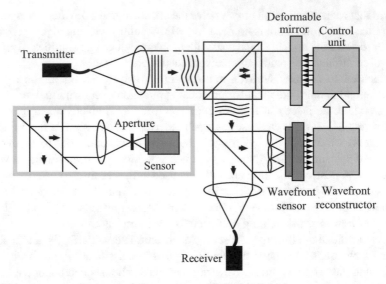

Fig. 10.17 Conventional adaptive optics system with wavefront sensor and reconfigurator

on microelectromechanical systems (MEMS). Wavefront sensors are implemented using detectors.

Conventional AO systems are built on the wavefront conjugation theory. Phase conjugation is achieved by an optoelectronic feedback loop system consisting of a wavefront sensor and reconfigurator, a control system, and a deformable mirror, as shown in Fig. 10.17.

To avoid wavefront measurements in strongly turbulent conditions, the wavefront corrector in an AO system can be controlled by model-free or blind optimization. A portion of the accepted signal is sent to the wavefront sensor, which generates a control signal and sends it to the deformable mirror, as shown in Fig. 10.18; however, this method of wavefront control imposes severe limitations on the control bandwidth. Model-free optimization in real-time AO system makes it difficult to control deformable mirrors with multiple controllable factors with a single scalar feedback signal. The information required to control individual mirror elements must be obtained from a single time-domain sensor rather than from parallel signals from a spatially distributed sensor array. Besides, adaptive optics requires the mirror shape to be adjusted at very high frequencies; thus, the deformable mirror size is generally small and the material requirements are very high. There have been incidents where deformable mirrors were unable to withstand high frequency adjustment and broke. Therefore, practical applications place very high demands on the deformable mirror and controller bandwidth required for real-time correction of wavefront aberrations.

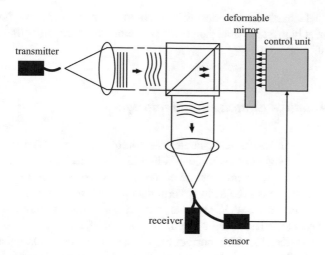

Fig. 10.18 Model-free adaptive optics system

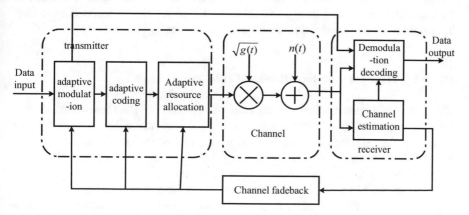

Fig. 10.19 Adaptive modem system

10.5.2   *Adaptive Modulation and Demodulation Techniques*

Underwater wireless optical channels are greatly affected by environmental factors, such as organic and inorganic particles in water, temperature, salinity, density, water depth, internal waves, water masses, reflections from the surface and the seafloor. Therefore, establishing a system that can sense the channel environment and perform adaptive modulation according to the communication distance and data volume is of great help for underwater optical communication.

The adaptive communication mechanism is established as shown in Fig. 10.19 [7] to track the channel changes through the feedback from the receiver, making the system adaptive modulation, coding, power allocation, etc. according to the tested channel characteristics during the communication process, thus enabling transmission at highest possible data rate at different distances and different channel environments.

10.5.3 Adaptive Equalization Techniques

The underwater wireless optical channel is a complex multipath channel that generates serious ISI in signal transmission. When ISI has a serious impact, the transfer function of the whole system must be corrected to make it close to distortion-free transmission conditions. The adaptive equalization technique involves inserting an adjustable filter into the UWOC systems to compensate for the amplitude and phase frequency characteristics of the system, which can effectively reduce the impact of ISI on the system [8]. This correction is carried out in the frequency domain and denoted as frequency-domain equalization. If the correction is carried out in the time domain, it is called time-domain equalization. With the development of digital signal processing theory and integrated circuits, time-domain equalization has become the main method used in most data transmissions.

10.5.3.1 Time-Domain Equalization Principle

The current common method of time-domain equalization is to insert a transverse filter (or cross-cut filter) after the baseband signal acceptance filter, which consists of a delay line with taps. The tap interval is equal to the period of the code element and the delayed signal of each tap is weighted and sent to a summing circuit for output in the same form as a finite impact response (FIR) filter, as shown in the diagram below. The summed output of the lateral filter is sampled and sent to the judgment circuit. The weighting factor for each tap is adjustable and is set to a value that eliminates ISI. Suppose there are $(2N + 1)$ taps with a weighting factor of $C_{-N}, C_{-N+1}, \ldots, C_N$. The sequence of sampled values of the input waveform is $\{x_k\}$ and the sequence of sampled values of the output waveform is $\{y_k\}$. Then we have

$$y_k = \sum_{i=-N}^{N} C_i x_{k-i}, k = -2N, \ldots, 0, \ldots, 2N. \tag{10.5}$$

The output matrix can be calculated using the matrix. Let

$$Y^T = [y_{-2N} \cdots y_0 \cdots y_{2N}]. \tag{10.6}$$

The output matrix can be calculated using the matrix. Let

$$C^T = [C_{-N} \cdots C_0 \cdots C_N].$$ (10.7)

The output matrix can be calculated using the matrix. Let

$$X = \begin{bmatrix} x_{-N} & U & 0 & \cdots & 0 & 0 \\ x_{-N+1} & x_{-N} & 0 & \cdots & 0 & 0 \\ \vdots & \vdots & \vdots & & \vdots & \vdots \\ x_N & x_{N-1} & x_{N-2} & \cdots & x_{-N+1} & x_{-N} \\ \vdots & \vdots & \vdots & & \vdots & \vdots \\ 0 & 0 & 0 & \cdots & x_N & x_{N-1} \\ 0 & 0 & 0 & \cdots & 0 & x_N \end{bmatrix}.$$ (10.8)

Then Eq. (10.5) can be expressed as

$$Y = XC.$$ (10.9)

In general, it is not possible to completely eliminate ISI with a finite number of taps, but when the number of taps is high, it is possible to reduce ISI to a fairly small level.

10.5.3.2   Adaptive Equalization Algorithm and Implementation

The transversal filter characteristics depend entirely on the individual tap coefficients, which are determined according to the effect of equalization. Just because of this, it is first necessary to establish a measure of the equalization effect. The commonly used measures are the peak aberration and the mean square aberration. The peak aberration is defined as

$$D = \frac{1}{y_0} \sum_{k=-\infty, \neq 0}^{\infty} |y_k|.$$ (10.10)

Its physical meaning is the ratio of the sum of the absolute values of the ISI at all sampling moments of the impulse response to the sampling values at $k = 0$ moments. The sum of the absolute values of the ISI reflects the maximum possible value, i.e., the peak value, of the interference of the preceding and following code elements at a given sampling moment in the actual information transmission. Clearly, for an impulse response without ISI, $D = 0$. When using the peak distortion as a criterion, the principle of choosing the tap coefficient should be to minimize the D of the impulse response after equalization.

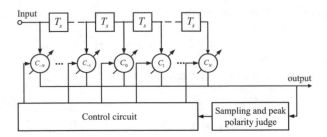

Fig. 10.20 Preset auto equalizer

Theoretical analysis shows that if the peak distortion before equalization is less than 1, then the minimum peak distortion after equalization must occur in the case where $y_k = 0(k = \pm 1, \pm 2, \ldots, \pm N)$. The number of taps in the transversal filter is considered here to be $2N + 1$. To determine the coefficients of the taps that force the inter-code crosstalk y_k to be zero, it is necessary to solve $2N + 1$ joint cubic equations. The most accurate and easy to implement method for automatically solving the joint cubic equations is the iterative method. The mathematical derivation shows that the peak distortion D is a convex function of the tap coefficient, so any iterative technique for adjusting the tap coefficient will minimize the peak distortion. An equalization algorithm based on forcing the ISI to zero is called a zero-forcing algorithm.

There are various options for the specific implementation of the zero-forcing algorithm. One of the simplest methods is preset auto-equalization, the block diagram of which is shown in Fig. 10.20. In a preset auto-equalizer, a single pulse signal with a low repetition frequency is sent before the actual information is transmitted, and a fixed increment is added to the original tap coefficient in the opposite polarity direction, i.e., the tap coefficient is increased or decreased once, according to the positive or negative polarity of each sample value in the equalized output impulse response (except y_0). When y_k is positive, the corresponding C_k decreases by one increment, and vice versa. For fast adjustment, the $2N + 1$ tap coefficients are usually adjusted simultaneously. The control circuit's role is to control the increment and decrement of the tap coefficients according to the outcome of each ISI judgment. It can be expected that the equalization accuracy achieved by this iterative method depends on the size of the increment, the smaller the increment, the higher the accuracy, but the longer the convergence time.

In UWOC systems, it is sometimes not possible to allow preset adjustment before transmitting the information, and even if preset adjustment is allowed, it is not guaranteed that the underwater optical channel remains unchanged during transmission. To automatically adjust the tap coefficients during the transmission of the information using the inter-code crosstalk information contained in the signal, adaptive equalization must be used. In adaptive equalization, each sample value's polarity cannot be used directly as control information, but rather the error information must be extracted from the sample value, the positive and negative

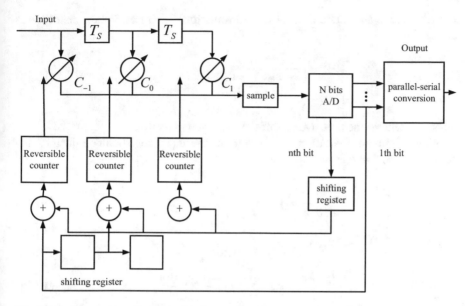

Fig. 10.21 Zero-forcing algorithm adaptive equalizer

polarity of the error is determined statistically, and then the direction of adjustment of the tap coefficients is controlled. It is not difficult to understand that in adaptive equalization using the zero-forcing algorithm, if the initial eye diagram is closed, errors in the error information and hence in the adjustment of the tapping coefficients occur, which may lead to non-convergence of the equalization process.

Figure 10.21 shows the block diagram of an adaptive equalizer with three taps. The parallel-to-serial conversion is of $\log_2 M$ bits, where M being the number of transmit signal levels, and the A/D converter is of $n = \log_2 M + 1$ bits. The first output code of the A/D converter indicates the sampled value's polarity, while the nth bit reflects the error information. The signal's polarity is correlated with the polarity of the error by a shift register using a modulo 2 sum and then fed to a reversible counter for counting, which controls the increase or decrease of the tap coefficient when the reversible counter overflows or recedes.

Another measure of the effect of equilibrium is the mean square aberration, which is defined as

$$e^2 = \frac{1}{y_0} \sum_{k=-\infty, \neq 0}^{\infty} |y_k|, \tag{10.11}$$

where y_k is the sampled value of the impulse response after equalization. In adaptive equalization, the output waveform of the equalizer is no longer a single impulse response, but the actual data signal, where the error signal is

$$e_k = y_k - \delta_k, \tag{10.12}$$

where δ_k is the amplitude level of the transmission. The mean square distortion is defined as

$$\overline{e^2} = \sum_{k=-N}^{N} (y_k - \delta_k)^2, \tag{10.13}$$

where $\overline{e^2}$ denotes the time average of the mean squared error.

With minimum aberration as a criterion, the equalizer should adjust its tap coefficients so that they satisfy

$$\frac{\partial \overline{e^2}}{\partial C_i} = 0, i = \pm 1, \pm 2, \ldots, \pm N. \tag{10.14}$$

From Eq. (10.13), we get

$$\frac{\partial \overline{e^2}}{\partial C_i} = 2 \sum_{k=-N}^{N} (y_k - \delta_k) \frac{\partial y_k}{\partial C_i}. \tag{10.15}$$

Substituting Eqs. (10.5) and (10.12) into Eq. (10.15) gives

$$\frac{\partial \overline{e^2}}{\partial C_i} = 2 \sum_{k=-N}^{N} e_k x_{k-i}, i = \pm 1, \pm 2, \ldots, \pm N. \tag{10.16}$$

The above equation shows that the head coefficient is optimal when the correlation between the error signal and the input sampling value is zero. As in the zero-forcing algorithm, the adjustment process of the tap coefficients in the minimum mean square error algorithm can also be iterative, where the tap coefficients can be refreshed once at each sampling moment, increasing or decreasing by one step. The minimum mean squared aberration algorithm can be used for preset equalizers as well as for adaptive equalizers. A block diagram of a three-tap minimum mean square aberration algorithm for adaptive equalizers is given in Fig. 10.22, where the reversible counter is used as a statistical average.

Theoretical analysis and experiments show that the minimum mean square error algorithm has better convergence and shorter adjustment time than the minimum mean distortion algorithm (i.e., the zero-forcing algorithm).

Preset equalizers are often mixed with adaptive equalizers in UWOC systems. This is because, in the above adaptive equalizer, the error signal is obtained in the presence of crosstalk and noise, which deteriorates the convergence in case of poor channels. In this case, preset equalization can be performed first and then transferred to adaptive equalization. The preset equalization can be performed using a known training sequence. The entire equalizer block diagram is shown in Fig. 10.23.

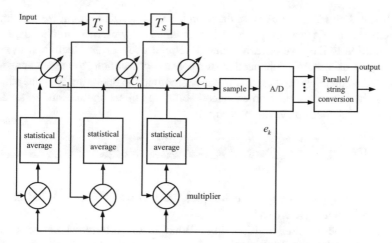

Fig. 10.22 Minimum mean square aberration algorithm adaptive equalizer

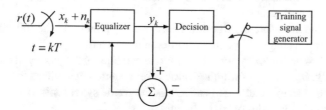

Fig. 10.23 Adaptive equalizer with preset equalization

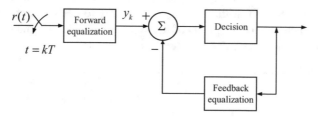

Fig. 10.24 Decision feedback equalizer

The adaptive equalizer technique described above is based on the use of linear filters, where the value of the error signal is obtained using a direct decision method. A nonlinear filter technique can be used to further improve performance, i.e., a decision feedback equalizer (DFE). In the DFE in Fig. 10.24, a lateral filter is used for linear forward filtering, and its decision is fed back to another lateral filter. If the previous decision is correct, the feedback equalizer removes the crosstalk caused by the previous code element. The feedback equalizer's tap factor is determined by the channel impulse response drag caused by including the forward equalizer. It is not difficult to understand that convergence is in principle guaranteed as long as the

BER is less than 1/2. Adaptive equalization is a complex and specialized technique, and, for reasons of space, only the basic principles are presented in this book.

In addition to the several techniques described above, other techniques exist to achieve the link system's enhanced performance. For example, the receiver aperture is adjusted so that the receiver aperture is larger than the operating wavelength, thus allowing the detector to have an enhanced ability to collect photons and mitigate the effects of turbulence on system performance.

? Questions

1. How many typical diversity techniques are there in UWOC? What are their similarities and differences?
2. What is time-division multiplexing? What is the difference between it and frequency division multiplexing?
3. What is the meaning of the term "channel" in FDMA, TDMA, and CDMA multiple access methods?
4. How are the zero-forcing algorithm adaptive equalizer and the minimum distortion algorithm adaptive equalizer implemented in a real system?
5. The PCM24 multiplex system is adopted, the sampling frequency of each channel is $f_s = 8\text{kHz}$, and each sampling value is represented by 8 bits. There are 24 time slots in each frame, and 1 bit is added as the frame synchronization signal, try to calculate the slot width of each channel.
6. The frequency bands of six independent sources are $W, W, 2W, 2W, 3W$, and $3W$, respectively. If time division multiplexing is used for transmission, each source adopts 8-bit logarithmic PCM coding. Try to design the frame structure and the total number of time slots of the system, and calculate the occupied time slot width and pulse width of each time slot.

References

1. J.A. Simpson, B.L. Hughes, J.F. Muth, A spatial diversity system to measure optical fading in an underwater communications channel, in *OCEANS 2009* (IEEE, 2009), pp. 1–6
2. X. Yi, Z. Li, Z. Liu, Underwater optical communication performance for laser beam propagation through weak oceanic turbulence. Applied Optics **54**(6), 1273–1278 (2015)
3. T.T. Nielsen, Pointing, acquisition, and tracking system for the free-space laser communication system SILEX, Free-space laser communication technologies VII. Int. Soc. Opt. Photon. **2381**, 194–205 (1995)
4. W.C. Cox, J.A. Simpson, J.F. Muth, Underwater optical communication using software defined radio over LED and laser based links, in *2011-MILCOM 2011 Military Communications Conference* (IEEE, 2011), pp. 2057–2062

5. G. Cossu, R. Corsini, A.M. Khalid, et al., Experimental demonstration of high speed underwater visible light communications, in *2013 2nd International Workshop on Optical Wireless Communications (IWOW)* (IEEE, 2013), pp. 11–15
6. M.L. Holohan, J.C. Dainty, Low-order adaptive optics: a possible use in underwater imaging? Opt. Laser Technol. **29**(1), 51–55 (1997)
7. Z. Wang, Y. Dong, X. Zhang, et al., Adaptive modulation schemes for underwater wireless optical communication systems, in *Proceedings of the Seventh ACM International Conference on Underwater Networks and Systems*, pp 1–2(2012)
8. T. Komine, J.H. Lee, S. Haruyama, et al., Adaptive equalization system for visible light wireless communication utilizing multiple white LED lighting equipment. IEEE Trans. Wireless Commun. **8**(6), 2892–2900 (2009)

Chapter 11
UWOC Networks

11.1 Introduction

In recent years, with the miniaturization and commercialization of advanced electronic equipment, underwater wireless networks have been increasingly used in military and commercial applications. The underwater wireless network has multiple access capabilities to meet the needs of communication among underwater users. Users include UUV, underwater sensors, submarines, ships, etc. Underwater wireless networks have played an important role in investigating climate change, monitoring marine ecological changes, detecting submarine pipelines, and maintaining offshore drilling platforms. The underwater wireless network can be realized through sound, light, and radio frequency communication links. Studies have shown that optical links have a high transmission rate and high bandwidth characteristics, which can meet UUV requirements and sensors to transmit high time-sensitive data. Chapter 10 of this book introduces a variety of multiple access technologies commonly used in wireless communications. Among these multiple access schemes, optical CDMA technology is the most widely used multiple access technologies in underwater wireless optical network (UWON) because of its fully asynchronous random access capability and the absence of a central network management protocol [1]. In addition, due to the limitation of the UWOC communication range, the relay-assisted UWOC is the key technology to realize UWOC. Relay technology can expand the system coverage, expand the communication distance, improve energy efficiency, and improve end-to-end system performance. An effective routing algorithm can fully consider the propagation characteristics of light underwater to give full play to the advantages of relay-assisted UWOCs. Therefore, this chapter first discusses serial relay and parallel relay technologies using amplify-and-forward (AF), decode-and-forward (DF), and bit detection-and-forward (BDF). Then, various routing protocols of the UWON are studied. Besides, the potential UWA network routing protocol that can be used in UWON is also studied. These potential routing protocols include hop-by-hop

© The Author(s), under exclusive license to Springer Nature Switzerland AG 2022
Y. Lou, N. Ahmed, *Underwater Communications and Networks*, Textbooks in Telecommunication Engineering, https://doi.org/10.1007/978-3-030-86649-5_11

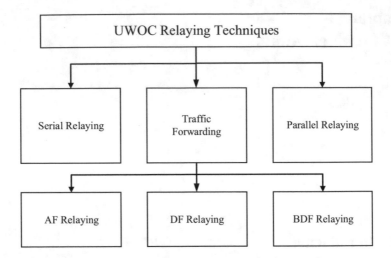

Fig. 11.1 Classification of underwater optical wireless relaying techniques

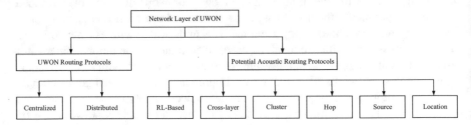

Fig. 11.2 Existing routing protocols for UWONs

routing, location-based routing, source-based routing, cluster routing, cross-layer routing, energy-time efficient routing, and reinforcement learning. Figures 11.1 and 11.2, respectively, classify the relay and routing technologies studied in this article [2].

11.2   Relay Technology

Relay technology is a technique to set up a relay node between the signal source and the target node so that the optical signal can be forwarded once or more to finally reach the target node. Relay technology can split a UWOC link into multiple UWOC links. The long distance and poor performance UWOC link can be replaced by multiple high-quality short distance UWOC links, so that the multi-hop UWOC link can obtain higher capacity and larger coverage area. Figure 11.3 shows the schematic diagram of UWOC through relay nodes are serial transmission and parallel transmission, respectively [3–5].

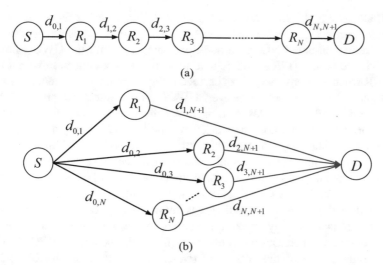

Fig. 11.3 Schematic diagram of serial transmission and parallel transmission network

11.2.1 Serial Transmission and PAT

Serial transmission (referred to as multi-hop transmission) means that the relay nodes in a link are configured serially along a fixed routing path. In this case, the relay nodes communicate with only a single receiver node at the same time. Compared to parallel transmission, serial transmission is more advantageous in extending the communication range and coverage of wireless networks. Serial transmission is often applied in ad hoc UWONs and cellular UWONs.

Since UWOCs are usually configured based on LOS, the beam energy can be focused by reducing the beam spread angle to increase the link distance per hop in serial transmission conditions. The use of narrow beam sources can effectively increase the link distance and reliability of the system at each hop. However, reducing the beam spread angle makes the link less tolerant to environmental interference. For example, random fluctuations in transceiver position caused by ocean turbulence may lead to UWOC link flicker and misalignment. Therefore UWOC links using serial transmission often require fast pointing, acquisition, and tracking (PAT) schemes to guarantee the robustness of the link. To the best of our knowledge, most of the current underwater serial UWOC applications follow the PAT scheme in the FSO domain. No dedicated PAT scheme has been proposed for the UWOC domain. As an alternative, several technologies that can reduce the PAT requirements of UWOC links have been proposed, including NLOS communication methods based on light scattering, omnidirectional fiber optic photodetectors, etc. Specifically

(1) The NLOS UWOC link is divided into two types: beam reflection by the water surface and scattering by the ocean component. Since the scattering-based

NLOS avoids the attenuation of the optical signal during reflection from the water surface, the scattering-based NLOS has strong robustness compared to the reflection-based NLOS. In addition, compared with the UWOC link with LOS, the NLOS UWOC link is able to achieve communication with a higher SNR when the transceiver is not perfectly aligned, effectively reducing the strict alignment requirements. However, UWOC link communication using NLOS method has disadvantages such as lower distance and higher confidentiality performance. In addition, the weaker received optical energy and more severe multipath scattering will also further the reliability of UWOC links. To address this issue, cutting-edge technologies such as adaptive beam spread angle, photon counting based mode, synchronization, channel estimation algorithms, and long-haul NLOS UWOC have been proposed.

(2) The detection area of photodiodes is influenced by the size of the aperture and can usually only reach the square millimeter magnitude. The small detection area increases the counterweight requirement of UWOC. To reduce the interference of the harsh underwater environment on the UWOC link, omnidirectional fiber optic detectors with large detection areas are proposed. The core component of the omnidirectional fiber optic detector is the scintillation fiber. The scintillation fiber works similarly to a luminescent solar concentrator in that it relies on molecules doped in the core of the fiber to absorb incident light and re-emit it at a longer wavelength. The outgoing light signal is then able to propagate along the core of the fiber toward the fiber end. The advantage of a scintillation fiber is that it can be made into a large-area optical receiver without significantly degrading the receiver's response speed. Figure 11.4 shows a large-area photodetector with a scintillation fiber title. The photodetector consists of about 90 strands of scintillation fiber, forming a flat detection area of about 5 cm^2. Compared to conventional photodiodes, the omnidirectional fiber photodetector has a larger detection area. It effectively reduces the PAT requirement and improves the robustness of UWOC in real marine environment.

In addition to the PAT mechanism, the accuracy of the positioning algorithm is also a key factor affecting the performance of UWONs. When the positioning accuracy is low, the link distance and beam diffusion angle can be traded off by adaptively adjusting the diffusion angle so as to guarantee that the receiver lens is always within the beam coverage area. In addition, fast closed-loop tracking and wavefront control techniques can also effectively reduce the probability of link disruption in UWONs.

11.2.2   Parallel Transmission and Relay Selection Protocol

The basic idea of parallel transmission is that the source node can communicate with multiple neighboring nodes, and these nodes can forward the information

Fig. 11.4 Commercial scintillation optical fiber photodetector

of the source node cooperatively. Figure 11.3b illustrates a typical parallel relay scheme. Because UWOC often uses LOS for communication and cannot achieve broadcasting, the light source in parallel-configured UWONs often uses multi-laser transmitters, and each laser points in a corresponding relay node direction. The source node sends signals to N relays. Moreover, based on different forwarding methods, the relay decodes and resends the signal, or the received signal is scaled and forwarded to the destination. Parallel transmission is an alternative way to realize the advantages of space diversity. It can effectively increase the UWONs system's diversity, resist the fading caused by turbulence, reduce the probability of interruption, and ensure the robustness of UWONs.

To achieve the theoretical maximum gain with parallel transmission is difficult due to the marine environment. This is because each channel of the UWON may be affected by turbulence. In addition, there are often a large number of relay nodes in UWON. The links composed of different nodes have different distances and transceiver pointing, resulting in a network where each hop of the link may have different path loss. Therefore, the complexity of UWON system is greatly increased. In order to guarantee the reliability and robustness of UWOC, a reasonable network design is very important. Relay selection is a key aspect of network design and is mainly a basic trade-off among beam divergence angle and received transmit power (or communication range for fixed power reception) and transmit-receive pointing. As an example, in the relay selection shown in Fig. 11.5a, relay node 6 is not involved in the communication link due to its poor performance. As shown in Fig. 11.5b, when different data links intersect at relay node 5, the node can be used for multiple data links simultaneously only if an effective multiple access scheme is used. However, in that case, node 5 tends to become a bottleneck for these data flows, and a critical node similar to node 5 determines the performance of the entire network. Therefore, it is very important to use an effective relay selection strategy in order to maintain and improve the network performance.

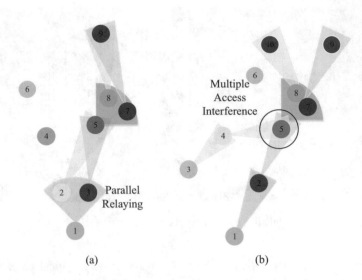

Multiple Access Interference

Parallel Relaying

(a)   (b)

Fig. 11.5 Relay selection method

11.2.3 Forwarding Methods

In UWONs, a relay either amplifies the information it receives or decodes it completely, re-encodes it, and re-transmits the source information. The two relay forwarding methods are named amplify-and-forward and decode-and-forward, respectively.

11.2.3.1 Amplify-and-Forward Relaying

Amplify-and-forward relaying means that the relay node does not modulate and decode the received signal, but directly amplifies the received signal and transmits it. In UWONs system, the common operation mode of amplify-and-forward relay is that: the node performs optical-electrical-optical (OEO) conversion of the received signal, amplifies the converted electrical signal, and finally converts the electrical signal into an optical signal for transmission to the next node. However, the OEO process often introduces additional noise. Amplify-and-forward relay can only take full advantage when nodes amplify optical energy directly without using OEO conversion. The main drawback of amplification and forward relaying is that noise is also amplified and forwarded at each node. In multi-hop links noise is constantly amplified and accumulated, degrading the reliability of the link.

11.2.3.2 Decode-and-Forward Relaying

The decode-and-forward relaying technique means that the relay will first demodulate and decode the received signal, and then the adjudicated data will be encoded and modulated for further transmission. Although decode-and-forward relaying improves system performance by limiting the propagation of noise in each of the links, it increases energy consumption of the nodes and significant delays because the signals undergo a complex decoding and encoding process in decode-and-forward relaying.

Decode-and-forward relaying is limited by the performance of the UWOC channel between the signal source and the relay. For channels with known physical characteristics such as turbulence and absorption scattering, the use of decode-and-forward relaying can effectively improve the performance of the total link. In more desirable water environment and in shorter links, the performance improvement of decode-and-forward relaying is not significant. In addition, considering the cost and the lifetime of UWONs, decode-and-forward relaying is not energy efficient compared to amplify-and-forward relaying. Therefore, in the practical application of UWONs, the designer can adjust the transmission method of the system according to the expected performance of the UWONs.

11.3 Underwater Routing Technology

The network layer technology of UWOC involves physical layer and link layer technologies that have been extensively studied. Section 7.2 of this book states that the propagation attenuation of optical signals depends on environmental characteristics, including salinity, temperature, pressure, and depth of the transceiver. In addition transmission loss equations as well as channel impulse response equations have been proposed. Although several research bases are available, the research on the network layer of UWONs is still in its infancy [6].

The design of UWONs is subject to many limitations, such as:

(1) UWOC link communication range is short, usually only up to 10 m magnitude.
(2) UWOC link is severely impaired by multipath dispersion and absorption scattering effects.
(3) UWOC link has extremely high directivity requirements. Affected by turbulence, the UWOC link is prone to flicker or interruption
(4) Due to scaling and corrosion, the lens of the underwater optical equipment may be obscured and prone to failure.
(5) Node life is limited by battery power. More energy-efficient communication methods (e.g., modulation and coding) and routing methods are to be studied.
(6) Some marine organisms are phototropic and may cause interference to UWOC communication.

11.3.1   Routing Protocols for UWONs

UWONs require the design of efficient and reliable routing protocols to address the channel parameters of light propagation in underwater media. Researchers have been working to provide solutions for the development of UWONs. Although many developed network protocols already exist in the field of underwater wireless communications. However, UWOC channels have unique turbulence-induced fading characteristics that require new efficient and reliable routing protocols. Only a few routing protocols have been developed for UWONs, divided into centralized and distributed routing protocols. Distributed solutions are suitable for situations where nodes have only local information about the state of their neighboring networks. In contrast, centralized routing relies on the availability of the global network topology and achieves better end-to-end (E2E) performance. However, collecting global network state information may incur additional communication overhead and energy costs [7, 8].

11.3.1.1   Centralized Routing

In centralized routing, a central node (CN) must be established which collects network information from all other nodes in the network. The CN is responsible for centralizing the overall information of the network. At the same time, the CN is able to design the best route based on the communication range of the nodes through a specific algorithm based on the information obtained, and then broadcast the updated route to each node. In other words, the routes are determined and shared by the CN to all nodes. Meanwhile, all the collected information and routing tables will be stored in a global database for easy network management.

Take a centralized UWON that has been proposed as an example. As shown in Fig. 11.6, the UWON consist of buoyed nodes, moored nodes, and seabed nodes. All three nodes carry optical transceivers. As shown in Fig. 11.6, the seabed node is fixed to the seabed to collect ocean information. moored nodes act as relay nodes and are responsible for receiving data collected from buoyed nodes and forwarding this information to the uplink nodes. The buoyed nodes near the sea surface need to transmit all the uplink information to the pooling point. In UWONs, the surface base is able to collect data measured by subsea sensors. In this network, buoyed nodes act as cluster-heads (CHs) and are responsible for collecting information from other sensors, designing routing protocols, and broadcasting it to other nodes.

Routing protocols are designed to optimize specific performance metrics of the UWON, e.g., power consumption, data rate, and BER, while satisfying constraints of some metrics. Depending on the metrics being optimized, routing protocols can be classified as

(1) Maximum end-to-end data rate routing. The goal of this routing protocol is to find a path that maximizes the end-to-end data rate while guaranteeing a target end-to-end BER.

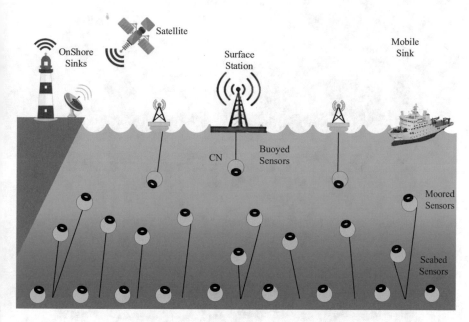

Fig. 11.6 Physical layer architecture of a UWON with centralized routing

(2) Minimum end-to-end BER routing. The goal of this routing scheme is to find a path that minimizes the end-to-end BER while satisfying a target data rate.

(3) Minimum power routing. The stored energy limit of the relay node has a great impact on the lifetime of the communication network. Considering the cost of battery replacement and engineering difficulties, an energy-efficient routing technique must be used to minimize the energy consumption of multi-hop communication.

11.3.1.2   Distributed Routing

In a distributed routing protocol, the network state is updated periodically. Each node in UWON has a local database. Each of these nodes is able to know the state of the network from other nodes and store the state information. The collected information is used to design the routing algorithm and to calculate their next destination node. In a new update cycle, all nodes will again broadcast their latest routing information to all neighboring nodes. In summary, with a distributed routing mechanism, all nodes are able to maintain and update their routing paths independently through some algorithms.

Recently a distributed routing algorithm called Light Path Routing Protocol (LiPaR) has been proposed. In LiPaR routing nodes only need to be informed of the location information of neighboring nodes. In LiPaR it is assumed that each node gives priority to communicate with the nearest node. This is shown in

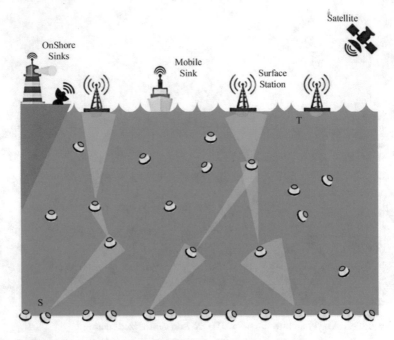

Fig. 11.7 Physical layer architecture of a UWON with distributed routing

Fig. 11.7. Since LiPaR considers the compromise between beamwidth and distance, LiPaR routing has a low requirement for alignment, does not require a strict PAT mechanism, and is able to provide better end-to-end performance in the case of pointing mismatch. Specifically, consider a two-dimensional UWON consisting of M ground stations/receivers (SS) and N nodes/sensors. The SS is responsible for propagating the data collected from the sensors to mobile devices. The underwater node is equipped with two optical transceivers to enable a bi-directional connection between the sensor and the SS. And assuming that the transmitter is capable of adaptively adjusting the beam divergence angle within a certain range, $\theta_{min} \leq \theta_n \leq \theta_{max}$. The position of the Mth surface station is estimated to be $\ell_m = [x_m, y_m]$ and the position of the Nth underwater relay node is estimated to be $\ell_n = [x_n, y_n]$ by the underwater positioning technique.

Thus, the set of feasible repeaters in the vicinity can be given by

$$\aleph_i = \left\{ j | D_i^j \geq \|\ell_j - \ell_i\|, \theta_{min} \leq \varphi_i^j \leq \theta_{max}, \forall j \right\} \quad \forall i. \tag{11.1}$$

The above equation is defined by the predetermined data transmission rate and BER for each hop. In the lower equation, the forwarder of the next hop is selected in the

$$h_{i+1} = \arg\max_{j} \left\{ (1 - \mathcal{P}_s^j) \times \left\| \ell_j - \ell_i \right\|, \forall j \in \aleph_i \right\}, \tag{11.2}$$

where each BER per hop $\mathcal{P}_s^j$ varies both with distance and with angle between the pointing vector. Let us consider two candidate Euclidean distances but different angles. If chosen as the next forwarder, the transmitter needs to reduce its divergence angle. Inversely proportional to the average propagation distance, it is minimized, so the average distance propagation distance is maximized. The current forwarder adaptively adjusts the beamwidth to cover the next node and sends the packet after informing the next node of its decision. This process is repeated along the path until one of the sink nodes is reached.

In addition, a sector-based opportunity routing protocol has been proposed for proper UWON operation in case the optical transceiver is not fully aligned. This protocol takes full advantage of the broadcast characteristics of the UWOC, as shown in Fig. 11.8. A node equipped with a large diffusion angle optical transmitter is linked to a set of candidate relay nodes. The UWOC communication with large diffusion angle is highly robust and can overcome to some extent the light intensity flicker and communication interruptions due to underwater channel corruption (e.g., mispointing, out-of-tune, turbulence, etc.) and marine life activities. The routing protocol also improves the data delivery rate because at least one of the many candidate relay nodes is able to perform high-performance single data routing with the node. The performance of the candidate relays can be ranked by means of distance progress (DP) and expectation metrics. Finally, to weigh the data rate and BER as well as the link distance and beamwidth, a candidate relay node selection

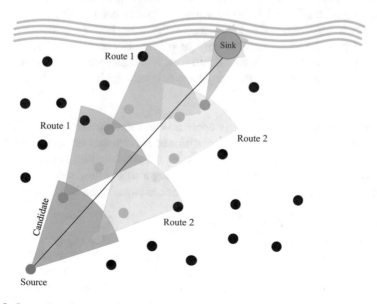

Fig. 11.8 Sector-based opportunity routing (sector) protocol

method and priority ranking (CSPA) algorithm has been proposed, which finds the coverage area of the best sector shape by scanning the feasible search space.

11.3.2   Potential Routing Protocols for UWONs

Some of the routing algorithms proposed for underwater acoustic networks can also be used in UWONs. These include:

(1) Location-based routing.
In location-based routing strategy, the location information of underwater sensors is utilized to seek the best path from the source to the destination. Each node should know its location, destination area and the location of its neighbors. The data is routed based on the location information.

(2) Source-based routing.
The source-based routing protocol picks the route that has the lowest transmission delay from the source node to the aggregation node. Once the route is specified, the nodes located along the route can also send data to the receiving node.

(3) Hop-by-hop routing. In hop-by-hop routing, relay nodes self-select the next hop. Hop-by-hop routing provides the network with both flexibility and scalability, but the choice of nodes may not always be optimal.

(4) Cross-layer routing.
Cross-layer routing protocols take into account the information that is available at different layers and provide solutions to several network problems, such as defining routing policies, scheduling, and power control. Cross-layer routing protocols also allow the selection of the next hop of a transmission by considering the transmission delay of potential candidate nodes, distance to the aggregation point, channel conditions and buffer size. Cross-layer policies improve the overall performance of the network and minimize the energy cost of the network.

(5) Clustered routing.
Clustered routing is particularly applicable to infrastructure-based UWONs. In clustered routing, the network is partitioned into clusters depending on the geoposition of the nodes. Once the network is partitioned into clusters, a cluster head is being selected for each cluster using a cluster head selection algorithm. The cluster head is used as a gateway for communication between clusters and between the aggregation nodes.

(6) Reinforcement learning based routing.
The reinforcement learning based routing protocol uses a Q-learning approach to the network state and adapts itself to changes in the topology. Nodes perform an analysis of the residual energy and energy of their neighboring nodes, apply a reinforcement function, and then pick the optimal node to forward the data.

(7) Energy-efficient routing.

Energy harvesting elements play an essential role in improving the lifetime of the network. Renewable energy can be harvested from the aquatic environment to extend the lifetime of UWONs. In terms of energy storage, the field of PV cell UWOCs with dual functions of signal acquisition and energy harvesting shows good promise.

11.4 Underwater Acoustic-Optical Hybrid Network

11.4.1 Introduction of Underwater Acoustic-Optical Hybrid Network

An ideal UWN scheme that provides high transmission speed with the advantages of robustness, high confidentiality performance, and low latency volume delay is the key to meet the growing demand for underwater high-speed communication. Section 6.2 of this book points out that the UWC technologies commonly used in UWN all have unique advantages and disadvantages, and a single UWC cannot achieve ideal high-performance underwater communication. Therefore, in order to maximize the advantages of different UWC technologies and achieve high-performance communication, mixing multiple communication technologies is an effective solution.

In early studies of acoustic-optical hybrid networks, UWN design solutions tended to select specific UWC technologies in advance for the needs of underwater applications in the network. To the best of our knowledge, the first acoustic-optical hybrid network was an underwater sensor network built by MIT in 2005 for detecting fisheries and coral reefs. The network structure is shown in Fig. 11.9 [9, 10]. The network consists of static and mobile underwater sensor nodes. The static underwater nodes are arranged equally spaced in a grid on the seabed. The static nodes carry a variety of sensors that are responsible for collecting ocean information such as temperature and pressure. The mobile underwater nodes are AUVs equipped with optical modems, which access all nodes in raster fashion. Two communication methods are used in this network, long-range low-speed UWAC between nodes and short-range UWOC communication between mobile nodes and static nodes. In this network, the mobile nodes act as data mules and are responsible for actively approaching the submerged static nodes and performing short-range UWOC to collect information. The researchers note that UWOC via data mules can reduce the energy consumption of the network by more than four orders of magnitude compared to direct acoustic dissemination of information between nodes. From an engineering point of view, the hybrid network has a lower cost and longer lifetime compared to a purely acoustic network.

Fig. 11.9 Underwater hybrid
network using mobile nodes
acting as data mules

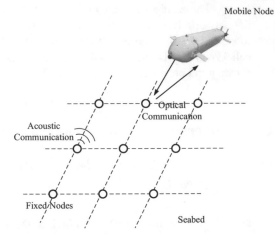

However, the network shown in Fig. 11.9 has two disadvantages.

(1) The design approach of selecting UWCs in advance locks the communication
methods between different underwater nodes and limits the ability of UWNs to
manage traffic adaptively.

(2) When the ocean environment changes over time or when there are nodes in
the UWN that can actively change their position (e.g., submarines and UUVs,
etc.), a single UWC technology may not be able to meet the practical needs of
underwater applications.

During the actual deployment of acoustic-optical hybrid networks, firstly, the
marine environment may change randomly, resulting in changes in the channel
characteristics of UWCs (including turbulence-induced fading, propagation loss,
etc.). Secondly, the distance of UWC links may change frequently in networks
containing moving nodes such as AUVs and submarines. These changes make the
underwater acoustic-optical hybrid network more demanding for the flexibility of
UWC. Therefore, in order to maximize the performance of hybrid networks and
integrate the advantages of different UWC technologies, the concept of multimodal
systems is proposed. In multimodal systems, the communication mode is not locked
in advance, but is switched according to the actual demand. hybrid nodes are
the key devices to build multimodal systems, and researchers generally consider
hybrid nodes as underwater nodes equipped with both acoustic and optical modems.
The nodes can use both communication systems or can select the communication system adaptively according to the needs of underwater applications. The
first simplex link multimodal system was implemented by a Hybrid Remotely
Operated Vehicle (HROV) called Nereus, which is capable of using both UWOC
and UWAC UWC technologies. As shown in Fig. 11.10, specifically, for remote
control and underwater positioning, the Nereus HROV can communicate over
long distances with low bandwidth through UWAC technology. In underwater
applications requiring fast response, the Nereus HROV can send optical signals

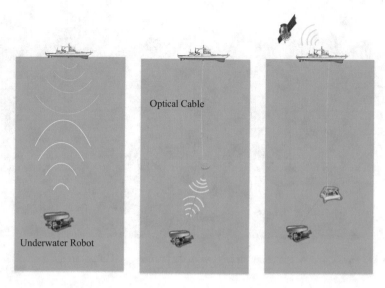

Fig. 11.10 Nereus HROV with simplex link multimodal system

via UWOC technology to underwater optoelectronic sensors connected by cable to a ship or surface base station. The development and successful testing of the Nereus HROV has laid the foundation for a multimodal system, demonstrating that the UWOC system can complement and integrate with the UWAC system, thus enabling the underwater node with the ability to perform short-range, high-speed, low latency communications and long-range, highly robust communications. Since Nereus HROV, multimodal systems have rapidly attracted the interest of researchers. A large number of multimodal network design solutions have been proposed.

11.4.2   Working Mode Switching Strategy

How the multimodal node adaptively selects its operating mode according to the actual requirements is the focus of deploying an underwater multimodal system. This change of operating mode is not limited to switching between UWAC and UWOC technologies, but also includes changing the implementation of a certain technology, for example, adaptively selecting acoustic modems operating at different carrier frequencies according to the communication distance, or changing the beam spread angle and FOV of the optical modem according to the environmental characteristics such as turbulence. From the above discussion, the underwater multimodal node should contain two subsystem technologies. One is the technology of different physical layers, and the other is the different ways to achieve wireless communication within the same physical layer. The selection of the working mode of the multimodal node should consider the actual demand of the underwater

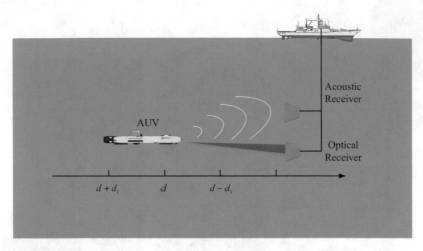

Fig. 11.11 Physical layer switching strategy with the primary goal of power enhancement

application, and this demand should be determined by considering the application scenario, system performance requirements, environmental factors, etc. This section focuses on the switching of physical layer technologies.

The problem of nodes switching between physical layer communication technologies involves several protocol stack layers, including the physical, network, and link layers. Among the many multimodal network schemes that have been proposed, physical layer metrics are the longest used to measure the switching of communication technologies. Physical layer metrics include received power, communication distance, beam coverage, etc.

In multimodal networks where boosting the received power is the main goal, the application layer often uses the constant bit rate configuration method in order to measure the received energy. In this configuration a fixed size packet is transmitted between nodes at a fixed rate. Figure 11.11 illustrates the basic idea of a typical physical layer switching strategy with the primary goal of boosting the received power. Assuming a multimodal system in which a free-motion AUV plays the role of a multimodal node, the AUV is able to detect the average power of the signal during the trigger cycle in real time, and the AUV decides whether to use UWAC or UWOC techniques for communication by judging the average power of the signal collected during a complete trigger cycle. During the trigger cycle, the multimodal nodes do not change their mode of operation. First, the minimum signal-to-noise ratio S_o at which the UWOC can operate properly is preset and the corresponding UWOC link distance d is calculated. P_{o-a} is defined as the received optical power of the link at a link distance of $d + d_1$m, and P_{a-o} is defined as the received optical power at a link distance of $d - d_1$m. The d_1 is set to provide a buffer for the operating mode switching process and to avoid frequent switching of the operating mode when the multimodal section is operating in a link with a signal-to-noise ratio of S_o. When $P_{rec} \geq P_{o-a}$, it indicates that the link distance of UWC is greater than the link distance corresponding to the preset signal-to-noise ratio S_o. At this time, it indicates that the optical signal-to-noise ratio is lower than the preset value and the UWOC link may not work properly, so the physical layer technology of the multimodal node should be switched from UWOC to UWAC. When $P_{rec} \leq P_{a-o}$, the optical signal-to-noise ratio is greater than the preset value, which indicates that the UWOC channel is excellent enough to communicate at high bandwidth and high speed at this time. Therefore, the physical layer technology of the multimodal node should be switched from UWAC to UWOC.

In the physical layer switching strategy, in addition to the received signal energy, the link distance can also be used as a basis for judging the switching physical layer technology. In the proposed multimodal system schemes, the measurement of link distance is usually achieved by underwater acoustic localization techniques. Take the acoustic-optical hybrid communication system proposed by Network Research Lab at University of California as an example. As shown in Fig. 11.12, the underwater contains two kinds of nodes, the stationary nodes for collecting information and completing the localization task, and the motion nodes including AUVs and submersibles. Both types of nodes carry acoustic and optical modems. The goal of the network is to communicate from the stationary node A to the moving AUV B.

The AUV is localized using Time Difference of Arrival (TDoA), which requires at least three nodes (B_1, B_2, B_3) with acoustic modems around the stationary node to assist in localization. The designer predetermines an upper limit of the link distance at which high-quality UWOC can be achieved. When the distance between nodes A and B is less than the preset distance, the UWOC technique can be used to communicate, and when the distance is too long, it switches to the UWAC technique. In the network, the acoustic-optical modem has a dual task. For one, the acoustic modem is responsible for locating the AUV. The acoustic device is also responsible for assisting the optical transceiver to achieve alignment when the AUV

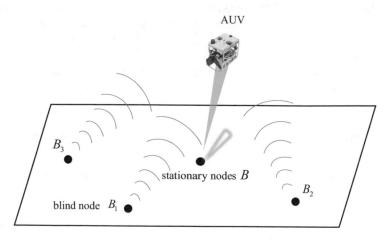

Fig. 11.12 Physical layer switching strategy using link distance as a criterion

is within the effective operating distance of the UWOC. Second, the acoustic device is responsible for achieving remote UWAC when the AUV is far from the stationary node. Influenced by the beam propagation characteristics, the beam coverage can also be used as a criterion for physical layer exchange in addition to the link distance. Specifically, the range of action of the UWOC is modeled as a cone-shaped area. It is assumed that within the cone-shaped area UWOC can provide communication quality that meets the requirements of underwater applications. Then, when the mobile nodes are able to reach the formulated conical range, the links are switched to UWOC technology between them. Outside the tapered range, the long-range UWAC technology is used. In addition, the marine environment has a huge impact on UWOC performance, so in some of the proposed multimodal system schemes, the effective range of UWOC is no longer a fixed value, but is adjusted in real time according to the actual physical phenomena, including temperature, salinity, composition changes, etc.

As the research on multimodal networks progresses, the selection of multimodal technologies based on physical layer metrics alone cannot meet the needs of complex underwater applications. To address this issue, researchers point out that the strategy to manage multimodal transmissions should be message driven. In a multimodal system, the switching of UWC technology should refer to a combination of physical layer metrics as well as link-level metrics. The physical layer metrics include link distance, signal energy, coverage area, ocean environment, etc., as defined above. Link-level metrics are the link performance including BER for a given physical layer. In the message driven switching strategy, the communication requirements can be divided into fast and robust requirements depending on the information to be transmitted. Fast communication is generally achieved through UWOC. In underwater applications with fast communication requirements, the node should be already in a usable UWOC link or have the ability to actively change its position and direction to achieve UWOC. The information to be transmitted includes large volume files such as marine environmental information or underwater images. On the other side, low-speed robust long-range communication is generally implemented through UWAC. In underwater applications of robust communication, data needs to be transmitted directly to the target user through a high availability link without prior detection, positioning, and alignment operations. The information transmitted through robust communication should have low data volume and high timeliness, such as control information and safety information.

11.4.3   Routing Protocol for Acoustic-Optical Hybrid Networks

Some routing protocols dedicated to underwater acoustic networks and underwater optical networks are presented in Sect. 11.3 of this chapter. Although these protocols have the potential to be applied to hybrid networks, they are not designed for the characteristics of UWOC and UWAC links and therefore may not maximize the performance of hybrid networks.

Fig. 11.13 Structure of the
MURAO routing

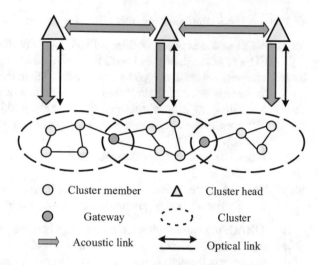

Only a very few routing algorithms dedicated to acoustic-optical hybrid networks
have been proposed. These include, the routing protocol MURAO, which is based
on multi-level Q-learning, the routing protocol where temporary networks are
periodically established by aggregation nodes, and the routing protocol CATPAIN,
which integrates data aggregation and data collection.

11.4.3.1   Routing Protocol MURAO

MURAO is the first routing protocol designed for acoustic-optical hybrid networks.
The physical layer structure provides the basis for the routing protocol. Therefore,
this section first describes the physical layer structure of the acoustic-optical hybrid
network to which MURAO applies. The network is logically divided into two parts
by means of by means of clusters. The upper layer of the network consists of
CHs and the lower layer consists of the members of the cluster. This is shown in
Fig. 11.13. To save energy, only the upper network nodes carry optical and acoustic
transceivers, and the members of the lower network, i.e., clusters, carry only optical
transceivers and acoustic receivers. The acoustic transceivers carried by the CHs are
mainly used to perform two functions; firstly, the CHs need to communicate with
the members of the clusters through the acoustic transceivers to plan the range of
each cluster in the network. Secondly, the acoustic transceiver carried by CHs is also
used for communication between CHs nodes. The UWAC between CHs nodes only
transmits control information to control the forwarding path of packets. In addition
the optical transceivers carried by CHs are used for communication in the upper
layer network along with the acoustic transceivers. While the acoustic receivers
carried by cluster members are used to identify the cluster they are in and to receive
control signals, the optical transceivers are used to deliver packets.

The specific routing process of MURAO is shown in Fig. 11.13.

Step1 A CH in the network notifies all CHs via UWAC broadcast signal that this CH will act as the central node to receive all packets.

Step2 The packets are acquired by the sensors. Then, the packets start to enter the lower network and spread among the members of the cluster.

Step3 According to cluster routing, the packets within the cluster are finally delivered to the egress gateway, which is also the entry gateway of the next cluster.

Step4 When the packet flows through the gateway, the egress gateway reports to the CH.

Step5 The new cluster gets the packet from the ingress gateway, and then the packet will continue to propagate within the new cluster.

In MURAO, the communication path design between clusters is implemented by Q-learning. In brief, Q-learning sets a value V for different CHs to measure the cost of communication through that path. when packets are successfully communicated between clusters, the communication gets a score of -1 and the value $V - 1$. when packet transmission fails, the value $-R_v$, R_v measures the degree of tolerance of Q-learning to link failure. The update of V value is done by two parts together. One is the score obtained by the egress gateway after successful communication in step 4 above. The second is the V-value of the next CH.

The UWAC between CHs in the MURAO scheme is much slower than the UWOC between cluster members. Therefore, in this network cluster members act as relay nodes to transmit packets, while CHs are only used to control data transmission. The MURAO scheme uses the scheme of cluster to decompose the complex routes into several parallel structured routes. And the Q-learning approach facilitates the network to adapt to the changing ocean environment and achieve a dynamic network. From the perspective of multimodal technology, the MURAO scheme controls the UWOC technology through UWAC technology and the optical communication paths between the cluster members through CHs, which effectively improves the network performance.

11.4.3.2 Routing Protocol Based on Reverse Route

Considering the changing ocean environment and the problem of random node movement due to turbulence, a routing protocol based on reverse route search is proposed. The physical layer structure of the network is shown in Fig. 11.14.

This acoustic-optical hybrid network is designed to monitor the marine environment. The sensors acquire information about the marine environment from the seafloor and pass the information through multiple relays to the surface base station. The network consists of three types of underwater nodes (fixed nodes, hover nodes and sink nodes), AUVs and surface base stations. Both fixed and hover nodes carry optical and acoustic transceivers. The fixed nodes are arranged on the seabed and carry image and pressure sensors, which are responsible for acquiring information

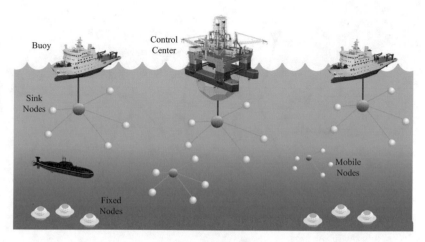

Fig. 11.14 Physical layer structure of marine environmental monitoring network

about the ocean environment and transmitting it to other nodes. The suspended nodes are suspended in the seawater and are responsible for collecting data and uploading to the upper nodes. To solve the problem of difficult networking of optical relay nodes, the designers implemented an approximate omnidirectional optical transceiver by using a spherical optical transceiver array. The sink node is connected to the surface base station by cable and carries acoustic transceivers and optical receivers, which are responsible for receiving data and sending control commands through the UWAC. The AUVs in the network act as data when data mules. when the marine environment data is highly time-sensitive and urgently needs to be uploaded to the base station quickly, the AUVs can collect data through the UWOC if the UWOC link between the nodes is not available. In the network, optical transceivers are responsible for transmitting marine environmental information and acoustic transceivers are responsible for transmitting control commands. When inter-node UWOC links are not available, information can be transmitted through low-speed UWAC technology or forwarded through AUVs.

Routing protocol based on reverse route features the ability to adapt the network to the randomly changing ocean environment, and can take full advantage of the high speed of UWOC technology to improve network performance. The routing structure is shown in Fig. 11.15. The specific process is shown below

Step1 The surface base station controls the sink node to send a fixed size of data called HELLO packet.

Step2 The levitating node that receives the HELLO packet returns the node status data (including the SNR of the received signal, the remaining energy and the information of the depth it is at) to the sink node.

Step3 The levitating node receives HELLO packets transmitted by multiple sink nodes, sorts the sink nodes according to the collected node information, and selects the highest ranked sink node as the relay node for the next hop.

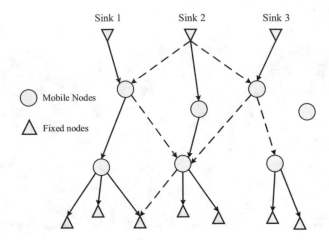

Fig. 11.15 Structure of routing protocol based on reverse route

Step4 When the packet flows through the egress gateway, the gateway reports to the CH.

Step5 The sink node transmits HELLO packets and repeats the process of the second three steps so that deeper nodes can adjudicate the optimal uplink data link. Until all nodes have been added to the network.

When a node fails to establish a link with another node for a period of time, it uses UWOC to send sensor status data. If the node is unable to communicate with the highest ranked node in the list due to environmental changes, the node will try to communicate with lower ranked nodes in the list. If the node is unable to communicate with all the nodes in the list through UWOC, the node switches to UWAC mode and tries to re-establish the link with other nodes and transmits the marine environmental information data at a lower rate. The underwater acoustic-optical hybrid network using routing protocol based on reverse route scheme can achieve wireless transmission of marine real-time data with lower energy consumption and higher data transmission speed. In addition, the network can choose the topology flexibly according to the environmental changes, which effectively improves the working life of the network.

11.4.3.3   Routing Protocol CAPTAIN

CAPTAIN routing protocol is designed for acoustic-optical hybrid networks performing data collection tasks. The physical layer structure of the acoustic-optical hybrid network under CAPTAIN routing protocol is shown in Fig. 11.16. In the network convergence nodes are arranged on the water surface. The suspended nodes are suspended in the water and the node positions are assumed not to change with time. The network contains multiple clusters, and the information collected by the clusters is transmitted to the receiving nodes through a tree route. All nodes in

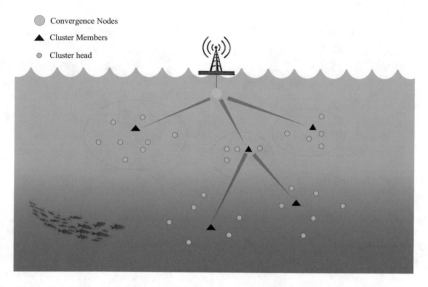

Fig. 11.16 Physical layer architecture of acoustic-optical hybrid network using CAPTAIN routing scheme

the network carry optical and acoustic transceivers. Similar to the network using routing protocol based on reverse route, omnidirectional communication is achieved by using an array of underwater optical transceivers. Unlike the MURAO routing protocol, in the network applying CAPTAIN, all nodes carry both optical and acoustic modems. Due to the difference in the propagation distance between optical and acoustic carriers, the number of neighboring nodes that can communicate with the nodes differs when the two different techniques are applied. This is shown in Fig. 11.17. In general, acoustic neighbors are more widely distributed, and optical neighbor nodes are usually also acoustic neighbors. In the network, the list of optical neighbor nodes is a more important reference for routing design in order to ensure high-performance communication between nodes.

In networks applying CAPTAIN, CHs are not predetermined but selected based on the actual communication effect. The campaign for CHs is all implemented through UWAC and the algorithm is divided into four key steps.

Step1   A node broadcasts a packet.
Step2   The node accepts a packet broadcasted by other nodes, calculates the distance between nodes based on acoustic localization technique, creates a list of optical neighbor nodes and acoustic neighbor nodes, and scores the neighbor nodes.
Step3   The node broadcasts the list information and receives the list information broadcasted by other nodes. If the node itself has the highest score in the list of optical neighbor nodes, the node becomes a CHs, otherwise the node becomes a member of the cluster.
Step4   Broadcast other nodes, whether this node becomes CHs.

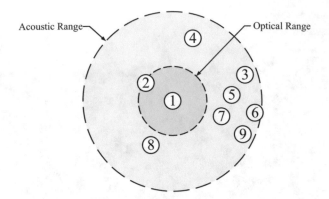

Fig. 11.17 Schematic diagram of optical and acoustic neighbor nodes

Fig. 11.18 Paths for
intra-cluster routing

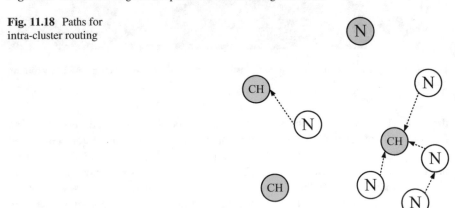

Similar to the MURAO routing protocol, the network shown in Fig. 11.16 is
also logically divided into two parts, one of which is the route for the cluster to
deliver information to the CHs, and the other is the route for the CHs to aggregate
information to the aggregation nodes.

First, Fig. 11.18 illustrates the selection of routes within the cluster. In the
process of campaigning CHs, the cluster has already rated the optical neighboring
nodes. Therefore, in the cluster, the node only needs to deliver the message to the
optical neighbor node with the highest score. Second, in the tree route where CHs
propagate messages to the aggregation node, the aggregation node first broadcasts
the information about the tree route. The CHs that receive this information join the
tree structure and record the distance from itself to the aggregation node as 1 hop.
Then these 1-hop CHs continue to broadcast the tree routing information, and the
CHs that receive the information immediately join the network and mark themselves
as 2-hops, and so on, and eventually all CHs join the network.

CAPTAIN routing algorithm can effectively reduce the energy consumption of
acoustic-optical hybrid networks, especially suitable for cluster networks with high

node density. In CAPTAIN, data transmission is mainly done through UWOC, and UWAC is only used for routing design, so the network has a lower average delay and higher data rate.

? Questions

1. UWON requires end-to-end performance with high reliability and high data rates. However, reliability and high data are contradictory to each other. To solve this problem, what routing protocols have been proposed to guarantee high reliability and high data rates for UWON?
2. What are the advantages and disadvantages of AF and DF relays?
3. What methods can be used to reduce the alignment requirements of the UWOC?
4. What are the advantages and disadvantages of distributed routing protocols and centralized routing protocols?
5. What are the advantages of an hybrid acoustic-optical network compared to UWAN and UWON?
6. Given that an all-optical amplified forwarding protocol is used for UWOC routing, the transmit power of the previous node is $P_{n-1}^t = 30$ dBm, the receive efficiency $\eta^r = 0.9$, the transmit efficiency $\eta^t = 0.8$, the receiver local noise variance $v_{n_l} = 0.1$ dBm, the channel gain instantaneous value $h = 0.12$, the amplification gain $G = 0.12$, and the variance of amplified spontaneous emission noise $v_{n_{ase}} = 0.2$ dBm, then find the average transmit power of this node P_n^t.

References

1. N. Saeed, A. Celik, T.Y. Al-Naffouri, et al., Underwater optical wireless communications, networking, and localization: A survey. Ad Hoc Netw. **94**, 101935 (2019)
2. M.A. Khalighi, M. Uysal, Survey on free space optical communication: A communication theory perspective. IEEE Commun. Surv. Tutor. **16**(4), 2231–2258 (2014)
3. M.V. Jamali, P. Nabavi, J.A. Salehi, MIMO underwater visible light communications: Comprehensive channel study, performance analysis, and multiple-symbol detection. IEEE Trans. Veh. Technol. **67**(9), 8223–8237 (2018)
4. A. Celik, N. Saeed, T.Y. Al-Naffouri, et al., Modeling and performance analysis of multihop underwater optical wireless sensor networks, in *2018 IEEE Wireless Communications and Networking Conference (WCNC)* (IEEE, 2018), pp. 1–6
5. G. Baiden, Y. Bissiri, High bandwidth optical networking for underwater untethered telerobotic operation, in *OCEANS 2007* (IEEE, 2007), pp. 1–9
6. M. Ayaz, I. Baig, A. Abdullah, et al., A survey on routing techniques in underwater wireless sensor networks. J. Netw. Comput. Appl. **34**(6), 1908–1927 (2011)
7. A. Celik, N. Saeed, T.Y. Al-Naffouri, et al., Modeling and performance analysis of multihop underwater optical wireless sensor networks, in *2018 IEEE Wireless Communications and Networking Conference (WCNC)* (IEEE, 2018), pp. 1–6

8. R. Alghamdi, N. Saeed, H. Dahrouj, et al., On distributed routing in underwater optical wireless sensor networks. Preprint (2018). arXiv:1811.05308
9. N. Saeed, A. Celik, T.Y. Al-Naffouri, et al., Energy harvesting hybrid acoustic-optical underwater wireless sensor networks localization. Sensors **18**(1), 51 (2018)
10. L.J. Johnson, R.J. Green, M.S. Leeson, Hybrid underwater optical/acoustic link design, in *2014 16th International Conference on Transparent Optical Networks (ICTON)* (IEEE, 2014), pp. 1–4

Part III
Underwater MI Communication and Networks

Chapter 12
Fundamental Principles of Magnetic Induction

12.1 Brief History of Magnetism and Electromagnetism

The history of magnetism is as old as the ancient Greeks when they discovered a mineral that would attract iron materials or similar minerals, and that mineral was later called as a magnet. However, the experimental and scientific research to study the phenomena of magnetism was first started in the sixteenth century by William Gilbert (1540–1603). It was William who found that Earth itself acts as a magnet. Charles Coulomb (1736–1806) then in the eighteenth century began studying magnetism quantitatively and presented his famous inverse square law. The link between electricity and magnetism was first explored by Christian Oersted (1777–1851) and then further investigated by Ampere (1775–1836) and Faraday (1791–1869) who carried out a detailed experimental study. In the nineteenth century, Maxwell (1831–1869) further demonstrated that both electricity and magnetism are indistinguishably related and laid the basis for electromagnetic physics. He developed the famous four equations that linked electricity and magnetism and are called Maxwell's equation as given in Table 12.1 [1].

These equations are often considered as quite complex, but to explain in simpler terms, these equations unify the phenomena of electricity and magnetism. The first equation, known as the Gauss's law, establishes a link between electric charges and the field created by these electric charges. The second equation explains the phenomena of magnetic fields. It implies that magnets always have two (north and south) poles regardless of how many parts a magnet is broken into. The third equation is commonly known as Faraday's law, which explains how electric fields are created by changing magnetic fields. This equation is the basis for the electrical generators and also the basis for our magneto-inductive (MI) communication. The final equation is Ampère's law, which was significantly modified by Maxwell. Ampère demonstrated that an electric current can produce a magnetic field but in the meanwhile Maxwell had introduced a refinement that a *varying* electric field produces a magnetic field.

© The Author(s), under exclusive license to Springer Nature Switzerland AG 2022   297
Y. Lou, N. Ahmed, *Underwater Communications and Networks*, Textbooks in Telecommunication Engineering, https://doi.org/10.1007/978-3-030-86649-5_12

Table 12.1 Maxwell's equation in differential and integral form

Name	Differential Form	Integral Form
Gauss's law	$\nabla \cdot \mathbf{E} = \frac{\rho}{\epsilon_0}$	$\oiint_{\delta(V)} \mathbf{E} \cdot d\mathbf{A} = \frac{Q(V)}{\epsilon_0}$
Gauss's law of magnetism	$\nabla \cdot \mathbf{B} = 0$	$\oiint_{\delta(V)} \mathbf{B} \cdot d\mathbf{A} = 0$
Maxwell–Faraday's equation	$\nabla \times \mathbf{E} = -\frac{\delta \mathbf{B}}{\delta t}$	$\oint_{\delta S} \mathbf{E} \cdot dl = -\frac{\delta \Phi_{B,S}}{\delta t}$
Ampere's circuital law (with Maxwell's correction)	$\nabla \times \mathbf{B} = \mu_0 \mathbf{J} + \mu_o \epsilon_0 \frac{\delta \mathbf{E}}{\delta t}$	$\oint_{\delta S} \mathbf{B} \cdot dl = \mu_0 I_S + \mu_0 \epsilon_0 \frac{\delta \Phi_{E,S}}{\delta t}$

Fig. 12.1 Magnetic field of a circular coil

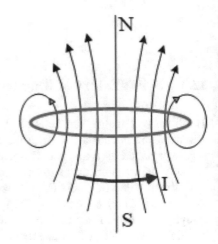

12.2 Basic Elements of Magnetism

12.2.1 Magnetic Fields

Magnetic fields can simply be understood as the area around a magnet where another magnet or electric charges can experience magnetic force. Magnetic fields are generally illustrated by the vector lines that generate from north/south pole and end in south/north pole. These vector fields are also known as magnetic flux lines and they can be a measure of magnetic strength too. The closer the flux lines, the stronger is the magnetic field and vice versa. Figure 12.1 shows magnetic field lines of a coil [2].

12.2.2 Magnetic Flux

The part of the magnetic field B moving through a surface A is called the magnetic flux Φ_B through that surface and is given by

$$\Phi_B = B.A = BA \cos \theta. \tag{12.1}$$

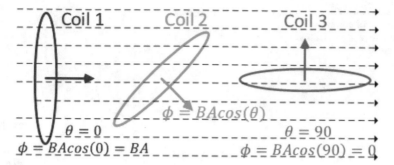

Fig. 12.2 Magnetic flux through coils of different orientations

Figure 12.2 shows a pictorial representation of magnetic flux where three circular coils with different orientations are placed in the presence of same magnetic field. Since the maximum number of field lines passes through Coil 1, magnetic flux through Coil 1 is maximum. On the other hand, the number of field lines passing through Coil 3 is none and therefore the magnetic flux through Coil 3 is zero. Magnetic flux through Coil 2 is $BA \cos \theta$.

12.3 Sources of Magnetic Field

The three main sources of magnetic fields are magnets, ferromagnets, and electric charges. Magnets were already known for the magnetism and their field lines were known for longer time. But notice of magnetism due to electric charges was first noticed by Danish Physicist Oersted in 1820. Oersted in a demonstration noticed a deflection in magnetic compass due to a nearby electric circuit. This discovery for the first time established a link between magnetism and electricity and paved way for other researchers to realize that electric charges also produce magnetic fields [3].

12.3.1 Ampere's and Biot–Savart's Law

After Oersted discovery, Ampere soon investigated the effect and performed a series of experiments. Ampere took a long straight wire and applied electric current to the wire. He then placed a magnetic compass near the current carrying wire to learn the magnetic field direction. Besides the compass, he used a test wire (with current carrying along the same direction) and observed that the two wires would repel each other (Fig. 12.3). Similarly, the two wires would attract each other when the current is along the opposite direction. Based on the experiments, Ampere then formulated his law known as the famous Ampere's Law stating that

Fig. 12.3 Ampère's
experiment of current
carrying wire along with a
test wire

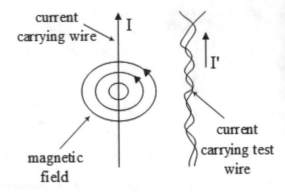

*"For a closed path with electric current, the integral of short lengths times the
magnetic field along its direction is equal to the magnetic permeability times
the electric current enclosed in that closed loop".*

$$\oint \mathbf{B} \cdot d\mathbf{l} = \mu_0 I_{enc}. \tag{12.2}$$

Ampere's law is thus generally used to find the magnetic field of closed loops
that are symmetrical and have steady currents flowing through them. However, a
more formal equation that is used to calculate magnetic field is known as Biot–
Savart's law, named after two mathematicians: Jean Baptiste Biot and Felix Savart.
They also performed experiments similar to Ampere and derived their mathematical
expression by observing the deflections of a magnetic compass. They also concluded
that any current element projects a magnetic field into the space around it. Biot–
Savart's law determines the net magnetic field using the superposition principle and
states that

*"Magnetic fields due to individual small current element are independent of
each other and that the net magnetic field at a point is obtained by vector sum
of individual magnetic field vectors"*

$$\mathbf{B} = \frac{\mu_0}{4\pi} \int_{wire} \frac{I d\mathbf{l} \times \hat{r}}{r^2}. \tag{12.3}$$

? Example

A cylindrical wire with radius R carries an electric current I. Let $I = 60\,\text{A}$ and
$R = 2\,\text{mm}$. Find the magnetic field at the wire surface at a distance of 5 mm using
Ampere's law (Fig. 12.4).

Fig. 12.4 A cylindrical wire carrying current

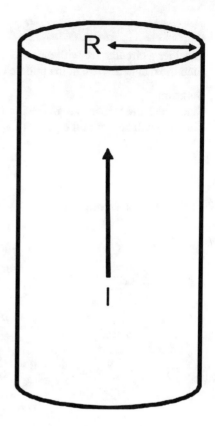

Solution

$$\oint B \cdot dl = \mu_0 I_{enc}$$

$$\int_0^{2\pi R} B dl \cos(\theta) = \mu_0 I_{enc}$$

$$B(2\pi R) = \mu_0 I_{enc}$$

$$B = \frac{\mu_0 I_{enc}}{2\pi R}$$

$$B = \frac{4\pi \times 10^{-7}\,\text{T} \cdot \frac{m}{A} \times 60\,\text{A}}{2\pi \times 2\,\text{mm}} = 6\,\text{mT}.$$

? Example

Consider a circular arc with radius R carrying a current I as illustrated in Fig. 12.5. Find the magnetic field at the middle of the arc using Biot–Savart's Law.

Solution

Since the current in the arc is flowing in anti-clockwise direction, the magnetic field direction will be out of the page. To apply Biot–Savart's law we start with

$$\mathbf{B} = \frac{\mu_0}{4\pi} \int_{wire} \frac{I\mathbf{dl} \times \hat{r}}{r^2}.$$

We can replace $\mathbf{dl} \times \hat{r}$ by $r d\theta$

$$\mathbf{B} = \frac{\mu_0}{4\pi} \int_{wire} \frac{I\mathbf{dl} \times \hat{r}}{r^2}$$

$$\mathbf{B} = \frac{\mu_0}{4\pi} \int_{wire} \frac{I r d\theta}{r^2}$$

$$\mathbf{B} = \frac{\mu_0 I}{4\pi r^2} \int_{wire} d\theta$$

$$\mathbf{B} = \frac{\mu_0 I \theta}{4\pi r^2}.$$

Fig. 12.5 Circular arc carrying current

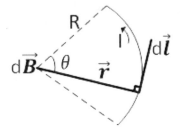

? Example

Figure 12.6 shows a filamentary wire of finite length from $z = a$ to $z = b$. Find the expression for the magnetic field in the xy plane. Next, find the magnetic field when $a \to -\infty$ and $b \to \infty$?

Solution As $I\overrightarrow{dl} = Idz\overrightarrow{a_z}$ and $\overrightarrow{R} = p\overrightarrow{a_p} - z\overrightarrow{a_z}$, so

$$I\overrightarrow{dl} \times \mathbf{R} = I_p dz \overrightarrow{a_\phi}.$$

Substituting in $\mathbf{B} = \frac{\mu_0}{4\pi} \int_c \frac{I\overrightarrow{dl} \times \overrightarrow{R}}{R^3}$, we have

$$\overrightarrow{B} = \frac{\mu_0 I_p}{4\pi} \int_a^b \frac{\overrightarrow{dz}}{(p^2 + b^2)^{3/2}} \overrightarrow{a_\phi}$$

$$= \frac{\mu_0 I_p}{4\pi} \left[\frac{b}{\sqrt{p^2 + b^2}} - \frac{a}{\sqrt{p^2 + a^2}} \right] \overrightarrow{a_\phi}.$$

This result shows that $\overrightarrow{B}$ has a non-zero component only in the $\overrightarrow{a_\phi}$ direction. This is expected because the current is in the z direction, and $\overrightarrow{B}$ must be normal to it.

By setting $a = -\infty$ and $b = \infty$ in the preceding expression, we obtain the $\overrightarrow{B}$ field produced at a point by a wire of infinite extent as

$$\overrightarrow{B} = \frac{\mu_0 I_p}{2\pi p} \overrightarrow{a_\phi}. \tag{12.4}$$

Fig. 12.6 Magnetic field due to a current carrying conductor in a xyz plane

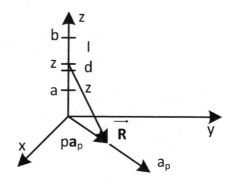

Fig. 12.7 A coaxial cable carrying current

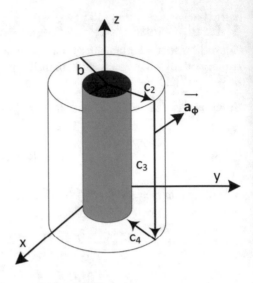

? Example

The inner conductor of a 100-m long coaxial cable of radius 1 cm with current 80 A in the $z-axis$. The outer conductor is very thin and has a radius of 10 cm. Calculate the total flux enclosed within the conductors.

Solution

Since both the inner and outer conductors are closer to each other, we can use Fig. 12.7 for the magnetic vector potential calculation at any point within the cable. The total flux enclosed can be found from $\Phi = \oint_c \overrightarrow{A} \cdot \overrightarrow{dl}$ as

$$
\begin{aligned}
\Phi &= \int_{c1} \overrightarrow{A} \cdot \overrightarrow{dl} + \int_{c2} \overrightarrow{A} \cdot \overrightarrow{dl} \\
&= \frac{\mu_0 I}{2\pi} \int_{-L}^{L} \ln(2L/a) dz - \frac{\mu_0 I}{2\pi} \int_{-L}^{L} \ln(2L/b) dz \\
&= \frac{\mu_0 I L}{\pi} \ln\frac{b}{a}.
\end{aligned}
\tag{12.5}
$$

Substituting the values $I = 80\,\text{A}$, $L = 50\,\text{m}$, $a = 1\,\text{cm}$, and $b = 10\,\text{cm}$ in Eq. (12.5), we get

$$
\Phi = 3.68\,\text{mWb}.
$$

12.4 Magnetic Induction

The discovery in 1820 established that electric current has the tendency to create magnetic field, there started a curiosity among the research community to find whether electric current could similar be created using magnetic field. It however took a decade to prove that magnetic field can also generate electric current. The process was called magneto-induction and the key ingredient that led to the discovery was a change in magnetic field. This was first discovered by Michael Faraday in 1830 during his experiments [4].

12.4.1 Faraday's Law of Magnetic Induction

Michael Faraday performed a couple of experiments to illustrate the phenomena of magnetic induction. In his first experiment, Faraday used a magnet and a coil connected with galvanometer as shown in Fig. 12.8a. He tried to move the magnet closer to the coil and away from the coil while looking for a deflection in the galvanometer. Surprising to his expectation, he did not notice any deflections in the presence or absence of the magnet near the coil. However, by close observation he found that, at the very moment when the magnet is brought closer to the coil, the galvanometer showed a slight deflection. Similarly, at the instance, when the magnet was removed, the deflection was observed in an opposite direction.

Figure 12.8b shows setup for his second experiment, where he took two coils (not connected to each other) wounded around an iron. He connected one coil to a DC battery and tried to observe the current flow on the other end with the help of a galvanometer. The galvanometer would not show any deflection, while the coil

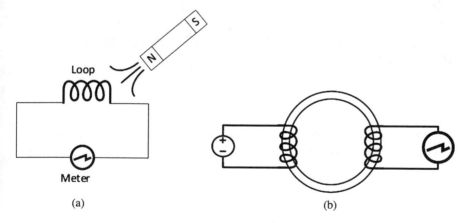

(a)   (b)

Fig. 12.8 Faraday's experiments to observe induction phenomena due to magnetic field. (**a**) Faraday's experiment with a magnet bar. (**b**) Faraday's experiment with a DC circuit

was connected to the battery. But as soon as the battery was disconnected from the coil, galvanometer showed some deflection. Similarly at the instance the battery was connected to the circuit, a deflection in the opposite direction was observed. Based on his experiments, Faraday discovered that a static magnetic flux has no effect on the secondary coil, but a varying magnetic flux does induce an electromotive force (EMF). He further concluded that this induced EMF due to varying magnetic flux allows the current to flow in the circuit. Faraday incorporated all his observations into a single law known as Faraday's law of magnetic induction that states that

"The induced emf in a circuit due to varying magnetic flux is directly proportional to the rate of change of the magnetic flux through that circuit" and is given by

$$EMF = \frac{d\Phi_B}{dt}. \tag{12.6}$$

To conclude, EMF induced in a coil depends on both the magnetic flux through the coil and the rate of change of the magnetic flux. A stronger magnetic field will create more magnetic flux resulting in an increased induced EMF. Similarly, increasing the frequency of the varying magnetic flux will also result in an increased induced EMF.

? Example

Calculate the induced EMF in a solenoid coil with a radius of 0.13 m and 10 number of turns when exposed to a varying magnetic field (ranging from 6 to 3 T in 10 s) at an angle of 60°.

Solution
From Faraday's law we know that EMF induced in a solenoid with N number of turns is given by

$$Emf = N\frac{d\Phi_B}{dt}.$$

Since the magnetic flux varies between $\phi_1 = 3\,\text{T}$ and $\phi_2 = 6\,\text{T}$ we can further simplify the equation as

$$Emf = N\frac{\phi_2 - \phi_1}{dt}.$$

Since $\phi = BA\cos\theta$, replacing and substituting the values

$$Emf = N\frac{B_2 A \cos\theta - B_1 A \cos\theta}{dt}$$

$$= 10\left(\frac{[6\,\text{T} \times \pi(0.13\,\text{m})^2 \times \cos 60] - [3\,\text{T} \times \pi(0.13\,\text{m})^2 \times \cos 60]}{10}\right)$$

$$= 0.06\,\text{V}.$$

? Example

Consider a circular coil ($radius = 40$ cm) lying in a xy plane. If the magnetic field in that area is $\overrightarrow{B} = 0.2\cos 500t\,\overrightarrow{a_x} + 0.75\sin 400t\,\overrightarrow{a_y} + 1.2\cos 314t\,\overrightarrow{a_z}\,T$, find the EMF induced in the coil.

Solution

As the coil is in the xy plane, the differential surface area of the coil is given by

$$\overrightarrow{ds} = \rho d\rho d\phi \overrightarrow{a_z}.$$

The flux passing through this area is

$$d\phi = \overrightarrow{B} \cdot \overrightarrow{ds} = 1.2\rho d\rho d\phi \cos 314t.$$

The total flux linking the loop at any time is

$$\Phi = 1.2\cos 314t \int_0^{0.4} \rho d\rho \int_0^{2\pi} d\phi$$

$$= 0.603\cos 314t\, Wb.$$

Because the flux varies sinusoidally with $\Omega = 314$ rad/s, the frequency of the induced emf is 50 Hz. The maximum value of the flux is 0.603 Wb. Hence the effective value of the induced emf, from $E = 4.44\,f\,N\Phi_m$, is

$$E = 4.44 \times 50 \times 1 \times 0.603$$

$$= 133.866\,\text{V}.$$

? Example

A tightly wound rectangular having N turns is rotating with the varying magnetic field as $B_m \sin \Omega t$, find the induced emf using Faraday's law of induction.

Solution

From Faraday's law we know that EMF induced in a coil with N number of turns is given by

$$Emf = N\frac{d\Phi_B}{dt}.$$

Since magnetic flux can be expressed in the following form

$$\Phi = \int_c \overrightarrow{B} \cdot \overrightarrow{ds}$$

$$= B_m \sin \omega t \cos \omega t \int_s ds$$

$$= \frac{1}{2} B_m A \sin 2\omega t.$$

The induced EMF can then be given by

$$EMF = \frac{1}{2} B_m AN \frac{d}{dt}(\sin 2\omega t)$$

$$= -B_m AN\omega \cos 2\omega t.$$

12.4.2   Lenz's Law

Faraday established the relationship between the flux and the induced emf; however, he did not mention what direction the induced current (due to the induced emf) will flow in the circuit. Lenz (1804–1865) was also independently investigating the induction aspects and he was the first one who predicted the induced current's direction. It is believed that Faraday knew about the current direction too, but it was Lenz who first stated it.

> *"The flow of the induced current in a conductor due to varying magnetic flux will follow the direction that resists a change in magnetic flux"*

(continued)

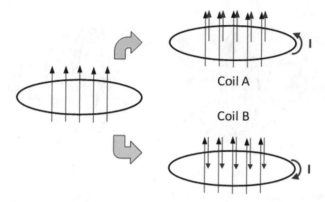

Fig. 12.9 Lenz's law explaining current direction

$$Emf = -\frac{d\Phi_B}{dt}. \tag{12.7}$$

Lenz's law contribution does not only add polarity to Faraday's law but it also illustrates law of conservation of energy as shown in Fig. 12.9. A coil with magnetic flux lines would induce a current in either a clockwise or anti-clockwise direction. Let us assume that for coil A the induced current follows an anti-clockwise direction, while for Coil B the induced current flows in a clockwise direction. Due to the flow of the induced current in Coil A, magnetic field with flux lines pointing in the same direction is produced. These lines add up to the total magnetic flux resulting in more induced current leading to more flux lines in the same direction. An infinite cycle thus continues going against the law of conservation of energy. On the other hand, the induced current in Coil B generates magnetic field with flux lines pointing in the opposite direction. The opposite flux lines cancel each other, reducing the amount of current flow in the coil. Eventually, the circuit becomes static following the law of conservation of energy and this is why Lenz's law to learn about the direction of the induced current is important.

12.5 Mutual and Self Induction

From Ampere's and Faraday's law we know that current flowing in a conductor generates magnetic field, while varying magnetic field induces current. To further extend these concepts and establish quantitative study of relationship between the current and magnetic flux, a study of mutual induction and self induction is presented in the following subsections [5].

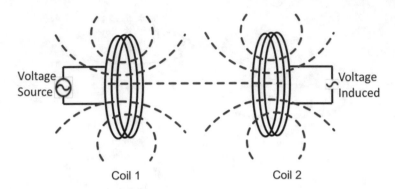

Fig. 12.10 Mutual induction between two coils

12.5.1 Mutual Inductance

Let us take two loops: L_1 and L_2 as shown in Fig. 12.10. If we excite L_1 with an alternating current I_1, a time-varying magnetic field B_1 is produced. The magnetic flux lines Φ_{21} of B_1 passing through L_2 thus induce an emf in L_2. This can help us establish a relationship between magnetic flux and the current that when there is no current in L_1, there is no flux in L_2. Similarly, if the amount of the current in L_1 is increased, the flux in L_2 is also increased. We thus can write the following expression:

$$\Phi_{21} = M_{21} I_1, \tag{12.8}$$

where M_{21} is the proportionality constant and called as the mutual inductance between L_2 and L_1. Similarly, the current flowing in L_2, due to induced EMF, generates magnetic field B_{12}, and proportional to the magnetic flux lines passing through L_1.

$$\Phi_{12} = M_{12} I_2, \tag{12.9}$$

where M_{12} is the proportionality constant and called as the mutual inductance of L_1 with respect to L_2. Furthermore, it has been demonstrated that the magnetic flux linkage between the two loops L_1, L_2 is the same and holds true regardless of the size and relative position of the two loops. Thus $M_{12} = M_{21} = M$, and we can write a general expression for M as

$$M = M_{ij} = \frac{\Phi_{ij}}{I_j}. \tag{12.10}$$

Mutual inductance M can thus be understood as the ratio of linking the magnetic flux as a result of the current flowing through the two loops. The mutual inductance can mathematically be expressed as

$$M = \frac{\mu_0 N_1 N_2 A}{l},\tag{12.11}$$

where μ_o is the magnetic permeability of free space, N_1, N_2 are the number of turns of L_1, L_2, A is the cross sectional area, and l is the length of the loop coils. It can also be understood intuitively that if the two coils are close to each other, almost all the magnetic flux will interact with the other coil and if the two coils are farther away from each other or at different angles, the magnetic flux will be lot lesser. Likewise, if there a bigger cross sectional area is available, more flux will pass through and vice versa.

12.5.2   Self-Inductance

A single coil can also have an inductive effect when a current flows through it. The flow of current I in the coil produces a magnetic field B creating a magnetic flux Φ. The magnitude of the magnetic flux is directly related to the amount of current flowing through it and is given by

$$\Phi = LI,\tag{12.12}$$

where L represents the self-inductance of the coil. Like mutual inductance, self-inductance can also be written as

$$L = \frac{\mu_0 N^2 A}{l},\tag{12.13}$$

where μ_o is the magnetic permeability of free space, A is the cross sectional area, l is the length of the loop, and N_1, N_2 are the number of turns of L_1, L_2.

? Example

If a current of 100 mA flowing through a coil (with 100 turns) produces a magnetic flux of 1.33×10^{-7} Wb, calculate the inductance of the coil. Moreover, if the flux takes 75 ms to reach the maximum level, find the emf induced in the coil.

Solution

$$L = N\frac{\Delta \Phi}{\Delta I}$$
$$= 100\frac{1.33 \times 10^{-7}}{100 \times 10^{-3}}$$
$$= 0.13\,\text{mH}$$

$$EMF = N\frac{\Delta\Phi}{\Delta t}$$

$$= 100\frac{1.33 \times 10^{-7}}{75 \times 10^{-3}}$$

$$= 0.17\,\text{mV}.$$

12.5.3 Inductive and Capacitive Reactance

The coil or loop described in our discussion is nothing but inductors. Following the AC circuit analysis, we can learn in detail about active and passive circuit components, but here we can briefly describe the inductive reactance and its dependence on the frequency. Inductive reactance is given by

$$X_L = 2\pi f L. \tag{12.14}$$

The inductive reactance of an inductor thus is highly dependent on frequency and increases with increase in frequency. This increase in the reactance then reduces the current flowing through it. On the other hand capacitive reactance is the reverse of inductive reactance and is given by (Fig. 12.11):

$$X_C = \frac{1}{2\pi f C}. \tag{12.15}$$

Thus with increase in frequency the capacitive reactance would decrease and allowing more current to flow through it. This therefore can lead to a combination

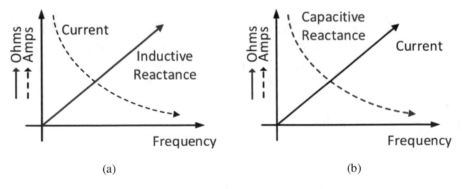

Fig. 12.11 Capacitive and inductive reactance response with frequency. (**a**) Inductive reactance. (**b**) Capacitive reactance

of inductive and capacitive circuits that cancels each other effect and help to resonate at a given frequency. The details will be covered in the next chapter.

Summary

- Magnetic field is the area around a magnet where other magnetic or electric charges experience some magnetic force.
- Magnetic flux is the measure of magnetic field strength.
- The common sources of magnetic field are magnets, ferromagnets, and electric charges.
- Ampere's law is used to calculate magnetic field for current carrying symmetric conductors.
- Biot–Savart's law determines the net magnetic field using the superposition principle and is used for both symmetric and non-symmetric structures.
- Faraday's law states that the induced emf in a circuit due to varying magnetic flux is directly proportional to the rate of change of the magnetic flux through that circuit.
- Lenz's law states that the flow of the induced current in a conductor due to varying magnetic flux will follow the direction that resists a change in magnetic flux.
- Mutual inductance M is the ratio of linking the magnetic flux as a result of the current flowing through the two loops.
- Self-inductance L is the ratio of linking the magnetic flux due to the current flowing in that loop.

? Questions

1. Explain the direction of magnetic field lines.
2. If a thin and long wire ($x = 0$ to $x = \infty$) carries a current I, express the magnetic flux at any point in the $x = 0$ plane.
3. A cylindrical wire with radius R carries an electric current I. Let $I = 60$ A and $R = 2$ mm. Apply Ampere's or Biot–Savart's law to determine the magnetic field at the surface of wire and at a distance of 5 mm above the surface.
4. Derive a magnetic field expression for a solenoid using both Ampere's law and Biot–Savart's Law.

5. Calculate the induced EMF in a solenoid coil with a radius of 0.13 m and 30 number of turns, when exposed to a varying magnetic field (ranging from 6 to 3 T in 10 s) at an angle of 90°.

References

1. X. Che, I. Wells, G. Dickers, et al., Re-evaluation of RF electromagnetic communication in underwater sensor networks. IEEE Commun. Mag. **48**(12), 143–151 (2010)
2. Y. Li, K. Cai, Y. Zhang, et al., Localization and tracking for AUVs in marine information networks: research directions, recent advances, and challenges. IEEE Netw. **33**(6), 78–85 (2019)
3. Z. Sun, I.F. Akyildiz, Magnetic induction communications for wireless underground sensor networks. IEEE Trans. Antennas Propagat. **58**(7), 2426–2435 (2010)
4. S.C. Lin, I.F. Akyildiz, P. Wang, et al., Distributed cross-layer protocol design for magnetic induction communication in wireless underground sensor networks. IEEE Trans. Wirel. Commun. **14**(7), 4006–4019 (2015)
5. N. Golestani, M. Moghaddam, Theoretical modeling and analysis of magnetic induction communication in wireless body area networks (WBANs). IEEE J. Electromag. RF Microwaves Med. Biol. **2**(1), 48–55 (2018)

Chapter 13
MI Communication System

13.1 Introduction

Figure 13.1 shows a block diagram of how MI communication is achieved between transmitter (Tx) and receiver (Rx) [1]. The data to be sent is first input to a communication block where necessary functions are applied. The prepared data is then forwarded to a matching network part, where electronic circuits generate the required varying waveform. As the varying waveform is applied to the MI-Tx coil, a varying magnetic field is generated around it. This varying magnetic field then induces an emf in the Rx coil that is translated back to a pattern with the help of a matching network. The communication block at Rx then recovers the transmitted data from this pattern. We thus define three building blocks (Coils, Matching Network, and Communication Block) for our MI communication system and explain them in detail in the following sections.

13.2 First Part: Coils

An alternating current flowing in a closed loop creates a varying magnetic field. This closed loop can be any conducting wire forming a complete electric circuit. In EM communication systems, we usually call the transmit and receive front end devices as *antennas* [2]. Antennas come in different sizes and shapes with different radiating patterns. Similarly, in acoustic communication, *transducers* are used to transmit and receive the acoustic signal. In case of MI communication we use a simple wounded copper wire (in circular or rectangular pattern) at both Tx and Rx ends. At the transmit side the coil generates a varying magnetic field and at the receive side an emf is induced in the coil after detecting the generated varying field. Both Tx and Rx phenomena are thus presented next.

© The Author(s), under exclusive license to Springer Nature Switzerland AG 2022   315
Y. Lou, N. Ahmed, *Underwater Communications and Networks*, Textbooks in
Telecommunication Engineering, https://doi.org/10.1007/978-3-030-86649-5_13

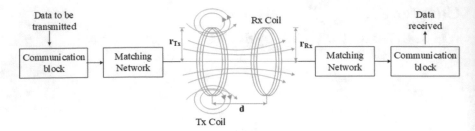

Fig. 13.1 This is an end-to-end communication system that clearly shows three main parts, starting with coils, matching network, and the communication blocks

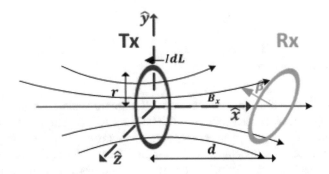

Fig. 13.2 Magnetic field generated by Tx coil and induced in Rx coil

13.2.1   Magnetic Field Generated by Circular Coils

When a current is applied to a coil, a magnetic field is generated around it (Fig. 13.2). To calculate the magnetic field on the axis, we apply Biot–Savart's law as

$$d\mathbf{B}_d = \frac{\mu_o I d\mathbf{L} \times \mathbf{y}}{4\pi a^2} = \frac{\mu_o I d\mathbf{L} \sin\theta}{4\pi a^2} \tag{13.1}$$

$$\sin\theta = \sqrt{d^2 + r^2} \tag{13.2}$$

$$d\mathbf{B}_d = \frac{\mu_o I d\mathbf{L} r}{4\pi (d^2 + r^2)^{\frac{3}{2}}}. \tag{13.3}$$

Similarly after integrating along dL, the magnetic field becomes

$$\mathbf{B}_d = \frac{\mu_o I 2\pi r^2}{4\pi (d^2 + r^2)^{\frac{3}{2}}}, \tag{13.4}$$

where μ_o is the magnetic permeability constant, I is the current flowing through the coil, and r is the radius of the coil. With the increase in distance between the Tx and

Rx coils such that $d \gg r$, it can be shown that magnetic field decays with distance as $1/d^3$

$$\mathbf{B}_{d,x} = \frac{\mu_o I 2\pi r^2}{2\pi d^3}\hat{\mathbf{x}}.$$ (13.5)

Also, with N number of turns, magnetic field will increase by N times

$$\mathbf{B}_{d,x} = N\frac{\mu_o I 2\pi r^2}{2\pi d^3}\hat{\mathbf{x}}.$$

? Example

A coil of radius $r = 10cm$ has a magnetic field of $B = 50 \times 10^{-3}$T at its center, when an alternating current of $I = 30$ mA is applied across its ends. Find the distance from the coil along its axis when the magnetic field reduces to $B = 5 \times 10^{-3}$T

Solution

$$B = \frac{\mu_o I 2\pi r^2}{2\pi (x^2 + r^2)^{\frac{3}{2}}}$$

$$(x^2 + r^2)^{\frac{3}{2}} = \frac{\mu_o I 2\pi r^2}{2\pi B}$$

$$(x^2 + (0.1\text{m})^2)^{\frac{3}{2}} = \frac{4\pi \times 10^{-7}\text{T}\frac{\text{m}}{\text{A}} \times 0.030\text{A} \times 2\pi \times (0.1\text{m})^2}{2 \times \pi \times 5 \times 10^{-3}\text{T}}$$

$$(x^2 + (0.1\text{m})^2)^{\frac{3}{2}} = 7.53 \times 10^{-8}\text{m}^3$$

$$x = 90\,\text{mm}.$$

13.2.2   Magnetic Moment

While discussing the transmitting magnetic field of a coil, it is also important to learn the concept of magnetic moment. Magnetic moment is an important and simpler measure of magnetic field strength and is given by

$$M_M = NIA,$$ (13.6)

where N is the number of turns, I is the current flowing through the coil, and A is the area of the coil. By easily modifying these simple parameters of the coil, the magnetic field strength can either be increased or decreased.

13.2.3   Voltage Induced in the Receive Coil

When a receive coil is exposed to a varying magnetic field, induction takes place depending on the amount of flux passing through the receive coil. We know that magnetic flux is

$$\Phi = \int \mathbf{B}.d\mathbf{A} \tag{13.7}$$

Applying Faraday's law of induction, the voltage induced at Rx coil is

$$V = \frac{d\Phi}{dt} \tag{13.8}$$

Since $\frac{d\Phi}{dt}$ is the rate of change of flux, we may also write it as

$$V = \omega\Phi, \tag{13.9}$$

where ω defines the change in Φ. Substituting Eq. (13.7) in Eq. (13.9), we get

$$V = \omega \mid B \mid A \cos(\beta) \tag{13.10}$$

with β as the angle between the magnetic field ($\mathbf{B}$) and normal to the Rx coil plane (Fig. 13.2). With N number of turns the voltage will be increased by a factor of N.

$$V = N\omega \mid B \mid A \cos(\beta). \tag{13.11}$$

13.2.4   Directivity Pattern

Directivity is an important metric for antennas and is a measure of communication robustness and reliability in a three-dimensional space. The directivity pattern of a single coil is shown in Fig. 13.3. It can be seen that due to directional nature of magnetic fields, the directive pattern of a single coil is quite directional. The maximum gain is along the axis of the coil (at $0°$), and the minimum gain is along the normal (at $90°$).

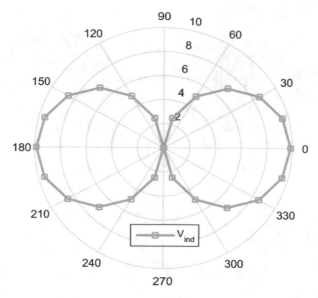

Fig. 13.3 Directivity pattern of a single dimensional coil

13.3 Second Part: Matching Network

We consider a matching network, an important part of our block diagram as it significantly influences the magnetic induction process. Depending on the electronic configuration used in a matching network, magnetic induction can either be (a) pure induction or (b) resonance induction [3].

Figure 13.4 shows a circuit equivalent representation of a pure induction mechanism where the coil alone (that acts as an inductor) is utilized on both transmit and receive sides. Pure induction strictly depends on the magnetic flux density produced by the transmit coil that enters the receive coil to establish a communication link.

Figure 13.5 shows a circuit equivalent representation of resonance induction, where the Tx/Rx coils are also loaded with capacitors to form a tuned LC circuit. Resonance induction depends on magnetic field density as well as the resonance frequency that resonates both the coils and is therefore more efficient.

Furthermore, to implement an efficient MI communication system with resonance induction, the transmit and receive sides use a series and parallel LC combination, respectively. The benefits of the two configurations are explained in the following sections.

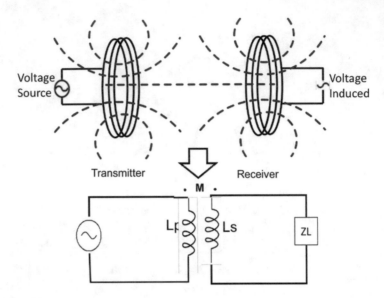

Fig. 13.4 Pure induction

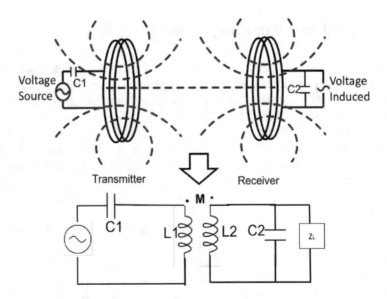

Fig. 13.5 Resonance induction where capacitors are introduced in the equivalent circuit

13.3.1 Transmit Coil with Series RLC Configuration

The transmit side uses the added capacitor in series to the coil and forms a series RLC circuit. In a series circuit, the current flows in only one path and the total impedance of the circuit is given by

$$Z = R + X_C + X_L. \tag{13.12}$$

To generate a stronger magnetic field in the Tx coil, a maximum amount of current is required. To allow maximum current to flow, the total impedance needs to be minimum. As inductors and capacitors are active components, their reactance (X_L and X_C) depends on the operating frequency. When frequency is higher, X_L is larger, while X_C is smaller, and when frequency is lower, X_L is smaller, while X_C is larger. Thus there exists a frequency point where X_L becomes equal to X_C and this point is called as the resonance frequency point. We can find this point by

$$X_L = X_C \tag{13.13}$$

$$2\pi f L = \frac{1}{2\pi f C} \tag{13.14}$$

$$f = \frac{1}{2\pi \sqrt{LC}}. \tag{13.15}$$

At this resonance frequency inductive and capacitive reactance cancels each other's effect leaving only the resistive element behind in the circuit as shown in Fig. 13.6a. The circuit impedance at this point is minimum, allowing maximum current to flow in the series circuit as shown in Fig. 13.6b. If the frequency is lower than the resonance frequency, the circuit will act as pure capacitive circuit. Similarly, if the frequency is higher than the resonance frequency, the circuit will act as pure inductive circuit.

13.3.2 Receive Coil with Parallel RLC Configuration

The receive side uses the capacitor in parallel with the coil to form a parallel RLC circuit, where the voltage induced will be similar in all the branches. In a parallel circuit, the combined impedance is given by (Fig. 13.7)

$$Z = \frac{1}{R} + \frac{1}{X_C} + \frac{1}{X_R}. \tag{13.16}$$

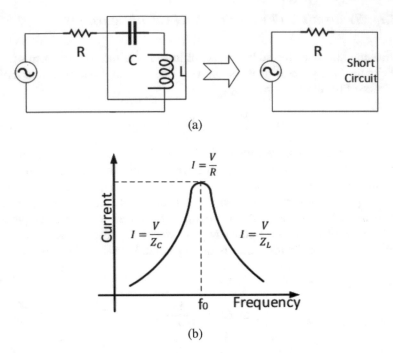

Fig. 13.6 Behavior of series resonance circuit at resonance frequency. (**a**) Series RLC circuit at resonance point. (**b**) Frequency response curve of series RLC circuit

Unlike the series RLC circuit, the impedance in parallel RLC circuit is maximum at resonance point $(X_L = X_C)$ allowing the receiver coil to sense a slight change in voltage.

? Example

An experimental setup with a resonance inductive mechanism is made to test a communication link. The coils on both transmit side and receive side have same radius and number of turns. The inductance of both coils is measured with the help of a LCR meter as $L_{tx} = L_{rx} = 340\,\text{mH}$. To establish the communication between the two coils at 125 KHz, what value of capacitor will be added on both transmit and receive side?

Solution
We can start with the resonance frequency as

$$f = \frac{1}{2\pi\sqrt{LC}}$$

$$\sqrt{C} = \frac{1}{2\pi f\sqrt{L}}$$

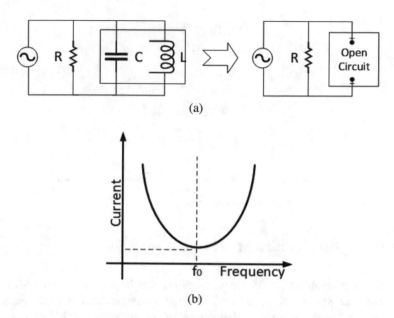

Fig. 13.7 Behavior of parallel resonance circuit at resonance frequency. (a) Parallel RLC circuit at resonance point. (b) Frequency response curve of parallel RLC circuit

$$C = \frac{1}{4\pi^2 f^2 L}$$

$$C = 4.7\,pF.$$

13.4 Third Part: Communication Block

Communication block is an essential part in ensuring a smooth transfer of data from one end (Tx) to the other end (Rx). Data can either be in the form of analog or digital waveform as shown in Fig. 13.8. An analog signal continuously varies with time, while a digital signal maintains its state for a given time period. Implementation and performance of MI system are different for both analog and digital communication and are explained in the following subsections [4, 5].

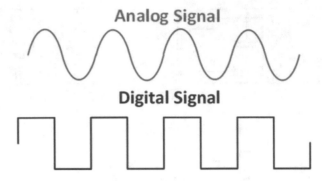

Fig. 13.8 Analog and digital communication signals

13.4.1   Digital Signals and MI Communication

This section presents an intuitive understanding and implementation of MI digital communication system. To understand MI digital communication, recall Faraday's experiment of induction when he first used a DC power supply to induce current in a neighboring circuit. From his experiment, he concluded that while the DC power supply creates a magnetic field that does not induce any current in the neighboring circuit. However, he observed that at the instant when the circuit is turned ON/OFF, current (for that short instant) is induced in the circuit. Same happens in case of MI digital communication as shown in Fig. 13.9. As a digital signal retains its level ("High for 1" and "Low for 0") for the entire bit duration, no current will be induced during this time. However, the change of level during the transition from a High/Low level to a Low/High level can be detected by the receiver end.

As shown in Fig. 13.9, the receiver can recover the data if transmitting sequence is alternating such as 101010; however, the data cannot be recovered successfully if the transmitting sequence contains consecutive ones (11) or zeros (00). This can lead to a serious faulty communication system, and in order to avoid such a scenario, implementation of an encoding (Manchester or Differential Manchester encoding) scheme is must to replace any consecutive sequence.

13.4.1.1   Manchester and Differential Manchester Encoding

Figure 13.10 shows representation of Manchester and differential Manchester encoding schemes. In both coding schemes, the transition is done with the transition of the clock signal. In case of Manchester coding, for input bit 1, the transition is made from High to Low, whereas for input bit 0, the transition is made from Low to High. On the receive side, decoder detects the transition direction and decodes the transmitted data. However, in case of magnetic induction it is not possible to

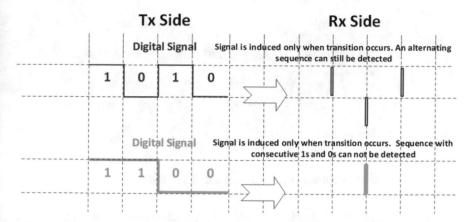

Fig. 13.9 MI digital communication

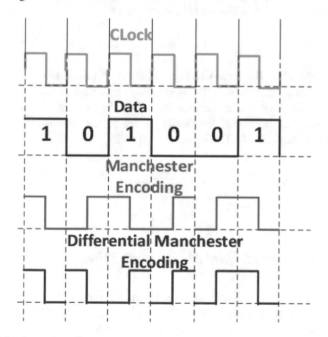

Fig. 13.10 Manchester encoding

simply detect the transition direction so the decoder will need to apply some novel techniques. Differential Manchester encoding on the other hand is modified form of Manchester encoding. In differential Manchester encoding, for each input bit 1, the transition is changed, while for each input bit 0, the previous transition is retained.

Tx Side   **Rx Side**

Analog Signal   Signal is induced as the signal at transmitter is varied

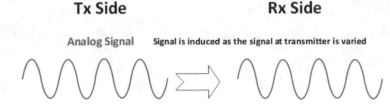

Fig. 13.11 MI analog communication

13.4.2   Analog Signals and MI Communication

Unlike the digital MI communication, analog communication is simpler and easier to implement as analog signal continuously varies with time. An analog signal is commonly represented by a sinusoidal signal such as

$$x(t) = A \sin 2\pi f t,$$

where $x(t)$ represents the time-varying signal, A is the amplitude of the signal, and f is the operating frequency. As previously discussed, MI communication can be both pure inductive and resonance inductive. In case of resonance inductive communication, the operating frequency is set to the resonance frequency for maximum efficiency (Fig. 13.11).

$$f = \frac{1}{2\pi\sqrt{LC}}. \tag{13.17}$$

13.4.3   Analog and Digital Modulation Schemes

As analog baseband communication is band-limited, analog modulation is required for better bandwidth and high data rates utilization. Analog modulation uses a carrier wave to modulate the analog signal that is given by

$$x_c(t) = A_c \sin(2\pi f_c t + \phi_c), \tag{13.18}$$

where A_c is the amplitude of the carrier, f_c is the carrier frequency (set to the resonance frequency), and ϕ_c is the phase of the carrier. Depending on variation of these three parameters three types of modulations can be performed (Fig. 13.12). Furthermore, it can be noted that in case of frequency modulation, significant changes in the matching network would be required to match C with the coil inductance L and resonance frequency.

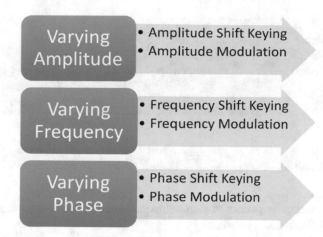

Fig. 13.12 Basic communication block diagram

Like analog modulation, digital modulation also requires the use of a carrier wave to represent the digital data (1 and 0). By using different amplitude, frequency, or phase for one and zero, the three modulation schemes can be implemented. The most common digital modulation scheme used in MI communication systems is On–Off Keying (OOK), which is a simpler form of amplitude shift keying.

Summary

- The three main parts of MI communication system are coils, matching network, and communication block.
- MI coils are made of simple copper wire wounded in a circular or rectangular structure.
- The directivity patter of MI coils is not omnidirectional.
- The matching network can either be pure inductive or be resonance inductive.
- Pure induction uses MI coil only at Tx and Rx end.
- Resonance induction MI coil with added capacitors at both Tx and Rx end.
- At Tx end, resonance induction uses the capacitor in series, while at Rx end, resonance induction uses the capacitor in parallel.
- MI digital communication system is challenging and will require encoding/decoding schemes.
- MI analog communication system is simple and easy to implement.

(continued)

- Both analog and digital modulation schemes are easy to implement in a MI communication system.
- OOK is the most common and simple modulation scheme for MI systems.

? Questions

1. Show the directivity pattern of single dimensional coil and also explain its effect in a three-dimensional network?
2. Consider a coil producing a magnetic moment of $10\,\mathrm{Am^2}$ with 5 A flowing in the coil. How would you modify the coil to get magnetic moment of $20\,\mathrm{Am^2}$?
3. Explain the difference between pure induction and resonance induction. Also enlist the advantages and disadvantages of both induction mechanisms?
4. Explain the advantage of series resonance circuit for MI transmitter and parallel resonance circuit for MI receiver.
5. Explain why digital communication in MI communication system is challenging and how can it be achieved?

References

1. I.F. Akyildiz, P. Wang, Z. Sun, Realizing underwater communication through magnetic induction. IEEE Commun. Mag. **53**(11), 42–48 (2015)
2. L. Yan, D. Wei, M. Pan, et al., Downhole wireless communication using magnetic induction technique, in *2018 United States National Committee of URSI National Radio Science Meeting (USNC-URSI NRSM)* (IEEE, Piscataway, 2018), pp. 1–2
3. Z. Sun, I.F. Akyildiz, Magnetic induction communications for wireless underground sensor networks. IEEE Trans. Antennas Propag. **58**(7), 2426–2435 (2010)
4. A.K. Sharma, S. Yadav, S.N. Dandu, et al., Magnetic induction-based non-conventional media communications: a review. IEEE Sensors J. **17**(4), 926–940 (2016)
5. M.C. Domingo, Magnetic induction for underwater wireless communication networks. IEEE Trans. Antennas Propag. **6**(60), 2929–2939 (2012)

Chapter 14
MI Channel Characteristics

14.1 Uniqueness of MI Channel

MI transmitter is surrounded by magnetic field in the form of a bubble. To establish a communication link with the MI transmitter, MI receiver has to be inside this magnetic field bubble, otherwise the communication link cannot be established. Thus to understand magnetic channel it is important to realize characteristics of the magnetic field bubble first.

We define magnetic field bubble as the *region where the magnetic field generated by the transmitter coil exists*. Magnetic fields do not propagate in space and are quasi static fields. Intuitively magnetic fields can be thought of eternal part of the magnetic body which do not travel. Moreover, strength of the magnetic field in the bubble is not uniform, rather it depends on how close are the magnetic flux lines. Since the flux lines are closer near the transmitter, the magnetic field is stronger near the transmitter, and as we move away from the transmitter, the flux lines have more spacing resulting in a weaker field. This uniqueness of magnetic fields leads to following interesting facts [1, 2].

14.1.1 Propagation Speed

Propagation speed is a property associated with propagation waves and is defined as the ratio of the distance traveled in a given time as:

$$Speed = \frac{Distance}{Time} \tag{14.1}$$

The propagation speed of EM waves is 3×10^8 m/s, and propagation speed of acoustic waves is 343 m/s in air, while 1480 m/s in water. On the other hand,

© The Author(s), under exclusive license to Springer Nature Switzerland AG 2022
Y. Lou, N. Ahmed, *Underwater Communications and Networks*, Textbooks in Telecommunication Engineering, https://doi.org/10.1007/978-3-030-86649-5_14

since magnetic fields are non propagating waves, it does not have any propagation speed. Once a magnetic field is generated, it co-exists in the entire bubble. The establishment of communication link for MI systems is thus instant, however, there may be a processing delay at both transmitter and receiver end.

14.1.2   Multipath and Doppler Effect

Multipath and Doppler effects are two important challenges of a communication channel. Like propagation speed, multipath, and Doppler effects are also associated with wave propagation. Multipath is a phenomenon when a wave reaches the receiver end after traveling through different routes, because of reflection, refraction, or scattering. The number of paths and the time a wave takes to reach the receiver reflects on complexity of a communication system. Since magnetic fields do not propagate, there are no multipaths and therefore requires a simple communication system.

Doppler effect in a communication system is the change of frequency that arises due to relative movement of either the transmitter or the receiver. In case of MI communication system, since frequency is the measure of how much a magnetic field is varying, it is not affected with the movement of either the transmitter or receiver. Hence there is no Doppler effect in MI communication system.

14.2   MI Channel Performance

The behavior of MI channel depends on the type of magnetic induction phenomenon utilized in MI communication system. This section thus presents the performance of MI channel with respect to both pure induction and resonance induction phenomenon [3–5].

14.2.1   Pure Inductive Communication Channel

In case of pure induction, the performance of MI communication channel is analyzed by the coupling coefficient. Coupling coefficient is the measure of how strong transmit and receive coils are linked with each other. A stronger coupling coefficient means that the transmitted information is less effected by the channel, while weak coupling coefficient means that the transmitted information is more effected by the channel. The coupling coefficient (κ) can be expressed as

$$\kappa = \frac{M_{Tx,Rx}}{L_{Tx}L_{Rx}}, \tag{14.2}$$

where $M_{Tx,Rxj}$ is the mutual inductance between the Tx and Rx coils, while L_{Tx} and L_{Rx} are the self-inductance of Tx and Rx coils, respectively. The value of coupling coefficient ranges between 1 -1, that is, $\kappa = [-1, 1]$. A value of $\kappa = 1$, implies that all the magnetic flux produced by the Tx coil passes through the Rx coil and thus induces the maximum current in the Rx coil. A lesser value of κ, implies that a smaller part of magnetic flux passes through the Rx coil while other part of the magnetic flux is not utilized by the Rx coil, reducing the overall efficiency. Thus in case of pure induction, the Tx and Rx coils need to be closer and perfectly aligned with each other for better efficiency.

14.2.2 Resonance Inductive Communication Channel

In case of resonance induction, the performance of MI communication channel is analyzed by the coupling coefficient as well as quality factor (Q). Quality factor defines the resonance behavior of MI systems. A high Q means resonating with higher amplitude, and a lower Q means resonating with a lower amplitude. Moreover, the Q factor is inversely related with the bandwidth of the system. A higher Q refers to a narrow bandwidth system while a lower Q refers to a wide bandwidth system. Since MI communication utilizes series resonance circuit on transmitter side and a parallel resonance circuit on receiver side, we therefore explain them separately.

14.2.2.1 Q Factor and Bandwidth of MI Transmitter

MI transmitter uses a series resonance circuit with resonance frequency

$$\omega_0 = \frac{1}{\sqrt{LC}}. \tag{14.3}$$

A series resonance circuit has minimum impedance and maximum current in the circuit as

$$I_{max} = \frac{V}{Z_{min}} = \frac{V}{\sqrt{R^2 + (\omega_0 L - \frac{1}{\omega_0 C})^2}}. \tag{14.4}$$

To find the bandwidth, we need to find the lower and higher cut off frequencies at which the power cuts half as compared to the power at the resonance frequency.

$$\omega_{lower} = -\frac{R}{2L} + \sqrt{\left(\frac{R}{2L}\right)^2 + \frac{1}{\omega_0^2}} \tag{14.5}$$

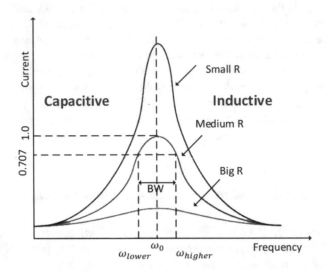

Fig. 14.1 The relation of bandwidth and quality factor for series resonance circuit

$$\omega_{higher} = \frac{R}{2L} + \sqrt{\left(\frac{R}{2L}\right)^2 + \frac{1}{\omega_0^2}}. \tag{14.6}$$

The transmitter bandwidth can then be written as

$$B_{series} = \omega_{higher} - \omega_{lower} = \frac{R}{L}. \tag{14.7}$$

The Q factor of the transmitter circuit is thus

$$Q = \frac{\omega_0}{B_{series}} = \frac{\omega_0 L}{R}. \tag{14.8}$$

Figure 14.1 illustrates the relationship of the Q factor, bandwidth, and impedance of a series resonance circuit. It can be seen that the curve with high Q (more amplitude) is sharper (narrow bandwidth) when the resistance of the circuit is smaller. Similarly, as the resistance increases, the curve gets wider (more bandwidth) and Q gets lower.

14.2.2.2   Q Factor and Bandwidth of MI Receiver

MI receiver uses a parallel resonance circuit with resonance frequency

$$\omega_0 = \frac{1}{\sqrt{LC}}. \tag{14.9}$$

Like the calculation for series resonance circuit, to find the bandwidth of parallel resonance circuit, we need to find the lower and higher cut off frequencies at which the power cuts half as compared to the power at the resonance frequency.

$$\omega_{lower} = -\frac{1}{2RC} + \sqrt{\left(\frac{1}{2RC}\right)^2 + \frac{1}{\omega_0^2}} \tag{14.10}$$

$$\omega_{higher} = \frac{1}{2RC} + \sqrt{\left(\frac{1}{2RC}\right)^2 + \frac{1}{\omega_0^2}} \tag{14.11}$$

The receiver bandwidth can then be written as

$$B_{parallel} = \omega_{higher} - \omega_{lower} = \frac{1}{RC}. \tag{14.12}$$

The Q factor of the receiver circuit is thus

$$Q = \frac{\omega_0}{B_{parallel}} = \omega_0 RC. \tag{14.13}$$

Like the series resonance circuit, the parallel resonance circuit also have similar response. A higher Q has less bandwidth and a lower Q has more bandwidth.

? Example

For a resonance inductive channel, calculate the bandwidth and Q factor of MI transmitter and receiver when a 50 Ω circuit is loaded with a capacitance of 5.9 nF. Assume that both the Tx and Rx coils have same radius of 17 cm and resonate at $\omega_0 = 100$ KHz.

Solution
For calculation of MI transmitter bandwidth (B_{series}), first calculate the inductance L from Eq. (14.3)

$$L = \frac{1}{\omega^2 C} = \frac{1}{(100 \times 10^3)^2 \times (5.9 \times 10^{-9})} = 16.9 \, \text{mH}.$$

For MI transmitter bandwidth and Q factor:

$$B_{series} = \frac{R}{L} = 2.9 \, \text{kHz}$$

and

$$Q = \frac{B_{series}}{\omega_0} = 0.029.$$

For MI receiver bandwidth and Q factor:

$$B_{parallel} = \frac{1}{RC} = 3.3\,\text{MHz}$$

and

$$Q = \frac{B_{parallel}}{\omega_0} = 33.$$

14.3   Factors Affecting MI Communication

MI communication channel is stable and predictable as compared to acoustic and optical communication channels. The instant speed enables MI signals with no delay, no Doppler effect and no multipath. Moreover possibility of high operating frequency enables MI communication system to transmit information at a higher data rate. However, there are still some factors that affect MI communication such as skin effect, background noises and existence of ferromagnetic materials between MI transceiver effect the MI communication [6, 7].

14.3.1   Skin Effect

Skin effect is frequency dependent and happens because of eddy currents. The eddy current arises due to the time-varying magnetic flux generated by the MI transmitter. Skin effect is often measured in terms of skin depth which is given by δ

$$\delta = \sqrt{\frac{2}{\omega\mu_0\sigma}}, \tag{14.14}$$

where ω is the operating frequency, and σ is media conductivity. Since the conductivity of sea water is higher, therefore, skin effect will be dominant in an ocean environment. To avoid skin effect, the operating frequency needs to be reduced for the MI systems which in-turn can decrease the overall data rate of the system.

14.3.2 Background Noise

Like every system, MI systems are also affected from background noises. Background noise can be natural or man-made (due to switching power supplies and generators etc.). To avoid background noise in MI communication, use of higher frequencies is recommended as noise distribution is inversely related to frequency. Higher frequency refers to lower noise and lower frequency refers to higher background noise.

14.3.3 Presence of Ferromagnetic Materials

EM waves depend on permittivity and permeability of the medium, and since permittivity of water is higher than air medium, EM waves attenuate at faster rate in water. On the other hand, magnetic field depends on the permeability of the medium only. Since permeability of water is the same as of air, MI performs similar in air as well as water. However, presence of materials with higher permeability may have significant effect on the communication. Thus it is important to consider the presence of ferromagnetic materials near MI transceiver while deployment in practical applications.

14.4 Path Loss Calculation

As MI systems follow transformers like principle, we therefore use transformer (Fig. 14.2a) and two-port network (Fig. 14.2b) models to calculate the power losses between a transmitter and receiver.

In case of transformer model, M represents the mutual induction between the Tx and Rx coils; V_s and V_{ind} denotes the input voltage and induced voltage, respectively; Z_{Tx} is the self-impedance of Tx coil, while Z_L represents load impedance; Z'_{Tx} is the effect of Rx coil on the Tx coil; and Z'_{Rx} is the effect of Tx coil on the Rx coil.

In case of two-port network model, M represents the mutual induction between the transmit and receive coil; V_s and V_{ind} denotes the input voltage and induced voltage, respectively; I_1 and I_2 are the input and output currents, respectively, $Z_{Tx} = R - Tx + jX_{Tx}$ is the internal impedance of transmit coil, while $Z_L = R_L + jX_L$ represents load impedance; Z_{11} and Z_{22} are the self-impedance; and Z_{12} and Z_{21} represents mutual impedance.

To derive the path loss expression, it is assumed that MI transceiver antennas/coils have no losses and transmit and receive coils are separated by distance d from each other. The Tx power and the Rx power can be written as

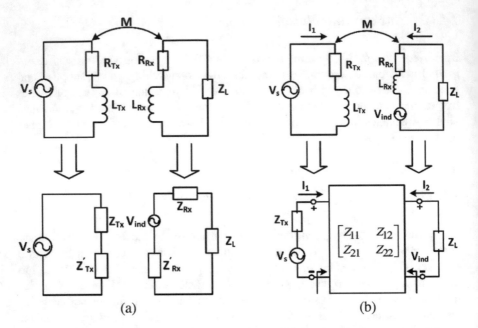

Fig. 14.2 MI equivalent circuit model. (**a**) Transformer model. (**b**) Two-port network model

$$P_{Tx}(d) = \text{Re}\left(V_s I_s^*\right)$$

$$= \text{Re}\left(Z_{11} - \frac{Z_{12}^2}{Z_L + Z_{22}}\right)|I_s|^2 P_{Rx}(d)$$

$$= |I_{ind}|^2 \text{Re}(Z_L)$$

$$= \text{Re}(Z_L)\frac{|Z_{12}|^2}{|Z_L + Z_{22}|^2}|I_s|^2. \tag{14.15}$$

For a very short transmission range d_0 between the Tx and Rx coil, the transmit power is $P_{T_x}(d_0) = \text{Re}(Z_{11})|I_s|^2$. Consequently

$$PL_{MI} = -10\log\frac{P_{Rx}(d)}{P_{Tx}(d)}$$

$$= -10\log\frac{R_L\left(R_{12}^2 + X_{12}^2\right)}{R_{11}\left[(R_L + R_{22})^2 + (X_L + X_{22})^2\right]}. \tag{14.16}$$

The received power can be maximized by carefully designing the load impedance Z_L. For high received power, the value of Z_L can be set equal to the complex conjugate of the input impedance at the secondary port. Furthermore, to lower the power, the value of transmit and receive coil resistances should be kept low.

Assuming that the internal impedance is approximately equal to zero, i.e., $Z_{Tx} \approx 0$, the load impedance can be written as

$$Z_L = R_{R_x} + \frac{\omega^2 M^2 R_{Tx}}{R_{Tx}^2 + \omega^2 L_{Tx}^2} + j \left(\frac{\omega^3 M^2 L_{Tx}}{R_{Tx}^2 + \omega^2 L_{Tx}^2} - \omega L_{Rx} \right), \qquad (14.17)$$

where M is the mutual inductance between Tx and Rx coils, which can be expressed as

$$M = \frac{\mu \pi N_{Tx} r_{Tx}^2 N_{Rx} r_{Rx}^2}{2 \sqrt{\left(r_{Tx}^2 + d^2 \right)^3}} \qquad (14.18)$$

with $\mu = \mu_o \times \mu_r$ is the magnetic permeability, while $\mu_o = 4\pi \times 10^7$ H/m is the magnetic permeability constant and $\mu_r = 1$ is the relative permeability of water. N_{Tx}, N_{Rx}, and r_{Tx}, r_{Rx} represent the number of turns and radius of the Tx and Rx coils, while d is the distance between Tx and Rx coil. The path loss of an MI communication system model can be re-expressed as

$$PL_{MI} = -10 \log \frac{P_{Rx}}{P_{Tx}}$$

$$= -10 \log \frac{R_L \omega^2 M^2}{R_{Tx}(R_L + R_{Rx})^2 + R_{Tx}(X_L + \omega L_{Rx}^2)}. \qquad (14.19)$$

14.4.1   Effect of Underwater Channel on Path Loss

MI communication in air and water is similar due to the same permeability of both the mediums, hence, path loss calculations can also be used for underwater channel. However, in case of sea water, eddy current contributes to more losses because of the conductive nature of sea water. Therefore, for sea water the path loss calculations incorporate losses due to eddy current as

$$PL_{EddyCurrent} = 20 \log(e^{\alpha d}) = 8.69 \alpha d, \qquad (14.20)$$

where $\alpha = 1/\delta = \sqrt{\pi f \mu \sigma}$ is the attenuation and is inverse of the skin depth δ, while σ represents the conductivity of sea water. Hence, the total path loss in case of sea water is the combination of the MI path loss (Eq. (14.19)) and attenuation path loss due to eddy current (Eq. (14.20)).

$$PL_{UMIC-total} = PL_{MI} + PL_{MI-EC}. \qquad (14.21)$$

Figures 14.3 and 14.4 plot the path loss for both fresh and sea water with varying area and operating frequency for the MI coils. An obvious increase in path loss with distance can be seen for both fresh and sea water, however, the path loss for sea water is more. An improvement in path loss can also be seen with bigger radii and low frequencies.

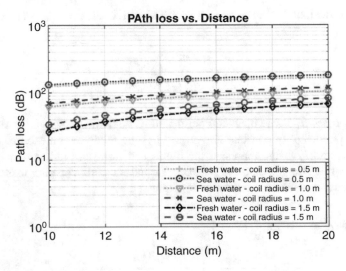

Fig. 14.3 Path loss vs. communication distance with varying coil radius

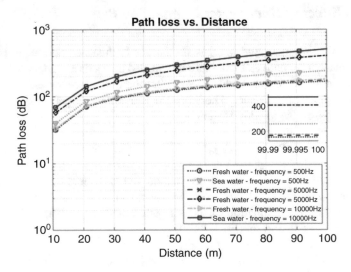

Fig. 14.4 Path loss vs. communication distance with varying frequency

Summary

- In pure inductive communication, the performance of the MI system is mainly analyzed by the coupling coefficient: stronger the coupling coefficient, less path loss will be experienced, hence, strong magnetic field strength will be achieved at the receiver.
- In resonance inductive communication channel, both coupling coefficient as well as quality factor are considered for the performance evaluation of the MI system.
- Higher quality factor means lower bandwidth whereas lower quality factor means higher bandwidth.
- The factors effecting MI communication considered are skin effect, background noises and existence of ferromagnetic materials.

? Questions

1. Consider a resonance induction circuit where a capacitor of 4.7 nF is added to the coil with self-inductance 580 μH. Assume the circuit resistance as 50 Ω on both transmit and receive side. If both circuits resonate at $\omega = 125$ KHz, calculate quality factor for both transmit and receive configurations.
2. Calculate the mutual inductance between two coils of radius 0.50 m and 1.5 m, respectively. The number of turns for Tx and Rx coils are 500 and 1000, respectively, while the communication range is 1 km.
3. Find the maximum communication range if the mutual inductance/coupling between the MI Tx and Rx is 5 dB. The radius of Tx and Rx coils is 1.5 m and 2.5 m, respectively, with same number of turns (500).
4. The conductivity of fresh water and sea water is 0.01 S/m and 4 S/m, respectively. Find out the total path loss for a communication range of 200 m when operating frequency of 1000 Hz is used. The number of turns and radius of the transmit and receive coils are kept same as 1000 and 1.5 m, respectively. Also assume the coil resistance in both Tx and Rx as 50Ω, while the self-inductance in both Tx and Rx is equal to 0.086 H.

References

1. M.C. Domingo, Magnetic induction for underwater wireless communication networks. IEEE Trans. Antennas Propag. **6**(60), 2929–2939 (2012)

2. Z. Sun, I.F. Akyildiz, Magnetic induction communications for wireless underground sensor networks. IEEE Trans. Antennas Propag. **58**(7), 2426–2435 (2010)
3. H. Guo, Z. Sun, P. Wang, Multiple frequency band channel modeling and analysis for magnetic induction communication in practical underwater environments. IEEE Trans. Vehic. Technol. **66**(8), 6619–6632 (2017)
4. Z. Sun, I.F. Akyildiz, On capacity of magnetic induction-based wireless underground sensor networks, in *2012 Proceedings IEEE INFOCOM* (IEEE, Piscataway, 2012), pp. 370–378
5. Z. Zhang, E. Liu, X. Qu, et al.. Connectivity of magnetic induction-based ad hoc networks. IEEE Trans. Wirel. Commun. **16**(7), 4181–4191 (2017)
6. S. Kisseleff, W. Gerstacker, R. Schober, et al., Channel capacity of magnetic induction based wireless underground sensor networks under practical constraints, in *2013 IEEE Wireless Communications and Networking Conference (WCNC)* (IEEE, Piscataway, 2013), pp. 2603–2608
7. J.I. Agbinya, M. Masihpour, Power equations and capacity performance of Magnetic induction body area network nodes, in *2010 Fifth International Conference on Broadband and Biomedical Communications* (IEEE, Piscataway, 2010), pp. 1–6

Chapter 15
Challenges and Advancements in MI Communication

15.1 Directionality Challenge and Multi-Directional Coils

MI coils play an important role in MI communication system. In-fact, the magnetic field generated by any single-directional coil is directional and offer only line of sight communication. In addition to the line of sight communication, the coils (Tx and Rx) also need perfect alignment in order to capture the maximum magnetic flux. Moreover, because of the continuous tidal movement in an underwater environment, it is not possible to fix orientation of the coils, and therefore this directional nature of magnetic field/flux lines is a bigger challenge in MI communication. To improve robustness, reliability and provide an omnidirectional MI communication, researchers have proposed the use of three directional coils, meta-material and spherical enclosed coils.

15.1.1 Tri-Directional Coil

Tri-directional (TD) coil refers to the use of three coils instead of a single coil at both Tx and Rx end. The three coils share the same center point but are perpendicular to each other as shown in Fig. 15.1a [1, 2]. This three directional coil offers a simple and low-cost solution to the directionality challenge, while significantly improve the robustness and reliability of MI communication.

Figure 15.1 shows the three possible combinations of how the TD coils can be connected together:

1. The three coils are connected in series with a single source. Once the source is connected to the circuit, all the three coils are excited in series (Fig. 15.1b).
2. Each of the three coils is attached to three independent sources allowing each coil to transmit three independent signals simultaneously (Fig. 15.1c).

© The Author(s), under exclusive license to Springer Nature Switzerland AG 2022   341
Y. Lou, N. Ahmed, *Underwater Communications and Networks*, Textbooks in Telecommunication Engineering, https://doi.org/10.1007/978-3-030-86649-5_15

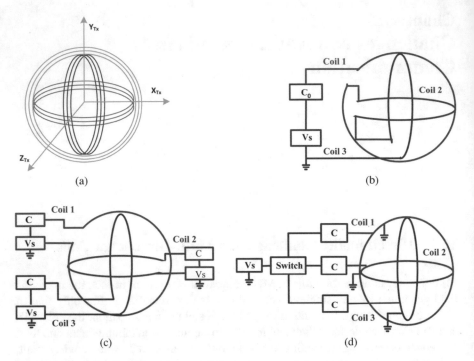

Fig. 15.1 Various types of multi-directional antennas. (**a**) Tri-directional antenna. (**b**) Configuration 1. (**c**) Configuration 2. (**d**) Configuration 3

3. The three coils are connected to a single source via a switch. The switch then choose the coil to be excited by the source (Fig. 15.1d).

Each of these configurations has pros and cons with respect to energy consumption and hardware complexity, however, they provide a reasonable omnidirectional pattern as shown in Fig. 15.2.

15.1.2   Meta-Materials and Spherical Array Enclosed Coils

Besides the multi-directional coils, the concept of meta-material based coils and spherically enclosed coils to enhance the magnetic field strength in an omnidirectional pattern are also proposed.

Figure 15.3a shows the meta-materials (M^2I) enhanced coils where a coil is enclosed by a meta-material shell. The meta-material shell is composed of periodic artificial or dielectric atoms. These periodic artificial or dielectric atoms produce special phenomena of backward waves and negative refraction indexes which boost the magnetic field strength. Furthermore, the mutual coupling is maximized between any two M^2I coils by matching the negative-permeability of the meta-material layer with the positive-permeability underwater environment.

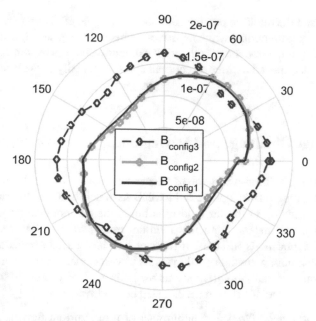

Fig. 15.2 Directivity pattern of the three configurations of multi-directional coil

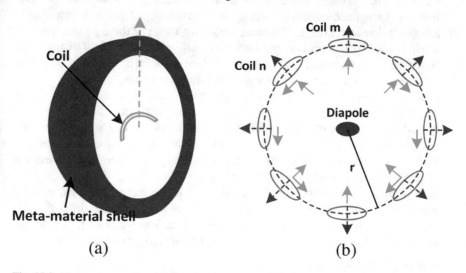

Fig. 15.3 Various types of multi-directional antennas. (**a**) MI Meta-material enhanced antenna. (**b**) MI spherical coil array enclosed loop antenna

Figure 15.3b shows a spherical coil array enclosed loop (SCENL), where multiple passive coils are uniformly surrounded by a central dipole. The central dipole generates a magnetic field which is further radiated by the surrounded passive coils. The total magnetic field strength is the sum of magnetic field generated by the central loop and the multiple passive coils. Moreover, the effective permeability on the sphere is omnidirectional while negative-permeability can be achieved by proper selection of the signal frequency and circuit elements. Hence, a reasonable magnetic field strength is achieved in three directional space, especially in the direction of the coils axes.

15.2   Range Challenge and Waveguides

The signal strength of radio/EM waves decays with square of the distance as $1/d^2$, however, the strength of magnetic field weakens much faster at the rate of $1/d^3$. This rapid weakening in the far field thus limits the communication range of MI systems to tens of meters. With this medium range, MI can be used for fewer ocean based applications in addition to river and lake based applications, but to utilize the maximum potential of MI systems in vast ocean, methods are needed to extend the communication range. Various methods are thus presented to extend the communication range of MI systems as listed below [3]:

- *Maximize Magnetic Moment*. The number of turns, current flowing through the transmit coil, and the coil area can all be used to increase the magnetic moment. Increasing magnetic moment, leads to an increased signal strength, resulting in a larger transmission range. However, modifying either of these parameters may result in more power and copper losses as well as heavier weight of the coils. Based on the application, the parameters values can thus be carefully chosen to achieve maximum transmission range with minimum losses.
- *Better coil design*. Another approach is to improve the coil design to (a) increase the magnetic field strength at Tx coil, and (b) allow maximum flux lines to pass through Rx coil. Use of meta-materials is one example, where artificial periodic or dielectric atom enhance magnetic field strength in an omnidirectional pattern. Such efforts help to increase the transmission range, however, the increase in range by this approach is not significant.
- *Waveguides*. A better and effective approach is to use waveguides. Waveguides exploit the already present sensor coils in the network and extend the communication range without additional power consumption. Waveguides are thus widely adopted in MI communication for applications with longer distances.

A detailed overview of waveguides is thus presented in this section.

15.2.1   Introduction to Waveguides

The term waveguide refers to structures that are capable of transmitting a wave between a transmitter and receiver. Figure 15.4 shows a simple MI waveguide implementation where multiple nodes/coils forming a finite array are placed next

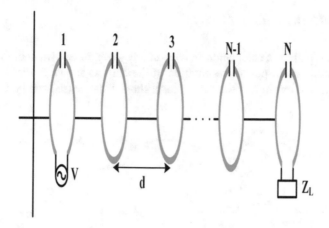

Fig. 15.4 Schematic diagram of MI waveguide structure

to each other separated by distance d. The first node (Tx) is excited with varying voltage V, and the receiver node N is loaded with impedance of $Z_L = R_L + j\omega_L$. The intermediate $(N - 2)$ nodes can be just passive coils with loaded capacitors.

The *principle* of MI waveguide communication is based on resonance induction principle where "the fist coil (Tx) generates a varying magnetic field around its region after excited with a sinusoidal signal. This varying magnetic field couples with the adjacent coil that further resonates its neighboring coil and the process continues till the last node (Rx) is reached." The intermediate (relaying) coils act as waveguides that guide the MI field from source (Tx) to destination (Rx), with no additional power and processing consumption.

Waveguides are utilized in various real-world applications such as microwave bandpass filters, wireless power transfer, implantable biomedical devices and are recently applied in non-conventional medium communication too. Waveguide technique is capable of reducing the path loss and increasing the signal power to noise ratio, resulting in a significant increase in range. Some of the advantages of MI waveguide communication in underwater medium are listed below:

- By carefully choosing MI waveguide parameters the number of relaying coils can be significantly reduced in an underwater sensor network to save the cost and resources of the network.
- Since it is not easy to replace a sensor node in an underwater wireless network, these waveguides can also act as redundant coils making a network more robust.
- As network lifetime is critical in an underwater wireless sensor network, the use of waveguide coils at no additional power and processing unit is extremely beneficial.
- An added advantage of waveguides is their utilization to transfer power to coils that are running short on power in the network.

15.2.2 MI Waveguide Model

To model the path loss calculation in case of MI waveguides, we again refer to the transformer and multi-port representation of the coils as shown in Fig. 15.5a and b, respectively. The transmit power (P_{Tx}) for a short distance (d_0) can be written as

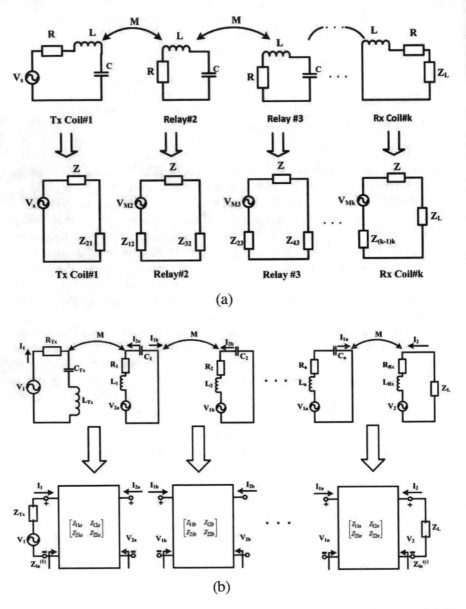

Fig. 15.5 Multi-stage transformer and multiport model. (**a**) Multistage transformer model. (**b**) Multi-port network model

$$P_{Tx}(d_0) = \text{Re}(V_1 I_1^*) = \text{Re}(Z_{11}a)|I_1|^2. \tag{15.1}$$

Similarly, at Rx (node N), the received power (P_{Rx}) expression with load impedance Z_L and current I_2 is written as

$$P_{Rx} = \text{Re}(Z_L)|I_2|^2, \tag{15.2}$$

where I_2 can further be split down to

$$I_2 = \prod_{g=a\ k=b}^{g=n-1\ k=n} \lambda_{gk} \frac{-Z_{21n}}{Z_L + Z_{22n}} I_1. \tag{15.3}$$

Here,

$$\lambda_{gk} = \frac{Z_{21g}}{\gamma_{gk}} \tag{15.4}$$

and γ_{gk} can be written as

$$\gamma_{gk} = Z_{22g} + Z_{11k-\frac{Z_{12n}\cdot Z_{21n}}{Z_L+Z_{22n}}} \tag{15.5}$$

with $g\epsilon\{a, b, \ldots, n-1\}$ and $k\epsilon\{b, \ldots, n\}$. With an assumption of $Z_{gk} = Z_{kg}$, the path loss can be expressed as

$$PL_{WG} = -10\log\frac{P_{Rx}(d)}{P_{Tx}(d_0)}$$

$$= -10\log\left(\frac{R_L\left(R_{21n}^2 + X_{21n}^2\right)}{R_{11a}\left((R_L + R_{22n})^2 + (X_L + X_{22n})^2\right)} \prod_{g=a\ k=b}^{g=n-1\ k=n} \frac{R_{21g}^2 + X_{21g}^2}{|\gamma_{gk}|^2}\right). \tag{15.6}$$

As the received power depends on the load impedance Z_L, to get the maximum gain, the load impedance Z_L needs to match with the input impedance Z_{in} as

$$Z_L = \overline{Z_{in}^{(y)}} = \overline{Z_{22n} - \frac{Z_{12n}^2}{\mu_{gk}}} \tag{15.7}$$

for $g = (n-1)$ and $k = n$, where μ_{gk} is given by:

$$\mu_{gk} = z_{22g} + z_{11k} - \frac{z_{12g}z_{21g}}{\mu_{g-1\ k-1}} \tag{15.8}$$

with $g \epsilon \{a, b, \ldots, n-1\}$ and $k \epsilon \{b \ldots, n\}$. Finally, for $g = a, k = b$, μ_{gk} is

$$\mu_{ab} = z_{22a} + z_{11b} - \frac{z_{12a} z_{21a}}{z_{TX} + z_{11a}}. \tag{15.9}$$

Since each relay node is coupled to its nearest relay node by mutual inductance M. The impedance in the ith relay node can be expressed as

$$Z_{ii} = R_{ii} + j\omega L_{ii} + \frac{1}{j\omega C_{ii}}. \tag{15.10}$$

We further assume that all nodes have similar impedance, i.e., $L_{ii} = L$, $R_{ii} = R$ and $C_{ii} = C$. The mutual impedance between nodes i and $i+1$ can be written as $Z_{i(i+1)} = Z_{(i+1)i} = j\omega M$, where $\omega_0 = 1/\sqrt{LC}$ denotes the resonance frequency and M is the mutual inductance:

$$M = \frac{\mu N_i r_j^2 N_i r_j^2 \pi}{2\sqrt{\left(r_i^2 + d^2\right)^3}}. \tag{15.11}$$

By assuming that $Z_{Tx} \approx 0$, the load impedance at the Rx can be calculated as

$$Z_L = R + \overline{\frac{\omega^2 M^2}{\mu_{gk}}} = R + \overline{\frac{\omega^2 M^2}{2R + \frac{\omega^2 M^2}{\mu_{g-1\,k-1}}}}. \tag{15.12}$$

Thus for resonating waveguides, the path loss can finally be written as

$$PL_{WG} = -10 \log \left(\frac{R_L \omega^2 M^2}{R(R_L + R)^2} \prod_{g=a\,k=b}^{g=n-1\,k=n} \frac{\omega^2 M^2}{\left(2R + \frac{\omega^2 M^2}{|\gamma_{g+1\,k+1}|}\right)^2} \right). \tag{15.13}$$

15.2.3 Waveguide Implementation Challenges

While waveguides increase the communication distance, it also offers a few challenges in real-time implementation. Two of the main challenges and their solutions are briefly listed here:

1. The minimum number of relaying node for a given network has always been a challenging question for waveguide implementation. Researchers have therefore investigated the use of optimal number of relaying nodes for MI-based network and established a relationship between number of optimal relaying nodes, the communication distance, and the required bandwidth.

2. Like the number of relaying nodes, another important question is to where these relaying nodes can be placed in a given network. In early study of MI waveguide technique, the relay nodes are always placed equidistantly between transmitter and receiver. This assumption proved significant in determining the advantage of achieving large transmission range, however, it was identified in later research that sometimes placing relay nodes equidistantly between transmitter and receiver may not perform better. Optimal relay position is different for different regions based on the various factors of the system such as coil radius, number of turns, resonance frequency.

15.3 MIMO and MI Communication

Multiple-input and multiple-output (MIMO) systems have been extensively used in wireless communication to enhance channel capacity. Underwater acoustic communication also utilizes MIMO to deal with the complex physical underwater acoustic channel. MI channel although is not as complex as acoustic channel, but researchers have still explored the implementation of MIMO to further increase the channel capacity [4, 5]. In a MIMO system, multiple transmission of data is carried out over available frequency bands by utilizing multiple transmitting and receiving coils. A simple 2×2 MI MIMO model is shown in Fig. 15.6, in which transmitter and receiver nodes are equipped with two coils each, respectively. For an $N \times N$ MI MIMO system (Fig. 15.7) the channel capacity is given by:

$$C = \log_2 \left[\det \left(I_N + \frac{\rho}{N} H H^* \right) \right] \text{bps/Hz,} \qquad (15.14)$$

where ρ is the SINR of the received signal, I is the identity matrix, H is the $N \times N$ channel matrix, and $*$ is the transpose-conjugate.

Though MIMO systems can be useful to increase the capacity, implementation of MI MIMO is challenging due to the lateral crosstalk and diagonal crosstalk as shown in Fig. 15.7.

1. The lateral crosstalks refer to the mutual coupling between neighboring (Tx/Rx) coils and with use of proper coil placement, it can easily be avoided. The proper coil placement refers to exploit the directionality of single coils. Since the orthogonal coils do not couple with each other, the idea is to place the two coils (with lateral crosstalks) orthogonal to each other to avoid the lateral crosstalks.
2. Diagonal crosstalks on the other hand refers to the mutual coupling between a Tx coils and unintended Rx coils. Precoding can be used to mitigate such type of crosstalks. In case of precoding, a suitable weight matrix is assigned to the transmitting signal that align the precoded signal to the intended links only and cancel the superimposed interference at each receiver. This precoding method is quite useful for MI-based systems because of the predictable and stable channel response. Precoding method for MI-based MIMO systems can further

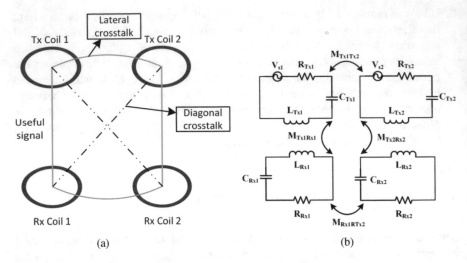

Fig. 15.6 MI MIMO 2×2 model. (**a**) MI MIMO 2×2 structure. (**b**) Equivalent circuit

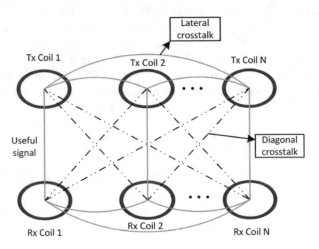

Fig. 15.7 MI MIMO $N \times N$ structure with various crosstalks

be classified into two types: linear and nonlinear precoding method. Linear precoding method usually follow zero forcing (ZF), maximum ratio transmission (MRT) and transmit Weiner precoding method, while dirty paper coding (DPC) concepts are used for the development of nonlinear precoding methods.

Summary

- Directionality and short range are the two main challenges in MI communication system.
- To cope the directionality challenge of single-directional coil, multi-directional coils are developed.
- Multi-directional coils proposed in literature are tri-directional (TD) coils, meta-materials enhanced coils and spherical coil array enclosed loop coils.
- In order to increase the range of MI communication system, a simple and easy approach is the use of waveguides in MI system.
- MI waveguides act as a passive relaying technique and help to significantly increase the transmission range.
- MIMO system can also be applied in MI-based systems to increase the channel capacity.

? Questions

1. Explain the directionality challenge and how much the proposed solution improve the directionality problem.
2. Explain the different configurations of tri-directional coils and comment on the energy consumption of these configurations.
3. Explain the principle of waveguides and comment about their effectivity in MI communications?
4. What are the implementation challenges associated with waveguides?
5. What is lateral and cross-talk in MI MIMO communications and how can they be avoided?
6. Magnetic field at the center of Tx coil (radius = 10) is measured as 50×10^{-3} T when a current of 10 A is applied across its ends. The Rx coil can detect the signal if the strength of magnetic field is at least 1×10^{-3} T. To achieve a total of 50 m range, find the minimum number of relays needed to achieve the communication range. Assume that each relay coil consumes the same power as the Tx coil to forward the signal.
7. Consider a relaying/waveguide network where there are two relaying coils (R1, R2) between the transmitter (Tx) coil and receiver (Rx) coil. Transmitter and Receiver are 100 m apart from each other, while R1 and R2 are placed at 35 m and 70 m from the Tx, respectively. All the coils have similar radius of $0.5m$, and number of turns 100. Calculate the (a) mutual inductance between R1 and R2, and (b) the mutual inductance between Tx and Rx.

References

1. I.F. Akyildiz, P. Wang, Z. Sun, Realizing underwater communication through magnetic induction. IEEE Commun. Mag. **53**, 42–48 (2015)
2. H. Guo, Z. Sun, M^2I communication: From theoretical modeling to practical design, in *2016 IEEE International Conference on Communications (ICC)* (2016), pp. 1–6
3. M.C. Domingo, Magnetic induction for underwater wireless communication networks. IEEE Trans. Antennas Propag. **6**(60), 2929–2939 (2012)
4. H.J. Kim, J. Park, K.S. Oh, et al., Near-field magnetic induction MIMO communication using heterogeneous multipole loop antenna array for higher data rate transmission. IEEE Trans. Antennas Propag. **64**(5), 1952–1962 (2016)
5. N. Tal, Y. Morag, Y. Levron, Magnetic induction antenna arrays for MIMO and multiple-frequency communication systems. Progr. Electromag. Res. **75**, 155–167 (2017)

Chapter 16
MI Wireless Sensor Networks

16.1 Underwater Wireless Sensor Network Applications and Architecture

With water comprising 70% of the earth, maritime offers a wide range of underwater wireless sensor network applications starting from military applications to commercial explorations as listed in Fig. 16.1. To better utilize the resources while designing a network for these applications, MI-based wireless sensor network may follow one of the following architectures [1].

- For an MI 1D-UWSN each MI sensor node itself is considered a network. This can be any MI sensor node deployed in any place to monitor specific task. Applications of AUVs can also be termed as 1D UWSN, where an AUV move in a region and collect the sensing data.
- A 2D-UWSN can be a cluster based network where there are more than one node, and each node can sense the data as well as communicate with a neighboring node.
- 3D-UWSN is also a cluster based network with three basic architecture as (1) Inter-cluster node communication at varying depths, (2) Intra-clusters connectivity between sensor/anchor nodes, and (3) Anchor-buoyant communication between nodes.
- 4D-UWSN is a combination of mobile UWSN and a fixed 3D-UWSN. The mobile UWSN comprises of remote underwater vehicles (ROVs) for collecting information from anchor nodes.

We next in this chapter present the design considerations and implementation challenges of localization, mac, and routing protocols for MI-based underwater wireless sensor network techniques.

Y. Lou, N. Ahmed, *Underwater Communications and Networks*, Textbooks in Telecommunication Engineering, https://doi.org/10.1007/978-3-030-86649-5_16

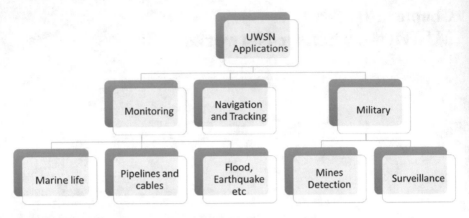

Fig. 16.1 Underwater wireless sensor network applications

Fig. 16.2 Main steps of any general localization process

16.2   Localization

16.2.1   Localization in Wireless Sensor Networks

A general localization process is shown in Fig. 16.2 where the information from anchor nodes (the input) combined with distance information and position estimation is used to locate a sensor node. Anchor nodes and sensor nodes are two types of nodes that are commonly used to localize a wireless sensor network. Nodes with known locations are called anchor nodes, while nodes with unknown locations are called sensor nodes. The scope of localization algorithm is to accurately locate the sensor nodes with the help of anchor nodes [2].

16.2.2   Distance Estimation

Inside a cluster, sensor nodes generally form a distributed wireless sensor network and therefore distributed localization algorithms are adopted to locate a sensor node in a cluster (Fig. 16.3). Distributed localization algorithms are classified into range free and range based algorithms. Range free algorithms consists of distance vector (DV)-hop, gradient algorithms, approximate point in triangulation (APIT), centroid systems and hop terrain. Range free algorithm are cost efficient but less accurate

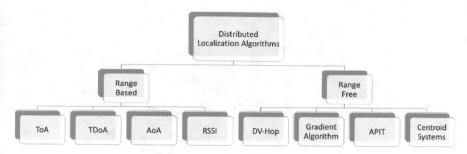

Fig. 16.3 Classification of distributed localization algorithms

and since hop count requires flooding the network, they are generally not suited for wireless sensor networks as they can reduce the overall network efficiency. On the other hand range based algorithms are widely adopted for wireless sensor network applications because of their better accuracy.

The common methods used to calculate the distance in a range based algorithm are (a) time of arrival (TOA), (b) time distance of arrival (TDoA), (c) angle of arrival (AoA), and (d) received signal strength indicator (RSSI). In MI communication since the communication link is established due to magnetic flux induction, it therefore is not possible to retrieve the time and angle information. The only parameter that a MI sensor node can use to find the distance is the amount of magnetic flux induced in it which is directly related to the strength of magnetic field. Thus to locate a MI sensor node, MI anchor node estimates the distance by measuring the induced voltage which is given by

$$V_{ind} = -\frac{d\Phi_B}{dt} = 2\pi f N B A \cos\theta. \tag{16.1}$$

It is also important to mention here that if the MI wireless sensor network consists of single dimensional coils only, the localization process will be extremely challenging because of the directionality issue as explained in the following section.

16.2.2.1 Localization Challenge with Single Dimensional Coil

When a one dimensional MI coil is excited with alternating current, magnetic field is created around the coil. These magnetic field lines are always directional where they are stronger on the axis of the coil and get weaker as move away from the axis. The receive coil therefore, if placed along the axis ($\theta = 0$) of the transmit coil, will get the maximum flux and hence the induced voltage will be maximum as given by the Eq. (16.1).

Figure 16.4 clearly illustrates the directionality issue of one dimensional coil, where four receiver nodes: Node 1 (N1), Node 2 (N2), Node 3 (N3) and Node 4 (N4) are placed at a distance of 10cm from the transmitter node N0. N1 is placed

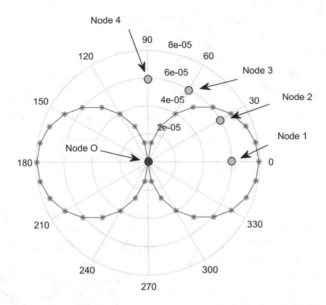

Fig. 16.4 Illustration of directionality challenge due to single dimensional coil

along the axis of the transmitter node at an angle of $\theta = 0°$, N2 is placed at an angle of $\theta = 30°$, N3 is placed at an angle of $\theta = 60°$ and N4 is placed at an angle of $\theta = 90°$.The induced voltage at the four receiver nodes is also shown in Fig. 16.4. When Tx node N0 and Rx node N1 are at $\theta = 0°$, maximum voltage is induced in the N1 node. V_{ind} will drop gradually as the angle of N2, N3 increases with N0, while it becomes zero for N4 where Tx and N4 nodes are at $\theta = 90°$. With distance point of view, it can be noted that all the four nodes are at the same distance from the transmitter node, however, the received signal strength at each node is quite different. This directional pattern of magnetic field thus offers severe challenge to apply RSSI with one dimensional coil. Thus for any MI-based wireless sensor network, if single dimensional coils are to be deployed, they should always be placed along the same axis.

16.2.3 Position Estimation

Once the distance information is obtained, the next step is to accurately find the position of the sensor node. The position estimation techniques utilizing distance information are called lateration. The simple lateration technique is *trilateration* where distance measurement from three anchor nodes is used to find the position of the sensor node. A pictorial illustration of trilateration is shown in Fig. 16.5, where the sensor node is located in the intersection region of the three anchor nodes. Based on the distance information acquired in the first step of *distance estimation*, a

Fig. 16.5 Trilateration
technique with three anchor
nodes and one sensor node

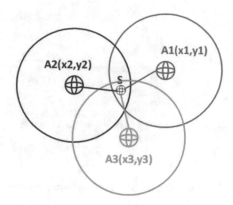

distance matrix is developed and used to find the exact position of the sensor node. To get more accurate position estimation, more anchor nodes can be utilized. A lateration technique that uses more than three anchor nodes is called *multilateration*.

16.3   Medium Access protocols

Traditional UWA communications face fundamental limitations in implementing standard WSN medium access (MAC) protocols, because of the long propagation delay associated with acoustic waves, however, for MI-WSNs there are no such obstacles. Because of the instant communication link, medium access techniques proposed for EM-based wireless sensor networks can be borrowed for MI-based networks. Nevertheless, to effectively and efficiently use the existing medium access techniques mainly developed for EM-based WSNs, there is a need to account MI channel characteristics such as, dependence on time-varying magnetic field. This means that, rather than considering the complex physical channel, the medium access techniques for underwater MI-based wireless sensor networks can use simple protocol models, in which interference from a neighboring node is only determined by whether or not a node lies within its range [3]. An overview of all the wsn protocols is presented in Fig. 16.6 that could easily be adopted with MI-based communication. The challenge in implementation of medium access techniques for MI-based network will again be with the sensor networks that involve single dimensional coils as presented next in case of hidden node problem.

16.3.1   Hidden Node Problem and MI Communication

Figure 16.7a shows a common hidden node problem that exists in contention based MAC protocols. There are three sensor nodes A, B, and C. Node A can establish

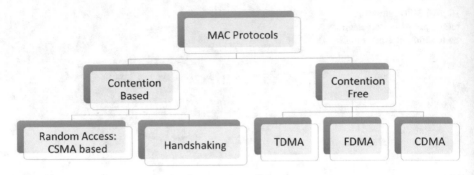

Fig. 16.6 Overview of wireless sensor network medium access protocols

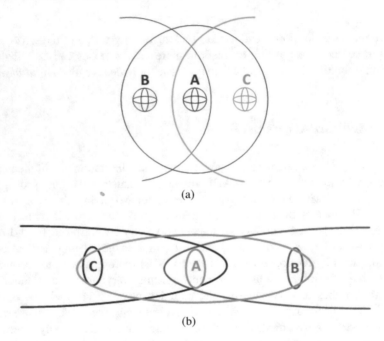

Fig. 16.7 Illustration of hidden node problem. (**a**) Illustration of hidden node problem in case of a multidimensional coil with omnidirectional pattern. (**b**) Illustration of hidden node problem in case of a single dimensional coil with directional pattern

communication link with both Node B and Node C, while Node B and Node C can only establish communication link with Node A. The issue arises when nodes B and C begin sending packets to access point A at the same time. Since nodes B and C are unable to detect each other's signals, they are unable to detect a collision before or during transmission.

Figure 16.7b shows scenario for MI-based network with single dimensional coils. In case of single coils, there are two main implementation challenges: (a) because of the directional behavior of single coils, a single node is accessible to fewer nodes only in a network, and (b) the hidden problem is more severe because a slight orientation of coil A can disconnect coil A from both coil B and C.

16.4   Routing Protocols

Like the medium access protocols, sending data over multi hops in an acoustic wireless sensor network is challenging due to the complex physical acoustic channel [4].

Sophisticated underwater routing protocols have been created to overcome the complexities of acoustic channels, however, the unique features of MI channels necessitate a rethinking of these underwater routing protocol design.

MI sensor nodes can easily forward the data to the next hop using waveguides. Waveguides act as passive relays, they forward the data without the use of additional power, and therefore limits the number of forwarding hops in a network. Consequently, the use of active and hybrid relaying can be more beneficial where the data is forwarded over longer distances by utilizing more hops.

16.4.1   Active Relaying

Active relaying technique requires the sensor nodes in the network to actively involve in forwarding a packet from source to destination. Figure 16.8 shows a mechanism of simple active relaying scheme where two sensor nodes actively relay the data to the receiver end in the transmit cycle and then relay back the feedback to the transmitter in the receive cycle. The sensor nodes in active relaying require additional power resources depending on their involvement in a given network. Moreover, the common relaying techniques are following

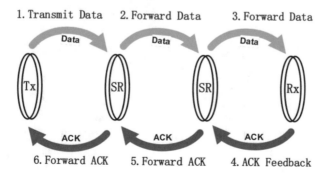

Fig. 16.8 A simple active relaying mechanism where two sensor nodes help to forward the data to the receiver end

- **Amplify-and-forward:** The simplest and easy to implement technique is amplify-and-forward technique (AF). In case of AF, a relay node receives, amplifies, and forwards the signal without decoding the signal.
  The downside of AF is that the noise also gets amplified with the signal amplification.
- **Decode-and-forward:** To overcome the noise amplification challenge in AF, decode-and-forward (DF) techniques are utilized.
  In case of DF, a signal is first received, decoded, then re-coded, and then transmitted to the next node. DF therefore requires the sensor nodes to be equipped with additional power and processing resources. Moreover, in case of false decoding, DF methods add an additional delay to the network.
- **Filter-and-forward:** A smart solution to the noise amplification challenge is the use of filter-and-forward (FF) technique. FF technique requires the sensor node to be equipped with dedicated hardware, that amplify the received signal after passing through an optimized bandpass filter to lower the noise level.

16.4.2 Hybrid Relaying

MI waveguides are passive relay nodes that forward data to neighboring nodes without additional power, however, the number of forwarding nodes is limited. Active relaying on the other hand allows each node in the network to work as active relaying node but on the cost of additional power. To utilize the networking resources in an optimal way, *hybrid relaying* are thus introduced. Hybrid relaying utilizes both passive and active relaying nodes to form an energy efficient and reliable network (Fig. 16.9). Hybrid relaying scheme, compared to waveguides and active relaying, consumes less power while transmitting same packets of information at a given distance, more specifically in long transmission range scenarios. Furthermore, hybrid relaying is useful in increasing the transmission range with less power consumption based on the optimal configuration.

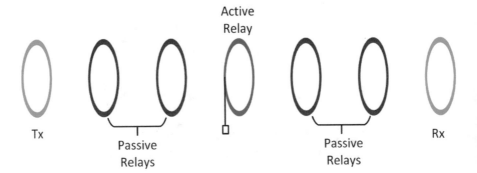

Fig. 16.9 Hybrid relaying: Combination of passive and active relaying nodes

16.5   Cross-Layer Protocols

Communication networks traditionally use a protocol stack of different layers to perform the different network functionalities. Each layer in a protocol stack has its own functionality which is then controlled by a subset of decision variables. Each layer functions independently of the others, and the entire structure is placed together using only a few interfaces between them. A typical protocol stack consists of the following five layers as shown in Fig. 16.10 [5].

Despite the fact that this layered communication architecture has been useful in many communication networks, researchers have begun to argue that in the case of wireless sensor networks, a cross-layer design might be necessary because in layered architecture, protocols are designed at different layers in isolation and operate without much interaction between functionalities.

The term "cross-layer architecture" refers to the sharing of information between layers in order to make the most effective use of network resources and achieve high adaptivity. Each layer in a cross-layer design is described by a few main parameters and control handles (Fig. 16.10). Layers use the parameters to decide the best adaptation rules for their control handles based on the network status. Cross-layer architecture is typically conceived as an optimization problem with multiple layers optimization variables and constraints. The optimal values for the control handles in the layers are determined by solving the optimization problem. MI wireless sensor networks also encourages the use of cross-layer architecture to better utilize the networking resources.

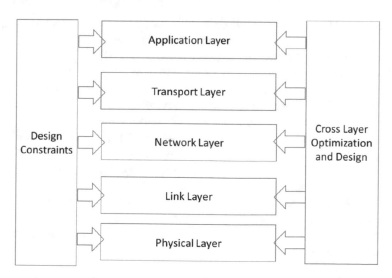

Fig. 16.10 Cross-layer design

Summary

- Underwater wireless sensor networks are classified into 1D, 2D, 3D, and 4D architecture.
- Localization in MI wireless sensor networks is challenging as the distance can only be found by the amount of voltage induced.
- Localization of MI wireless sensor network with single dimensional coils is severe challenging due to the directionality problem.
- MAC layer implementation for MI wireless sensor network is easy and simple. The existing mac protocols of EM-based networks can easily be applied to MI-based networks.
- Routing in MI is easily achieved with active and hybrid relaying techniques.
- Wireless sensor networks in general encourages the use of cross-layer protocols and can easily be implemented in MI-based networks.

? Questions

1. What is a hidden node problem in contention based MAC protocols? Also explain the hidden node problem for single directional and three directional coil.
2. Given that a multidimensional MI coil is used in a MI-based wireless sensor networks, what ranged based technique can then be adopted for localization?
3. Explain the advantages and disadvantages of active and hybrid routing protocols?
4. How is cross-layer design different from conventional OSI layer design? Explain the need of a cross-layer design in a wireless sensor network?
5. If magnetic field is 5×10^{-3} T at 1 m distance from the Tx coil (radius = 0.1 m) along its axis. Calculate the induced voltage in the Rx coil if (a) Rx coil is perfectly aligned to the Tx coil, and (b) Rx coil (radius = 1 m) is misaligned at $60°$ The operating frequency is 100 kHz.
6. If magnetic field at distance 100 m along the axis of Tx coil (radius = 10 m) is 50 T, when a current of 10 A is applied across its ends, and the voltage induced in Rx coil is 0.08 V. Find the alignment angle of the Rx coil with respect to the Tx coil. The operating frequency is 100 kHz.

References

1. S. Kisseleff, I.F. Akyildiz, W. Gerstacker, Disaster detection in magnetic induction based wireless sensor networks with limited feedback, in *2014 IFIP Wireless Days (WD)* (IEEE, Piscataway, 2014), pp. 1–7
2. H. Huang, Y.R. Zheng, Node localization in 3-D by magnetic-induction communications in wireless sensor networks, in *OCEANS 2017-Anchorage* (IEEE, Piscataway, 2017), pp. 1–6
3. I.F. Akyildiz, P. Wang, Z. Sun, Realizing underwater communication through magnetic induction. IEEE Commun. Mag. **53**(11), 42–48 (2015)
4. Z. Sun, I.F. Akyildiz, Deployment algorithms for wireless underground sensor networks using magnetic induction, in *2010 IEEE Global Telecommunications Conference GLOBECOM 2010* (IEEE, Piscataway, 2010), pp. 1–5
5. S.C. Lin, I.F. Akyildiz, P. Wang, et al., Distributed cross-layer protocol design for magnetic induction communication in wireless underground sensor networks. IEEE Trans. Wirel. Commun. **14**(7), 4006–4019 (2015)

Index

Printed in the United States
by Baker & Taylor Publisher Services